THE RISE OF NUCLEAR FEAR

THE RISE
—OF—
NUCLEAR
FEAR

SPENCER R. WEART

HARVARD UNIVERSITY PRESS
Cambridge, Massachusetts, and London, England
2012

An earlier version of this work was published by Harvard University Press as
Nuclear Fear: A History of Images (1988).

Library of Congress Cataloging-in-Publication Data

Weart, Spencer R., 1942–
The rise of nuclear fear / Spencer R. Weart.
p. cm.
Includes bibliographical references and index.
ISBN 978-0-674-05233-8 (alk. paper)
1. Nuclear energy—History. 2. Nuclear energy—
Psychological aspects. 3. Antinuclear movement—History.
4. Radiation—Public opinion—History. I. Title.
QC773.W44 2012
621.48—dc23   2011035170

CONTENTS

PREFACE

When I began studying the history of nuclear energy I did not think that images were important in themselves. I was wrong. Radioactive monsters, utopian atom-powered cities, weird ray devices, and many other images have crept into the way everyone thinks about nuclear weapons and power plants. The images, connecting with major social and psychological forces, have exerted a strange and powerful pressure on our history. This is no story of things locked away safely in the past: the images are as strong today as ever.

How can an investigator connect the colorful images in people's heads with the attitudes that affect historical events? For an answer I rely on commonsense ideas that have been validated by rigorous experiment. Minds form associations between separate things; the sound of a can opener brings my dog to the kitchen and may even make her salivate. This mechanism is fundamental to all thought, including raw perception. Experiments show that we must have a rough idea of what to look for, some previously learned mental picture, or we even have trouble recognizing an object in a drawing. In short, as a result of experience every simple image becomes connected with various other things in a web of associations. Such durable associations can be forged not only by repetition (as with Pavlov's dogs) but by a single traumatic experience. The process is inborn, automatic, and almost instantaneous. Visceral emotions like fear or disgust are especially important components of our mental clusters; our brains are designed so that such emotions can help us to respond to a situation quickly and efficiently, if not always correctly.

Our brains scarcely distinguish between direct experiences and vivid imaginary experiences, as seen, for example, in a movie. Thus some things become associated with a whole cluster of other things in a public image that may be shared by many people. Such a cluster may include, alongside the inarticulate feelings, various kinds of unconsidered beliefs and images.

In some cases we will overlay the tangle with carefully thought-out ideas (as I hope to do in this book). The result is the attitudes that finally determine action.

The history of images matters. For example, with two zigzag strokes of a pencil I can draw a design that most people will call a swastika. To a Hindu scholar in 1925 the design would recall symbols that he had seen carved on white temples, evoking feelings of religious devotion. To a Nazi youth in 1935 the associated material would include flag parades and national pride. To a Jew in 1945 the same simple design would evoke nightmares of mass death. As this example shows, such associations may come through age-old tradition (as with the Hindu), through conscious manipulation by propagandists (as with the Nazi), through historical events (as with the Jew), or from all of these together.

This book will mention only in passing the historical events and social forces concerned with nuclear energy, which have been analyzed in many other books. I focus on psychology and imagery, which are poorly understood. Indeed, nuclear imagery stands before us as a screen that prevents us from truly understanding how the other forces work.

My central argument is that images buried within our minds have larger consequences in history than has commonly been thought. I do not mean that our thoughts must respond helplessly to these pressures. In pointing out the power of imagery, I hope to make it easier to resist deliberate manipulation by propagandists and by our own unconscious biases: to make reason, not imagery, our guide.

In 1987 I finished writing a book on this subject, published the following year as *Nuclear Fear: A History of Images*. The present book is different in three ways. First, more than two decades have passed, bringing major historical developments that have altered public thinking. Second, advances in fields from historical scholarship to social psychology to neuropsychiatry have cast new light on how people respond to imagery; my original analysis, while nowhere incorrect, can be made deeper. Third, although there is more to cover, this book is much shorter. The earlier book was too long for an easy reading. Lose a sentence, gain a reader! Details on additional pre-1988 topics and a full scholarly apparatus remain available in the earlier book.

In addition to the people and institutions acknowledged there, I give my thanks to Bo Jacobs and the Hiroshima Peace Institute; the American Institute of Physics and its Center for History of Physics and Niels Bohr Library & Archives; and Ann Bisconti, Paul Boyer, John Canaday, Michael Edwards, Paul Forman, and Alain Michel.

THE RISE OF NUCLEAR FEAR

1

RADIOACTIVE HOPES

Once there was a man who sought after hidden knowledge. The story says that he hoped to make human civilization more noble, and if there was an ugly, mad streak in him, as in all of us, he controlled it strictly. This man arduously studied not only modern science but also alchemy, and it was after pondering the arcane philosophers' stone that he discovered the most prodigious secret of physics: the release of vast energy from within atoms. He knew at once that this energy would change the world. He feared vast explosions, but at the same time he hoped that atomic energy would save civilization, which he believed was otherwise destined to collapse when its fossil fuels ran out. A vision came to him of white towers rising from gardens, a peaceful and prosperous future city centered upon gleaming atomic power plants.

Up to here the story is historical fact, but the rest becomes increasingly like a dream. The man built a shining cylindrical device to project atomic rays. With delight he explored the astonishing effects of the radiation, finding it could cure cancer and other ills. However, in his experiments the rays sometimes did not cure people but gave them horrible cancers, or changed their very genes so that their children were monsters.

The destructive power of atomic rays might be useful, the man thought, for his nation was under deadly threat from enemies. If he could make an all-powerful weapon, surely nobody would dare to start a war. He went to a laboratory hidden down a shaft deep in the earth, and there he used his rays to construct a weird creature, a sort of living robot armed with irresistible energies.

In this story there was also a woman who might have been the scientist's lover. He had found little time to court her when all his efforts were focused on knowledge and power, but she nevertheless visited his workroom. Just then he had been thinking of a ray that might render living creatures immortal. As the woman approached he aimed a ray device toward her and proposed an experiment; she fled in horror. Rage exploded in the man's over-taxed brain, and he screamed that everyone had abandoned him, leaving him alone in the world. Climbing into a recess in his robotic creature, he rode it to the surface of the earth. But when he emerged the terrified authorities attacked him, and that automatically activated his weapons. Enormous clouds mushroomed into the sky; radioactive poisons swept life from the planet. In the ashen landscape lay the robot, blackened and deformed.

From an underground room where she had taken shelter, the woman emerged. When she tried to lift the ruined creature it cracked apart like a shell and the man crawled out, his madness purged away. The pair joined hands. A new world would rise on the ashes of the old, a purified and wiser race, perhaps with a white city after all . . .

There are some curious things about this legendary tale, which I have constructed as a composite of numberless familiar stories. The stories combine contradictory ideas, yet in an odd way the ideas fit neatly together. Still more remarkable, the images are plausible. Atom-powered city, potent ray, strange creature, blasted plain—each could happen, and to a degree each *has* happened. Images so plausible, and also so impressive, might have been expected to exert some kind of influence on the people who made the political, economic, and military decisions bound up with the history of nuclear energy.

The most curious and unsettling thing is that every theme in such tales was already at hand early in the twentieth century, decades before the discovery of nuclear fission showed how to actually release the energy within atoms. The imagery, then, did not come from experience with real bombs and power plants. It came from somewhere else.

Legends conceal grave truths, but not truths about nuclear physics. Such tales are about more important matters: the forces of human history, social structure, and psychology. In this book I explore these matters as seen in the interaction of imagery with the history of nuclear energy. In the first four chapters I deal with the years prior to the discovery of nuclear fission, focusing in turn on the white city, the destroyed planet, the transforming ray, and the monstrous creature—images that have exerted a surprising influence on the history of our times.

There really was a man who studied both science and alchemy, found the secret of atomic energy, and exclaimed that it would lead humanity to paradise or doomsday. His name was Frederick Soddy. As a youth at the turn of the century he had been bright, ambitious, quarrelsome, and lonely; photographs show his features set firmly and scornfully, like a gentleman boxer about to enter the ring. His chance to become famous came while he was teaching chemistry in Montreal and fell in with Ernest Rutherford. At thirty Rutherford was a hearty, well-liked, and well-established professor, all the things that Soddy was not. But the two men shared a gift for science, and they also shared an ambition: to crack the puzzle of radioactivity.

Radioactivity had attracted scant attention when it was discovered in 1896. It seemed only a curiosity that a few minerals such as thorium and uranium emitted feeble rays resembling a sort of invisible light. Then Marie Curie discovered the new metal radium, whose rays, compared with the whisper from uranium, were like a piercing shout. When the cream of the world's physicists gathered in Paris for a congress in 1900, Marie and her husband, Pierre, proudly displayed little vials of radium compounds so active that they glowed with a pearly light. The newspapers began to pay attention to radioactivity, and so did Soddy and Rutherford.

Late in 1901 the pair discovered that radioactivity is a sign of fundamental changes within matter. A pulse of radiation signals that an atom is changing into a different kind of atom, a different element with its own chemical properties. At the moment he realized this, Soddy recalled, "I was overwhelmed with something greater than joy—I cannot very well express it—a kind of exaltation." He blurted out, "Rutherford, this is transmutation!"

"For Mike's sake, Soddy," his companion shot back, "don't call it *transmutation*. They'll have our heads off as alchemists." Already at the instant the new science was born, it stirred both joy and anxiety.[1]

What did it mean, this word "transmutation," which elated Soddy and gave Rutherford pause? To most people the word was associated only with gold-making charlatans and crackpots. But in fact the concept of transmutation had once been the central strand of a far-reaching and ancient web of thought. It gives us a clue that can help to explain almost every strange image that would later appear in nuclear energy tales.

At first Soddy and Rutherford could not quite say why atomic transmutation was important, but during 1902 they found one of the reasons: energy.

The pair, and meanwhile Pierre Curie in Paris, showed that radioactivity released vastly more energy, atom for atom, than any other process known. Soddy explained the discovery to the public promptly in May 1903 in a British magazine read by cultivated ladies and gentlemen. Radioactivity, he said, pointed to "inexhaustible" power; henceforth matter must be considered not just as inert stuff but as a storehouse of energy. A year later, while taking a long journey by steamship to Australia where he was to lecture on radium, he made a more specific calculation, the kind of tangible example that an audience would not forget: a pint bottle of uranium contained enough energy to drive an ocean liner from London to Sydney and back!

Soddy had more in mind than such mundane tricks. Writing in an American magazine, he said that if we could manage to tap the energy within atoms, "the future would bear as little relation to the past as the life of a dragonfly does to that of its aquatic prototype." He summed up his ideas in 1908 in a widely read book, *The Interpretation of Radium.* "A race which could transmute matter would have little need to earn its bread by the sweat of its brow," he declared. "Such a race could transform a desert continent, thaw the frozen poles, and make the whole world one smiling Garden of Eden."[2]

Eden restored, the dragonfly springing from its pupa—where did Soddy find these extraordinary images? The few facts then known about radioactivity gave no support for such language. Rather, these were images that the solitary chemist had already cherished before nuclear energy was discovered.

At the start of the twentieth century, many felt that science would lead humanity to an abundance not only of material goods but of brotherhood and wisdom. The greatest enthusiasts were scientists themselves, such as the eminent French chemist Marcellin Berthelot, who fascinated people with his books. By the year 2000, he declared, the earth would be a garden where a kinder and happier humanity would live amid the abundance of a Golden Age. He explained that it was the discoveries of scientists like himself that would bring this to pass, for example, by providing a limitless source of energy. Many of Soddy's declarations could have been taken straight from Berthelot's discourses, or others like them. In 1893 a simulacrum of the perfected city had actually been constructed, if only for a summer. The fairgrounds of the Chicago International Exposition, dubbed the "White City," were a fairyland of broad avenues and sparkling fountains, incandescent at night under the new electric lamps, with steel dynamos gleaming alongside alabaster sculptures.

Reasons for enthusiasm in the scientific future were seen even in the planet Mars. A few astronomers convinced themselves that they saw spidery lines across the blurred disk in their telescopes. Camille Flammarion, a French astronomy popularizer, said in 1892 that the lines were probably canals, the grand engineering projects of an elder race. Percival Lowell of Boston explained that Mars must represent a later stage of evolution than the Earth, a planet grown dry over the ages. Thus the future of our planet was written upon our neighbor. Not only did the colossal scale promise astounding technical prowess, but the way the canals spanned the entire planet proved that the Martians had outgrown war. Although most professional astronomers scoffed at such ideas, the public was half convinced that the Martian engineers existed.

Of course Flammarion and Lowell had simply projected onto the ambiguous disk an image of future civilization that existed only in their heads. I say this to emphasize a mechanism that is central to any history of images: ideas already held in the mind tend to creep into the picture that people think they perceive.

By the start of the twentieth century the image of a White City, expanded to planetary scale and projected onto the blurred screen of the future, was at a peak of popularity, well positioned to become the first symbol associated with the energy of atoms. After all, modern civilization was founded on energy. Coal and electricity were visibly transforming nations, a change more rapid than has happened to any generation before or since. As a child Soddy had read by smoking oil lamps and traveled behind plodding horses, technologies little different from those used by the ancient Romans; by the time he joined Rutherford, electric trams were humming down brilliantly illuminated avenues.

Much remained to be done. Industry was carried on the stooped backs of coal miners, working in the dark until their health broke. Coal smoke choked the cities, yet millions preferred the soot and tuberculosis of industrial slums to the traditional rural life of ignorance, exhausting labor, and malnutrition. Progressive thinkers had good reason to say that civilization would lift itself far higher—provided it continued to replace human muscle with the energies manipulated by scientists.

Experts warned, however, that the planet's finite stock of fossil fuels must eventually be exhausted. Soddy was only repeating familiar ideas when he wrote that "the world's demand for energy is ever increasing and will continue to increase, while the available supply of fuel is ever diminishing." Other scientists promised that before the coal gave out they would find other sources of energy—most likely sunlight. Visionaries wrote of a sun-powered civilization in which slums would become true white cities. Or perhaps, since sunlight is everywhere, people would abandon cities altogether for a prospering countryside. Soddy, however, saw poor prospects for solar power (investors in schemes for solar boilers were in fact regularly disappointed). He warned that the "inevitable coming struggle" over dwindling fuel supplies might hurl the world into poverty sooner than people guessed. But if we could exploit the energy within atoms? Then all the wonders predicted for humanity's future would indeed unfold.[3]

There was ample reason for scientists to tell the public their dream for making a better future, as Soddy and other prominent scientists did in numerous lectures, writings, and newspaper interviews. The obvious motive was pride in their profession and in their personal discoveries. Beyond that, they could hope to find money for their laboratories, and students for their classes, only if the public joined in their belief in the value of science. What is more interesting is how avidly journalists took up the same belief.

Newspapers and magazines zealously praised science in general and atomic science in particular. After Rutherford and Soddy's discovery of what everyone soon called transmutation, every medium from newspapers to public lecturers exclaimed that it gave scientists a tool that could revolutionize civilization. Radium might be harnessed to illuminate cities, create new metals, and do almost anything else imaginable. By the 1920s, when one or another magazine remarked that a bottle of uranium might propel a steamship across an ocean, it was repeating a tired cliché known to schoolchildren around the world.

An obvious reason that journalists said so much about radium was that it was a fine source for sensational tales that could attract a paying audience. Besides, as the twentieth century wore on, not only Sunday supplement curiosities but more and more of the front page had a connection with science. The trend posed a problem for the rising mass press of the 1920s. It was no longer enough to snip the most striking remarks from an eminent scientist's pronouncements, or ignore them altogether in favor of sensational claims by some pseudoscientist. Specially trained journalists were needed to separate the wheat from the chaff.

Science journalists, a new breed of writers with a solid background of scientific training, rose to the challenge. In their work as reporters these young men (never women) spent many hours in the company of leading scientists, remarkable people whom they admired, and upon whom they relied for their information. These journalists had come to their specialty in the first place because they saw science as a force for progress, and saw themselves as missionaries for the scientific viewpoint. The public must be taught to admire science, for the support of society was needed to sustain the scientific endeavor (and, to be sure, science journalism).

The dean of these writers was Waldemar Kaempffert. In 1927, when he became science editor of the *New York Times*, Kaempffert had already been reporting for a quarter-century in clear, enticing prose on subjects like radium. Over the years he came to resemble the famous professors he reported on, robust and impressive, by turns charming and pompous. Agreeing with his scientist friends, Kaempffert wrote that they had brought amazing benefits to society, and we could look forward to yet greater marvels: rockets to the moon, precooked meals, and towns with "plenty of garden space . . . and a finer outlook on life." In 1934 he wrote, "Probably one building no larger than a small-town post office of our time will contain all the apparatus required to obtain enough atomic energy for the entire United States." With transmutation under control, gold might be a waste by-product of the new industry, used for roofing material.[4]

Buildings roofed with gold? The social forces that brought atomic physics into association with the utopian scientific White City may seem plain enough, but these moved alongside deeper forces of still greater power. The concept of "transmutation" was a doorway through which archaic images would make their way into nuclear energy tales.

Physicists and the press loved to compare atomic scientists with the medieval alchemists who had sought to turn lead into gold. By the 1930s even level-headed Rutherford titled his popularizing book on atomic physics *The Newer Alchemy*. It became a cliché that "the famous problem of the alchemists has been solved." But in reality the problem of the alchemists was a wholly different kind of transmutation, which remained unsolved.[5]

For some 2,000 years men and women from Egypt to China had devoted lifetimes of intense effort to blending chemicals and metals in their furnaces, producing kaleidoscopic displays, grotesque lumps and stenches, glittering crystal formations. Arduously studying these substances day and night, the alchemists were seized by a conviction that the chemical changes pointed to something tremendously significant. In fact they were projecting their most deeply held feelings onto the ambiguous transformations in their crucibles. Alchemy opened a window onto a secret inner landscape of images.[6]

The wiser adepts knew that they were working less with matter than with their own minds. Masters warned their followers not to aim merely at physically changing lead into gold, for such work should be only an aid and a symbol for something infinitely more important. They explained that the true philosophers' stone—the central goal and secret of transmutation— was a spiritual matter. To achieve transmutation meant the perfection of the soul, its transformation from a dull leaden state into the golden state of divine grace and mystical illumination.

Some writers went further, hinting that the true alchemical gold would be found in perfecting the spirit not just of one individual, but of everyone. Transmutation in the fullest sense meant achieving a world of justice, peace, and plenty, a world described in the legends of many peoples and known in Western culture, through no coincidence, as the Golden Age. That realm might be located in the past, like the Garden of Eden, but in Middle Eastern and European tradition the Golden Age might also lie in the future. Sometimes whole populations eagerly awaited the Jewish Messiah or the Mahdi of Islam with their kingdoms of righteousness, and repeatedly in the history of Christendom mass movements proclaimed that the millennium was at hand. When Soddy and others exclaimed that transmutation of the elements could mean a new Eden, a society emerging like a dragonfly from its pupa, they were reconstructing a legend out of ideas that had already been associated with one another centuries earlier.

Less obvious at first to Soddy, though familiar to alchemists, was the fact that transmutation also involved death. The adepts explained that in their crucibles matter must literally die before it could be refashioned into a golden state. In chemical practice this process meant turning substances into a black mass, inflicting a fall into corruption and putrefaction. That was a deliberate metaphor of a mental process that religious mystics insisted upon: humans must make an agonizing descent into darkness and chaos, purging themselves in the "divine furnace," painfully surrendering attachments, breaking ingrained patterns of thinking, before they could win inner peace and spiritual renewal. In the end transmutation became a symbol for the greatest of all human themes: the passage into death and beyond. This was the theme symbolized by the man in my nuclear energy legend when he survived his catastrophe to begin a new and better life. I will show him again in many other guises.[7]

This individual experience of passage through death sometimes expanded to a cosmic scale. People of many times and places believed the world was decaying into social collapse and warfare, but would emerge renewed from the chaos. The nadir might be the world conflagration that ethnographers have recorded in native tales from every continent. In Christian culture countless preachers insisted that humanity must pass through the fires and bloodshed of Armageddon before we could enjoy the millennial Golden Age and the shining City of God.

Alchemists themselves were convinced that the secret of transmutation held boundless perils, both spiritual and mundane. That was why the adepts kept it secret, hiding chemical formulas amid a confusion of arcane symbols to ward off misuse. In short, long before Soddy's day the idea of transmutation was tangled up with ideas about vast hidden forces, cosmic transformations, and apocalyptic perils.

The last of the great alchemists was Isaac Newton, who devoted more years to the occult quest than to his discoveries in physics. After he failed to craft the philosophers' stone the ideas began to unwind from one another, each to go its own way. By the nineteenth century the image of a perfected White City would not remind most people of the alchemist's dark secrets. Likewise taking separate historical paths were ideas about the end of the world, the concept of the transformation of the soul, and the stereotype of the alchemist himself.

Yet in an age of books no idea is altogether forgotten. The whole complex of transmutational symbols was known, for example, to the chemist Berthelot, who took many hours away from his laboratory to peruse antique manuscripts and write a multivolume history of alchemy. Soddy read it around 1900. As a university lecturer he had chosen to teach the history of chemistry, so he read up on alchemy and was fascinated by the imagery. From the moment in 1901 when he told Rutherford that what they had discovered was "transmutation," Soddy deliberately dealt in potent old symbols. In his popular writings throughout the following decade he hinted at tremendous long-hidden secrets, primordial powers of evolution, and cosmic cycles of creation and decay, which he now linked with the modern question of energy supply and with radium.

The meager collection of facts known about atomic energy in the early twentieth century was a blurred picture upon which scientists and their admirers could project their most urgent ideas. Atomic energy was coming to stand for things more important than atomic energy itself.

2

RADIOACTIVE FEARS

W ould the future hold a White City or a desert of ash? Put another way, was the man in the nuclear energy legend good or evil? These questions stood in for a larger one that sophisticated people would debate with increasing passion as the twentieth century advanced: could science be trusted to take civilization in the right direction? At the base of the debate lay divergent beliefs about the role of all expert authority in a technology-based society. That debate is the subject of this chapter, and to some degree of the entire book.

People only slowly came to see technological authority itself as a problem; they had more obvious concerns. One clear danger was new scientific weapons. Would atomic energy be used to make war more frightful? Here, as with other topics, talk of radioactivity had a remarkable propensity to call up primitive images and connect them.

The first well-publicized hint of a link between atomic energy and weapons came from Sir William Crookes, a scientist with a spectacular white goatee and mustachio, well known to the British public. In early 1903 he devised a startling image of atomic energy. Any physics teacher, asked to illustrate a quantity of energy, would be inclined to talk about lifting a weight. Crookes chose to lift the British Navy. The energy locked within one gram of radium, he calculated, could hoist the entire fleet several thousand feet into the air.[1]

The idea developed in a revealing way. From Crookes's scientific discourse the newspapers picked up the naval example as the stuff of headlines, adding helpful engravings of a cluster of battleships suspended in midair. The public quickly caught the image's latent meaning. Soddy, passing through Boston, wrote to Rutherford that people were repeating Crookes's remark that a gram of radium could "blow the British Navy sky high." Scientist, press, and public had together crafted a new thought.[2]

It was not news that scientists could invent weapons. After all, Crookes himself was on the British War Office's Explosives Committee. Typical of public thinking was an 1892 rumor that Thomas Edison was building an electrical device that could annihilate a city from a distance, followed by a newspaper satire about the great inventor destroying England with a push-button "doomsday machine." In 1903 Soddy became the first scientist to explicitly add atomic energy to the roster of possible weapons. The idea promptly became a cliché, with radium missiles adding spice to newspaper Sunday supplements and science-fiction books.[3]

In 1913 a more disturbing atomic weapons tale came from the pen of H. G. Wells—one of the most influential authors of the era, at least among science-minded young men. *The World Set Free*, one of the worst written yet best conceived of Wells's novels, was directly inspired by Soddy's writings, and explicitly dedicated to him. Wells described a world war of the 1950s in which aviators erased entire cities by dropping (coining a phrase) "atomic bombs." Colossal pillars of fire raged while "puffs of luminous, radio-active vapor" drifted downwind, "killing and scorching all they overtook." The near extinction of civilization taught the survivors a lesson. They created a world government that nurtured a brilliant new society, regulated by an elite of virtuous technocrats. At the story's end citizens could travel where they chose in atom-powered aircars, building atom-powered garden cities in deserts and arctic wastes, enjoying liberty and free love. Wells had neatly fitted together fragmentary notions about science and atomic energy to craft the first full-scale scientific legend of atomic Armageddon and millennium.[4]

It was a legend with special meaning for Wells himself. From a wretched beginning as a subservient clerk, where thoughts of destroying society had seemed almost attractive, he had saved himself through an education in science. Wells thought that society should be rearranged so that science-minded men (like himself) could rise still higher, taking charge of affairs with facts and logic.

Wells's book was barely in print when real warfare made the power of science an urgent problem. Soon everyone knew what could come from poison gas research and the like. After the war scientists worked to drive home the lesson, advising nations to spend more money on their laboratories or be defeated in the next war. The thought was taken to an extreme in many pulp-fiction stories. Whereas before 1914 two-thirds of fictional apocalypses had been due to natural causes, after 1914 two-thirds were caused by humans, and of these, three-quarters of the doomsdays came in world wars with scientific weapons. Worry about fantastic catastrophes was also accepted among serious thinkers. Winston Churchill remarked that humanity already had the tools to "pulverize" civilization; Sigmund Freud wrote that science had given us the power to kill one another "to the last man."[5]

What would pulverize civilization in the next war? The most common fear was of attack from the skies—not with atomic bombs, which were expected only in a distant future if at all, but with the weapons already at hand. Most widely influential was an Italian military theorist, General Giulio Douhet, who claimed that victory would come within weeks to whichever side struck with its bombers quickest and hardest. "The bomber will always get through," Prime Minister Stanley Baldwin told a dismayed House of Commons in 1932. "The only defense is in offense, which means that you have to kill more women and children more quickly than the enemy, if you wish to save yourselves."[6]

Many ideas that became central to debates over nuclear weapons in later decades got a trial run during the 1930s. For example, the "deterrent": a nation should develop bomber fleets and poison gas so that its enemies would fear to start a war. Science-fiction stories from the 1920s onward applied the same idea to atomic bombs. In both atomic fiction and sober strategic thinking, a related idea was the preemptive "knockout blow" (what nuclear theorists of the next generation would call a "first strike"): one side might be able to annihilate the other's forces by surprise the day the war started. It followed that a large bomber force might be a dangerous provocation that could trigger an attack simply out of fear of being attacked first. Therefore nations pursued disarmament negotiations, hoping to establish an international inspection system for bombers. The loudest public debate was over "civil defense." Some said that providing everyone with gas masks and bomb shelters would dissuade an enemy from launching an attack; others declared defense against air raids to be hopeless. Especially cherished was the idea that war had already become too horrible to consider in practice. General Douhet argued that citizens would force their leaders to bring any war to a halt. As soon as bombing began white flags would spring up over the cities, fluttering frantically.[7]

It was no new idea that terror of modern weapons would bring an age of peace. The prospect of atomic weapons in particular inspired as much hope as fear. "When we have discovered the secret of the atom," a journalist wrote in 1921, "it is likely that all nations will be ready and willing to lay down their arms and abolish their armies and navies." Belief in the virtues of science and technology could be so strong that even a threat of destruction might sound like a promise of peace.[8]

———

The secrets of the atom seemed to open possibilities on a cosmic scale. Within weeks after the idea of atomic weapons first excited the public, atomic energy was linked with a vastly greater threat: the end of the world.

When Soddy first told the public his ideas about atomic energy, in May 1903, he added that our planet is "a storehouse stuffed with explosives, inconceivably more powerful than any we know of, and possibly only awaiting a suitable detonator to cause the earth to revert to chaos." Here was an entirely new idea, that science might allow a person to deliberately destroy the world. A colleague of Rutherford noted in a magazine article that the scientist had "playfully" mentioned that "an explosive wave of atomic disintegration might be started through all matter which would transmute the whole mass of the globe into helium or similar gases." Rutherford later pointed out that such a disaster was not credible. Over billions of years, radioactive atoms had been randomly trying every conceivable reaction, so that if the planet were unstable it would long since have disintegrated. Yet Rutherford's little joke that "some fool in a laboratory might blow up the universe" slipped into the public consciousness with remarkable ease. One reason was that the image reflected rising anxieties about the entire technological future.[9]

It was in the early nineteenth century that ideas about the end of the world first began to separate from their ancient mythical and religious contexts, joining up instead with science. A first tentative step came in 1805, when a fantastic novel, *The Last Man*, was published in France. The author still referred to the Judgment Day of Christian tradition, but the bulk of his story showed the human race dwindling to a lonely end through natural processes. The author was himself a desperately lonely man—late one winter night after finishing his book, he drowned himself. His unhappy image of a Last Man survived him, inspiring dozens of poems, paintings, and stories during the next decades. The most substantial of these was *The Last Man* novel that Mary Shelley published in 1826. According to her biographers, this author, too, had reasons for writing about desolation, for she had grown up motherless, and before writing this novel she had lost her husband and other good friends through tragic early deaths.[10]

Meanwhile another traditional element in doomsday, the idea of future social breakdown, also separated from the religious context and forged ahead on its own toward a linkup with science. If technology was driving back natural perils such as plague and famine, it was also overturning accustomed social relationships. Some people began to feel that the chief danger to society came from headlong social change itself. Opposing the general self-confidence of the late nineteenth century, a few prominent authors predicted that insurrections and wars would bring down civilization. The most outspoken writers felt personally alienated from society, but for whatever reason, they vividly presented an image of doom brought by the advance of technology.

Soddy's atomic nightmare differed from all these catastrophes in one essential way. The exhaustion of resources or a rising tide of insurrection involved forces beyond the control of the individual. But atomic energy might bring disaster through the carelessness or malice of a single man.

The idea of an individual meddling with dangerous powers was far from new. The popular stereotype had an ancestry reaching back through medieval legends of the wizard or magician to prehistoric shamans. Such an individual might release pestilence, demons, and other evils, as witches did, or might simply propagate heretical ideas, as some protoscientists in fact did. People have always feared that an overweening individual, through evil thoughts or magical acts, might contaminate neighbors or the entire community.

This idea seemed to find confirmation in the lives of real people. Most notorious was Dr. Johann Faust, a sixteenth-century itinerant rakehell who boasted that he controlled demonic forces. Pious churchmen wrote pamphlets that combined Faust's actual career with legends of black magicians to craft a warning. Beware the rising breed of rationalist, humanist skeptics who rejected traditional Christian authority! Through the nineteenth century Faustian stories proliferated. Popular authors from newspaper hacks to Nathaniel Hawthorne modified the old tales of witch or sorcerer to create a new fictional figure: a man who combined tremendous pseudoscientific powers with stunted morality, endangering himself and those around him.

A few writers suggested that such a figure might desire to end the world personally, but the early authors did not suggest that such a deed could really be possible. As technology altered entire landscapes, sometimes for the worse, opinions began to shift. The change from a sorcerer contaminating his neighborhood to a technical expert who threatened all the world could be seen as early as 1817 in Mary Shelley's *Frankenstein*. The scientist was forced by his monstrous creation to prepare an artificial bride, but at the last moment he hesitated. Would the two fiends mate, the scientist wondered, and produce "a race of devils," polluting the earth and perhaps exterminating humanity?[11]

The emerging stereotype of the dangerous scientist-inventor was associated with surface characteristics of the legendary alchemist and wizard, such as a gloomy laboratory with esoteric instruments and glassware, and also with more essential features. Like the traditional reclusive who hunted forbidden secrets, he (never "she") was remote from everyday life; he was not swayed by women's love; he seized powers over life and death; he boasted an impious pride; and he tended to die an uncanny and violent death. I will come later to the meanings underlying each of these standard features.[12]

Jules Verne, whose writings were easily as influential as Wells's, did more than anyone to develop the new stereotype. A typical Verne novel told of a mad chemist who had invented a mighty explosive that he used to blow up, if not the planet, at least his island hideout and himself. The premises seemed plausible; Verne had apparently based his fictional character on an actual inventor of explosives, who sued for libel.[13]

Soddy and Rutherford's concept of a planetary chain reaction provided the first superficially scientific description of how a person might in fact destroy the entire world. Leading scientists passed along the idea. For example, a well-known textbook on atoms asked whether the exploding stars (novas) occasionally seen in the sky might be outbursts of atomic energy "brought about perhaps by the 'super-wisdom' of the unlucky inhabitants themselves." By the 1930s even schoolchildren had heard about the risk of a runaway atomic experiment destroying the planet.[14]

Still more chilling was the thought that someone might cause some kind of cataclysm on purpose. After all, bomb-throwing anarchists had caused notorious damage since the late nineteenth century. The public connected their powerful new explosives with science. Thus scientists were coming to be personally linked with the old idea that civilization would perish in social upheaval. More than one novel featured the cold-blooded Professor, a chemical expert who plotted to blast society apart, to make "a world like a shambles" with his infernal devices. In a popular French novel of 1908, physics-minded terrorists destroyed entire cities with pocket-sized atomic explosives.[15]

The legendary dangerous scientist was usually shown as more or less mad. An especially convincing portrayal was in *Wings over Europe*, a play that enjoyed a modest run on the stage in London and New York beginning in 1928, and frequent revivals among college theater groups during the 1930s. Center stage was held by Francis Lightfoot, a youthful scientist, brilliant but reclusive and a bit unbalanced. Lightfoot announced that he had found the secret of releasing atomic energy. Staring into space, he exclaimed hoarsely that he could bestow the power to make a Golden Age, "the power of . . . a god, to slay and to make alive." Losing his bearings, Lightfoot decided the world was wicked and made ready to blow it to dust.[16]

To be sure, the great majority of writings showed scientists in a less lurid light. Sometimes they were heroes like Pasteur (or a heroine, Marie Curie) working for the good of humanity. But throughout the nineteenth and the early twentieth centuries the most common stereotype showed the scientist as simply harmless—well-meaning, unworldly, and absentminded. For example, in 1908 a Sunday newspaper comic strip showed a "Doctor Radium" who was a scrawny, nearsighted old pedant, serving as the butt of small boys' pranks while he mumbled scientific gibberish to himself.

Over the years the scientist began to seem less of a joke. A 1941 *Batman* comic book featured a very different Professor Radium, a robust genius who meant to work wonders with radioactivity, but accidentally turned himself into a murderously insane "human radium ray." This was no isolated case. Pressure from somewhere was gradually altering the public image of scientists, and by implication the image of the radioactivity they worked with.

While some felt new misgivings that science was making wars worse and weakening social cohesion, many others insisted that science was not the problem but the solution. After all, scientists themselves had already formed their own international community—peaceful, cooperative, and dedicated to reason. "Religion may preach the brotherhood of man," said Kaempffert, but "Science practices it."[17] Since the 1890s juvenile fiction had swarmed with technological heroes like Frank Reade Jr., who rode in on his gigantic electric tricycle to break up the Nigerian slave trade. The play *Wings over Europe* went further. At the end a group of Wellsian virtuous technocrats who had learned the secret of atomic energy came soaring over the world's cities in huge green airplanes, bearing atomic bombs, to enforce the will of a "League of United Scientists of the World."

A good many young scientists and science journalists, mostly on the political left, agreed that the proper way to reshape society was to give a greater role to scientifically trained people (that is, to people like themselves). An example was J. B. S. Haldane, a fiery Marxist biologist who threw himself into writing for the public. He insisted that science was languishing under the thumb of capitalists, who used technology only to enrich themselves. Let affairs instead be efficiently organized by science-minded people, and progress would be amazingly swift; atomic energy was only one of many wonders he promised.

Applying science through rational planning implied a particular type of Golden Age. The gleaming cities of the planners, ranging from Plato's authoritarian Republic to the industrial society projected by Marxists, would fit individuals into their roles as harmoniously as the components of a machine. But many visionaries over the centuries had imagined an opposite sort of paradise—a rural Arcadia where couples would lie like flowers in the gentle breeze, blending into nature, ignorant of technology and social authority. Some writers, like Kaempffert, hoped for a compromise: a middle state between untouched nature and orderly civilization, where electrically powered villages stood amid landscapes gardened with the help of machinery. But if one had to choose between the two extremes, people like Kaempffert would choose the benefits of order and technology. "We may cry 'back to nature' as loudly as we please," he wrote, "but the scientist within us answers, 'forward to the laboratory and the machine.'"[18]

Not everyone agreed that science should set the pace. In the 1920s and early 1930s, popular German writers and essayists in American literary magazines declared that because science by definition had nothing to do with such ideas as beauty and holiness, it was destroying traditional values. Humanists recalled the spiritual horrors of the First World War, in which poison gas and artillery bombardment had rendered the old knightly virtues useless. A leader of the French Senate complained that technological breakthroughs had upset industry and brought on the Great Depression. A prominent British bishop suggested that there should be a ten-year moratorium on research to give society time to adjust.[19]

The problem with scientists was that their role in society was anomalous. Most of them were either poorly paid professors or underlings in a corner of some industrial organization, yet their discoveries were radically transforming civilization. Thus they had an incongruous sort of power over the future, nothing like the power normally granted to a business leader or government official. A few outspoken men like Haldane proclaimed that this situation must be rectified by honoring scientists and engineers above others, reorganizing society to conform to scientific ideals. The people who were used to shaping society's ideals—clergymen, humanists, popular writers, and the like—could well feel that their world was threatened by something from outside the established order of things. These were the groups that in fact helped to spread talk about blasphemous scientists.

The mechanism resembled one that was later described by anthropologists. In many societies, anyone who upset other people might be accused of attacking them with secret magic. These accused witches or sorcerers or heretics were not the decrepit hags of modern children's books, but sharp-tongued men and women, often taking a social or moral role that did not fit the established pattern. It was fitting to accuse incongruous "moral entrepreneurs" by saying that they wielded incongruous powers. And in practical terms, the attack could put them in their place. I believe that worries about the dangers from scientists, alongside whatever rational arguments might be presented, similarly reflected uneasiness about their disquieting social position.[20]

Atomic energy was especially apt for evoking anxiety about science; had not scientists themselves proclaimed it the most mysterious and potent of powers? Thus atomic energy became particularly closely associated with all the uneasiness that people felt over matters of science and technology.

Typical of the critics who invoked the atom was Raymond B. Fosdick, an idealistic American lawyer. Already deeply impressed by the destructive forces revealed in the First World War, Fosdick happened to meet Soddy, who warned him about the future of science in general and atomic weapons in particular. In 1928 Fosdick repeated the warnings in a widely read book, *The Old Savage in the New Civilization*. He declared that humanity, like a savage or a child playing with matches, could scarcely cope with the powers of technology. We needed to give less heed to scientists and more to humanists who would boldly question society (that is, to people like himself).[21]

Fosdick's challenge was answered by Robert Millikan, the leading spokesman of American science, famous for his work on atomic particles. When Millikan was photographed alongside his friend Herbert Hoover it was Millikan, with his handsome features and silver hair, who looked more like a president. Although his political views were at the opposite pole from Haldane's Marxism, Millikan was no less insistent that science was the key to progress. In a 1930 magazine article he scoffed at the critics who talked about diabolical scientists tinkering, "like a bad small boy," with atomic energy and blowing the earth to bits. On the contrary, studying whether atomic transmutation could be controlled would be worth spending a billion dollars (some of it, presumably, in his own laboratory). The by-products of the research by themselves would repay the expense— "And if it succeeded, a new world for man!"[22]

If sensational claims for atomic energy made people nervous, Millikan was more than willing to tone them down. Atomic apocalypse was a mere hobgoblin for the ignorant, he insisted. Still more widely heard in the 1930s were Rutherford's denials. He was more famous and self-confident than ever, for he had proved that the source of all radioactivity is not the atom as a whole but its nucleus, that tiny kernel buried amidst the cloud of electrons that fills most of an atom's volume. (Thus radioactivity did not really represent "atomic" energy but "nuclear" energy, although most people used the terms interchangeably.) Newspapers paid close attention when Rutherford explained that "Anyone who says that . . . with our present knowledge we can utilize atomic energy is talking moonshine."[23]

Skeptical pronouncements by eminent physicists did not stop the colorful newspaper stories. Kaempffert pointed out that Rutherford and Millikan were speaking only of the known techniques of physics, and that other methods might be invented one day. Rutherford and Millikan themselves, when they addressed not the public but fellow physicists, hinted that the release of nuclear energy was conceivable. Many other prominent physicists were openly optimistic, telling the public that an astonishingly beneficent atomic revolution might begin at any time.

Science journalists, too, and with them the public, paid less attention to possible dangers from science than to expected benefits. Corporate and government officials likewise promised technological progress. The public, these pronouncements implied, need only wait, passively and patiently, for authorities and experts to achieve the inevitable progress. Advertisements for "scientifically" improved products never ceased to present an image of a future world that delighted in advanced technology. C. P. Steinmetz of the General Electric Company suggested that electricity would become "so cheap that it is not going to pay to meter it."[24] The whole set of futuristic images was accepted by educated elites around the world, perhaps most of all in the Soviet Union (the writings of Soddy and H. G. Wells were translated into Russian and widely read). Occasionally a Russian author would describe the atom-powered Soviet White City of the future. Communist thinkers spoke frankly about the role of such propaganda: it mobilized the populace to sacrifice themselves in fulfilling the leadership's plans for industrial development.

Ceaseless technological advance was structurally embedded in the cultural, economic, and political institutions of modern society. The promise that technology would improve their followers' lives was indispensable to democratic politicians, Fascist and Communist dictators, labor union leaders, advertisers, and the writers of corporate reports to stockholders. The critics of technology, a halfhearted minority of humanities professors, clerics, and authors, swayed no other social groups to their concerns. By the late 1930s, the ripple of criticism had subsided. The evil scientist lived on only in lurid fiction. From the capitalists with their Millikans to the Marxists with their Haldanes there was only one ideology: Progress through Science.

3

RADIUM: ELIXIR OR POISON?

If there had been no thought of ever releasing energy wholesale from radioactive atoms, radioactivity would still have impressed the public as a symbol of the powers of science. For the rays could transform living flesh, for better and for worse. When Pierre Curie carried a bit of radium around in a pocket, it burned his skin. Curie killed a mouse with a dab of the element, and announced that in the hands of criminals, radium could be a great danger. But he and other scientists emphasized that in the hands of experts who took proper precautions, the prodigious power of radioactivity would be all to the good.

Already in 1903 Soddy suggested that people with tuberculosis, the most dreaded disease of the time, might benefit from inhaling radioactive gas. Physicians raced to try out such ideas. Radium did prove helpful in reducing certain skin cancers and tumors, for if the rays were harmful to ordinary flesh, they were far more harmful to some types of cancer cells. Newspapers quoted doctors who speculated that radium would cure every type of cancer and many other diseases. Meanwhile Pierre Curie and other scientists, studying the waters in health spas, found that the springs carried mild radioactivity up from the deep rocks. Did the rays somehow stimulate the body toward health?

In the absence of solid knowledge, fantasy had free play. Newspapers proclaimed: "Old Age May Be Stayed by Radium." A young scientist mixed sterile bouillon with radium in a test tube and created curious shapes—had radioactivity created altogether new life? Journalists reported that radioactivity might reflect a vitality stirring within every atom of matter itself, the very "Secret of Life." A chemist who did some research with Soddy on transmutation received a number of letters from spiritualists who declared that radioactivity was a supernatural agent, and solemnly warned scientists to keep their hands off this mystic force. Such extremes of enthusiasm faded as laboratories began to turn out solid facts. But even in the

1920s Millikan could propose that all matter was in a sense alive, continually at birth or in decay.[1]

In plain fact, of course, a rock is no more alive than ever if it is radioactive; nuclear physics contains no secret of immortality. Then where did these extraordinary ideas come from? Soddy gave a clue that can lead us through the maze. The philosophers' stone, he pointed out, "was accredited the power not only of transmuting the metals, but of acting *as the elixir of life.*"[2]

Ideas about physical forces had always been inseparable from ideas about life. Chinese, Middle Eastern, and European alchemists all elaborated traditional theories about a fundamental life-force: an occult power, perhaps like a fire or a fluid, underlying every process from the growth of plants to chemical change. By manipulating the life-force in their furnaces, the adepts hoped not just to make gold but to transmute human flesh itself into incorruptible perfection; they would achieve both spiritual rebirth and bodily immortality.

When alchemists said that gold would grow in their laboratories, they meant it literally. Some deliberately shaped their sealed vessels in the form of an egg or a womb, believing their transmutational work paralleled events within a mother's body. And just as human life begins with a union of man and woman, so to move matter toward rebirth the adepts aimed to bring about a marriage of "male" and "female" chemical substances. Manuscripts illustrated this concept with frank drawings of a naked couple in coitus. The life-force could be a specifically sexual energy.

Wiser philosophers used the marriage of chemicals as another symbol of processes leading to spiritual transformation, often within a religious context. From early times into the twentieth century, in religions from Catholicism to Tantric Buddhism, a "mystical marriage" symbolized spiritual renewal. The symbol evoked grand associations extending to cycles of cosmic creation. Long before Soddy and Millikan connected radioactive transmutation with the primeval evolution of matter, some alchemists had imagined that in their retorts they were reenacting the destruction and rebirth of the world.

Ethnographers have found among peoples on every continent the myth of an ancestral couple who survived world catastrophe to engender a new cycle of humanity. In modern Europe the theme remained visible in such places as *The Last Man* novel of 1805, in which the lonely protagonist fruit-lessly sought a mate. In a pioneering 1901 novel, after years of wandering a poisoned world the Last Man did find a Last Woman. The couple became the Adam and Eve of a new and better race, a triumph of the sexualized life-force.[3]

The cosmic life-force is a remarkably widespread feature of human culture and probably of human psychology. Anthropologists found that most native peoples tended toward animism, seeing purposeful life in everything that moved or changed. Jean Piaget discovered that in modern Europe, too, young children spontaneously tended to treat all things as somehow filled with living desires. More recent scholars speculate that human brains have evolved to be watchful for intentions: a useful trick not only in dealing with other humans, but in hunting animals and even in finding plants that "prefer" to grow in particular conditions. Projecting a concept of life into a shape in the dark could save you if the shape is a lion, and does little harm if it is a bush.

Children, questioned with care, said that they felt surrounded by a world of mysterious and sometimes frightening forces. To master their fears they fantasized about wielding magical forces themselves, like a fairy-tale hero. And after they won through the dangers, "they married and lived happily ever after": the philosophers' stone in its most primitive form. After all, children were especially fascinated by that genuine source of life, the parents' sexuality and the mysteries of birth. Curiosity about such matters was a childish anticipation of more mature aspirations to understand life, but the concepts that emerged in childhood tended to persist. A wide-ranging symbolism of procreative life-force was unequivocally observed in many cultures, shaping images that people projected onto all sorts of things.

In summary, the concept of life-force connected with speculations about the cosmic order, questions of procreation, and fantasies about magical powers capable of anything up to bodily immortality and world rebirth. How did these associations become bound up with ideas about nuclear energy? There were many strands in the tangle, but one was particularly strong: radiation.

Primitive peoples naturally connected rays of sunlight with the growth of crops, and hence with procreation. Sometimes light rays took an explicitly sexual form, as in folktales in which sunlight brought conception (the many variants of the Danaë myth that persisted in Europe into modern times, along with tales found in Siberia, Mexico, Fiji, and so forth). In Western culture from ancient Egypt onward, people saw light as a general symbol of life-energy. Respected medieval philosophers said that sunbeams, as well as the invisible rays that carried down astrological influences from the stars, conveyed occult forces. Light also seemed close to the divine "illumination" of ecstatic mystics—not to mention the radiant golden haloes of painted saints. Light was even central to the Bible's description of God's original Creation (Genesis 1:3; John 1:7–9). As alchemists and churchmen contemplated light, cosmic energies and cycles opened before their dazzled eyes.

Up to the time of the scientific revolution, the tangle of images associated with transmutation, life-force, and rays formed a coherent cluster; leading thinkers like Newton swallowed it whole. Later scientists struggled to sort out the true agents of physical change, for example, electricity. But the old associations were hard to shed. Into the nineteenth century people speculated that electrical forces could energize mineral slime into primitive life, and sick people flocked to quacks like the Englishman James Graham, in whose electrified bed couples were guaranteed to beget children.

During the nineteenth century, physicists defined "energy" in precise mathematical terms, but ordinary people continued to use the same term for spiritual energy, sexual energy, and so forth. The ancient and sacred symbol of thunderbolts now streamed out, raylike, in advertisements for electrical appliances, while quacks sold tens of thousands of battery-powered "electrical belts" guaranteed to cure everything from impotence to old age. Scientists might keep patiently working to unwind the strands of imagery, but each strand kept twining back around the others.

By 1910 the strands had knitted firmly together around ideas of radioactivity. The ancient transmutational imagery had included visions of primeval light and the creation of the world; as if in confirmation, excited physicists pointed out that for countless ages radium had been emitting light while transmuting, as though it contained the energies of the original creation. When doctors reported that atomic rays could heal, they seemed only to confirm that radioactivity meant life-force. Some people even suspected that, as a headline proclaimed, "Secret of Sex Found in Radium."

If electricity was becoming a humdrum household matter, then the mysteries once associated with it could be transferred to radioactivity, where they seemed to fit even better. A 1935 movie serial brought cowboy actor Gene Autry back from the dead in a "radium reviving room," although the apparatus also crackled with traditional electric sparks.[4]

A power great enough to transfigure flesh could also destroy it: the death ray. Like other radiation themes, this idea had a long history of emotive associations. The associations would one day appear in controversies over bomb tests and nuclear reactors, but they were already plain in ancient tales.

In their earliest form, malevolent forces were simply thrown from a distance. Native peoples around the world believed firmly in witches who caused disease by projecting some sort of magical contamination into their enemies. Nature, too, might project harmful influences; for example, European physicians well into the eighteenth century believed that people could be literally struck ill by astrological forces, sometimes depicted as explicit rays from an evil star.

The psychological meaning of radiating influences is plainest in the image of willpower sent out through the eyes. Pictures of a powerful, radiating eye can be found everywhere from ancient Egyptian papyri to the drawings of modern schizophrenics, and in Irish and Persian myths a divine eye could even hurl thunderbolts. Talismans against the "evil eye" remain common in many cultures. The ancient concept that eyes can emit a sort of luminous beam survived in modern figures of speech, as when an 1890s political leader was said to have a gaze as "penetrating" as X-rays. In all these cases the radiated power was associated with a dangerously strong person. After all, animals and people do tend to submit when they face a dominating stare.[5]

Stern eyes could also seem threatening when they saw "into things," for example, into a child's misdeeds or forbidden sexual matters. The discovery of X-rays called forth nervous jokes about people using gadgets to peep through closed doors; a brash London firm offered X-ray-proof underclothes for sale. Decades later, a psychologist reported the case of a male X-ray technician who had chosen the profession because, as he learned under analysis, he had a secret wish to see within his mother's body. A symmetrically opposite case was a female psychology student who fainted when being X-rayed, fearful of being "seen through." Thus baneful penetration and discovery of secrets were added to the many powers latent in the symbol of rays.

The whole cluster of symbols was most openly displayed in minds not restrained by logic. The insane of medieval times raved about demons, but by the early twentieth century well-read paranoiacs were insisting that machines projecting electricity or X-rays or radioactivity were injecting uncanny feelings and thoughts. Rays carrying sexual power or willpower, emanating, for example, from a "radium tube," were ruining the entire world or healing it.[6]

Ideas about rays that came up spontaneously in minds operating outside normal limits, ideas that could be traced far back in history, ideas that were common folklore among peoples around the world, whether primitive or civilized—such ideas must not be taken lightly. During the nineteenth century, death-ray imagery had been generally scattered and inconspicuous, part of the anonymous noisy background of popular culture. The discovery that radium rays could harm or heal flesh began to weave the diverse images together again.

Biological and medical speculations about radiation were joined by ideas about physical ray weapons, as when newspapers speculated that radium rays could smash battleships. Magazine fiction followed, such as a 1908 story by Jack London about a genius in an island hideout who used ray powers to kill people around the world. Unlike talk of atomic bombs or talk of medical applications, these death-ray stories had not a shred of scientific fact behind them, no relation to any of radium's actual properties. The stories were descended entirely from older myths.

The Romans had a legend about Archimedes blasting an invading fleet with giant mirrors that concentrated the rays of the sun, and talk of a philosopher's burning-glass destroying armies or cities had continued down the centuries. H. G. Wells modernized the ray weapon in his 1898 novel *The War of the Worlds*, in which Martian invaders blasted soldiers with beams of heat. As soon as Roentgen announced his discovery of X-rays, some people wrote him to express fear of his "death's rays." While most scientists scoffed, the press kept the pot boiling for decades by quoting a few respectable scientists and inventors who declared that ray weapons really could be devised.[7]

The pulp science-fiction magazines that became popular among boys and young men in the United States from the 1920s on, along with radio shows and comic books like the famous Buck Rogers stories, could hardly have made it through an issue without rays. Atomic or "radium rays" would incinerate enemies, transform living creatures, and enslave minds. The ancient heritage of such images was transparent in a 1940 Captain Marvel comic (made into a movie) in which the villain used rays not only to attack people but to make gold; as he rightly said, these were "powers such as men have dreamed of since the beginning of time." More than half a dozen movies about ray weapons appeared before 1930 and at least two dozen more between 1930 and 1940.[8]

Roughly half of these movies used the rays as a defense against airplanes, the most dreaded menace of the 1930s. Typical was the 1940 American film *Murder in the Air,* featuring the actor Ronald Reagan as an agent who guarded the secret of a projector that could bring down enemy bombers— "the greatest force for world peace ever discovered." The idea seemed so plausible that in the mid-1930s the British government assembled a secret panel of top scientists to study air defense, with ray weapons at the top of their agenda. The British committee concluded that they could not build a death ray—and went on to develop radar, a true defense ray device.[9]

By the 1930s many solid facts were known about atomic radiation. Educated people could shrug aside the fantasies about rays that struck people like thunderbolts or brought them back to life. But as verified research accumulated, it showed a surprising ability to keep reminding people of primitive life-force imagery.

At the turn of the twentieth century use of the new X-rays had brought a new malady: radiation injury. The victims felt weak, suffered vomiting and diarrhea, perhaps temporarily lost patches of hair. Months or years after the exposure to X-rays, cancers developed. Equally unsettling to the public was the news that men whose genitals were heavily X-rayed could be left sterile for weeks or longer. But public concern over X-rays was muted, for almost all the injured people were physicians or technicians. It was an honorable tradition that when a new therapy was plied against deadly disease, one would have to accept a certain number of accidental deaths, especially among the physicians themselves.[10]

When radium came along a few years after X-rays, people benefitted from the earlier experience. Although some doctors and technicians who used radium came down with radiation injury and cancer, the tragedies were a small price to pay for the benefits. Physicians devised ways to use radioactive substances to destroy various types of tumors without risky surgery; during the 1930s more than 100,000 people around the world were treated with radium every year, mostly to their advantage. Meanwhile doctors experimented by poking X-ray projectors and radioactive substances into every part of their patients' bodies. (Curiously, these men took a special interest in irradiating the vagina and uterus: gynecological experiments led the way in both X-ray and radium therapy.)

The greatest transmutational wonder of all was announced in 1927 by Hermann Muller. An independent and combative biologist, Muller had been convinced from his boyhood that humanity must seize control over its own evolution, advancing toward ever-better social organization and even improving our species' biological form. When Muller bombarded flies with heavy doses of X-rays, he found that the more X-rays were delivered, the more mutations showed up among later generations of flies. Mutations, such as oddly colored eyes or shrunken wings, were nothing new to fly breeders. They suspected (correctly, as it turned out) that mutant creatures could be manufactured on demand by exposing the parents or even the grandparents to certain common chemicals, or even by slightly raising their temperature. But Muller was the first to alter genes deliberately. It was particularly exciting that he did this with radiation. Consciously intending to make a connection with nuclear physics, he called his discovery the "Artificial Transmutation of the Gene." Muller attracted wide attention when he suggested that his discovery could help scientists guide the evolution of improved plants and animals, not to mention people.[11]

Making it all seem more plausible, scientists pointed out a genuine connection between radiation and evolution. The world has always been bathed in natural radiation, coming down from outer space in cosmic rays, coming up from the traces of uranium and other radioactive elements that are present in ordinary rocks, even coming from the radioactive minerals that all creatures carry within their bones. After Muller's announcement scientists realized that this radiation, along with natural chemicals and other sources of mutation, had provided evolution with its raw materials by altering genes at random. Writers for pulp science-fiction magazines imagined that a scientist who could concentrate cosmic rays might compress millions of years of evolution into hours, crafting a monster or a superman.

In every case, real or imaginary, the hopes for benefits far outran any worries. The most famous of radioactivity workers, Marie Curie, refused to take precautions that might slow down her work, and regularly got radioactive matter into her mouth and lungs. She refused to believe that her wonderful radium, which was saving so many lives, was killing her by inches. The public was little wiser.

A 1929 European pharmacopoeia listed eighty patent medicines whose active ingredients were radioactive. You could take radioactivity in a tablet, bath salts, liniment, inhalation, injection, or suppository. You could eat mildly radioactive chocolate candies and then brush your teeth with radioactive toothpaste. Manufacturers promised that their nostrums would give relief from any number of ailments, notably rheumatism, baldness, and "symptoms of old age," as if radioactivity were a genuine elixir of youth.[12] Even after a few people died from regularly drinking a radioactive "tonic," the public did not become fearful of radioactivity itself. Radium was only one among many kinds of patent medicines, some of them far more deadly. When journalists warned against such nostrums they took note that, in the hands of competent physicians, radiation remained a great force for health.

Radium had another large-scale use, for it could make certain ordinary chemicals fluoresce. The public was delighted with luminous paint, and everything from children's toys to doorknobs began to glow in the dark. At the center of the industry, in New Jersey, hundreds of young women applied delicate lines to watch dials, tipping the points of their brushes on their lips, unaware that they were ingesting radium from the paint. Radium that is eaten behaves much like a chemically similar mineral, calcium, tending to settle in the bones. Local dentists began to notice an epidemic of problems that they called "jaw rot." In 1925 radium was suggested as the culprit. Newspapers around the world exposed the young workers' travails in mawkish stories about the "Case of the Five Women Doomed to Die."[13]

To the ordinary reader this was just another industrial tragedy, and not the most striking one. For example, the world had heard a few years earlier about the horrible death throes of petroleum industry workers poisoned by the ubiquitous gasoline additive tetraethyl lead. By comparison, radium poisoning was not so much a problem as a curiosity.

During the 1930s nuclear physics research became more closely connected than ever with medical marvels. The medium was that universal connector of ideas, transmutation. Devices like Ernest Lawrence's cyclotrons could transmute many kinds of atoms by bombarding them with particles, and the changed atoms would often be artificially radioactive.

By the late 1930s physicists could serve up radioactive forms of sodium, iodine, and numerous other elements, which promised to surpass radium in treating cancer. To impress an audience, a scientist would make a "cocktail" of radioactive sodium that a volunteer or the lecturer himself would drink. A few minutes after drinking the elixir, the subject would wave his hand in front of a Geiger counter. The counter would chatter, and the audience would applaud.

Just how far did approval of radioactivity go? To get some facts, I looked over the titles of articles in American magazines listed in the *Readers' Guide to Periodical Literature* under such rubrics as "Radioactivity" and "Atomic Energy." Titles that would awaken little emotion, or that balanced positive and negative feelings, made up three-quarters of the total published between 1900 and 1940. Some of the remaining titles, such as "Curing Cancer with Radium," would evoke mainly positive, optimistic feelings; others were likely to evoke negative or even fearful feelings: "Radium Poisoning" and so forth. In articles from 1900 until the mid-1920s I found very little negative language. In the late 1920s the news of radium poisoning brought overt fears, yet positive titles still outnumbered the negative titles by nearly two to one. By the mid-1930s radium hazards were no longer news, and by the end of the decade there were again almost no anxious titles among the many hopeful ones.

Later in the century, when radioactivity frightened many citizens, experts suggested that it was feared particularly because people could not see it or feel it, or because it brought up instinctive concerns about sexual reproduction and deformed babies, or because it could cause the especially dreaded disease of cancer. However, the public in the 1930s already knew these features of radioactivity, yet showed no special anxiety about them. After all, there were other agents such as X-rays, common chemicals, and viruses that also could not be seen or felt and were known to cause sterility, birth defects, and cancer. In the 1930s anxiety about radioactivity existed on the same moderate level as anxiety about those other things.

Yet already there was a special anxiety connected with radioactive life-forces. It was separate from ordinary talk of health problems and scarcely appeared in nonfiction magazine titles. One could detect it only by looking closely into imagery, as a psychologist might notice a minute twitching in a seemingly confident patient with a hidden neurosis.

4

THE SECRET, THE MASTER, AND THE MONSTER

Now I come to the core of the nuclear energy tales that proliferated in the first half of the twentieth century. Almost every story had a tremendous forbidden secret; a powerful authority who mastered the secret; and a device, often personified in a robot or monster, whereby the secret power caused harm. By the 1930s this cluster of themes had a remarkable hold on the public's imagination.

It started with the secret. Dangerous secrets were important in native traditions around the world; Western culture had Adam and Eve, Prometheus, Lot's wife, Bluebeard's wife, the Sorcerer's Apprentice, and a thousand more who came to grief by grasping after forbidden knowledge. The Christian apocalypse itself would begin when the book of seven seals, which no man is worthy to open, was unsealed (Revelation 5:1–4). In this tradition the medieval Catholic church distrusted alchemy, thinking it sinful to pry too far into God's mysteries, while the alchemists themselves wrote endlessly about the perils they concealed.

More modern researchers had their own reasons to talk of mysteries. Typical was a radio talk by Ernest Rutherford in which he called himself an heir to the quest for the philosophers' stone. Like the medieval adepts, he denied that he sought mere mundane gold; his real interest was the "search into one of the deepest secrets of Nature." Science journalists repeated the theme, as in an article by Waldemar Kaempffert titled "Ultimate Truths Sought in the Atom." By boasting that research was close to unveiling cosmic mysteries, scientists and their admirers frankly hoped to increase public support for science. After all, humans are curious and honor those who can answer questions.[1]

But raw curiosity was only one of the psychological forces that talk of secrets might awaken. Fear of the unknown is an instinct in all creatures, a valuable aid to survival. The process is especially important in childhood, when many specific fears are learned. Robert Millikan had been on target when he said critics were accusing scientists of transgressing boundaries like a "bad small boy." Everyone knew from an early age that a secret is by its very nature a thing forbidden. It was presumably forbidden because it was dangerous to know. And if you peeked against the rules you risked direct punishment. In Western culture, children who probed into the mysteries of sex and birth in particular often encountered a boundary to their curiosity like a jolting electric fence. Twentieth-century psychologists noticed that the secret of the womb or of coitus was often the most fearful secret of all.

The theme of the cosmic secret was inseparable from another theme: an assault on the secret in a search for mastery. Physicists inadvertently fitted themselves into this pattern when they said they planned to attack or disintegrate atoms in order to unravel their mysteries. Such statements sounded much like the central work of alchemical transmutation, which also involved breaking down substances before reconstructing them.

The alchemists had specifically symbolized matter as female (the word "matter" itself derives from the same root as *mater*, mother); thus the basic alchemical operation of breaking down matter could sound like a symbolic assault on something female. Ever since the invention of the plow, people had spoken of how "man" used technology to exploit Mother Earth and all Nature. Around the time of the scientific revolution, both the actual exploitation of natural resources and the language used to describe it grew bolder. Enthusiasts for science like Francis Bacon described scientists as men who would "master," "disrobe," and "penetrate" a feminine Nature. By the nineteenth century it was commonplace to say that a scientist "interrogated Nature" in his laboratory, "unveiled" her, and "forced her to respond" to him. It all resembles the modern clinical material collected by the child psychologist Melanie Klein, who observed children with fantasies of attacking their parents, striking into their bodies in belligerent attempts to discover the secrets of the womb.[2]

Atomic scientists and journalists used this primitive metaphor as a matter of habit. Physicists investigated "the most intimate properties of matter," indeed "penetrated" hidden mysteries, "tore away the veils" to reveal inner secrets, and "laid bare" the structure of atoms. Millikan, for one, wrote of the "satisfaction in smashing a resistant atom."[3]

In the 1930s such talk came to a focus on "atom smashers," that is, cyclotrons and other apparatus built to study nuclear structures. Physicists worked to publicize the devices, if only because they needed considerable money to run them. The press and public were fascinated by images of shiny metal towers, massive and phallic, shooting powerful rays in what Kaempffert called "violent assaults" on the nucleus.

Of course there were also more lofty associations. The press rarely failed to mention that atom smashers were designed not just to smash atoms but to transmute matter. Once the atom was split, wrote Kaempffert, once the nucleus, its "holy-of-holies, its secret shrine," was laid bare, gold could be created on an industrial scale, and society would begin anew. Indeed he who penetrated the shrine would understand all "the secrets of the universe."[4]

Splitting the basic elements of matter did not sound attractive to everyone. The more enthusiastic atomic journalism became, the more it reminded some people of the bad small boy, recklessly poking into things best left alone. The most obvious problem raised by nuclear energy tales was that once the legendary scientist learned forbidden secrets, he might use his new powers unwisely. What was to stop him from building uncanny devices to master and alter people?

This typical assumption of a domineering tendency corresponded to another widespread human problem. From childhood conflicts with parents into adulthood, everyone struggled through relationships with authorities. Often the authority figure seemed dangerously knowing, powerful, and heedless of the wishes of others. These were threats that the scientist was especially apt to symbolize.

The arrogant villain who sought to rule the world, a stereotype as old as the first tyrants and kings, was increasingly associated with science. Especially memorable were American movie serials that captivated young people of many nations during the latter 1930s. Gene Autry descended into an underground atom-powered city to battle a prince armed with radium ray devices; Crash Corrigan overcame an undersea ruler who planned to enslave the world using "the atom—the most destructive force known to science"; and the incomparable Flash Gordon sabotaged an "atom furnace" to foil his archenemy Ming the Merciless, who had boasted that "Radio-activity will make me Emperor of the Universe!" To be sure, Ming and his ilk might not look much like scientists themselves. But the use of advanced science was becoming a key attribute of the totalitarian nightmare.[5]

Tales about technological dictatorships can be traced back to H. G. Wells's 1896 story *The Island of Dr. Moreau*. The brilliant doctor had strapped down and vivisected animals to turn them into half-humans, then controlled them with hypnosis and crude propaganda; his island held fiction's first society of scientifically dominated under-people. Political oppression was only part of the point. Dr. Moreau wielding a bloody knife upon a screaming animal-woman: that called up more primeval thoughts.

The most biologically primitive fear is a shrinking from pain and mutilation, but this is only an entry to the anxieties that psychologists have analyzed. Within mutilation they uncovered thoughts of castration, a fear more common among men than many supposed, while women had a similar and not unfounded fear of attack on their genitals. Both men and women particularly dreaded an attack that could rob them of the ability to procreate. A more general fear was that of losing all pleasure and power; psychologists found that the worst thing some people could imagine was to be immobilized as a frozen, helpless victim. Many imagined that death was just such a paralysis, while others believed in a more explicit hell, but either way the fear of death itself looked much like the fear of being made a victim. In short, not unlike the primitive peoples who believed that any harm was due to attack by witches, most people associated their basic human fears with the thought of being victimized. It was these interlinked fears that brought the evil scientist to life in countless tales.

Anxiety about what others might do was not limited to their attacks. According to evidence gathered by psychologists, at the root of many people's anxieties lay something almost opposite: the fear of being abandoned. Many people dreaded death itself mainly because it seemed like a total separation from other people. Such separation can be truly devastating, as in certain orphaned children who never laugh or cry, but sit by themselves uncomplaining and hopeless. In a lesser form everyone has felt if only for an hour the icy touch of that loneliness, the feeling of being cut off from humanity and from our own inner springs of emotion. And if adults have reason to fear abandonment by loved ones, children dread it still more, knowing by instinct how utterly their well-being depends on others. To the nursling, the mother is the world, and separation from such a caretaker can be as devastating as the loss of "Mother Earth" itself. Being abandoned is like being the last person alive in a ruined world.

Such associations may have had special meaning for Frederick Soddy. His mother had abandoned him by dying when he was near eighteen months old. Soddy himself believed it was this loss that made him solitary and indifferent to the people around him; modern studies of motherless children suggest that he may have been correct. Soddy had special reason to imagine how the world might be ruined.

But why should thoughts of separation be connected with thoughts of dangerous science in particular? For many people the association probably came through thoughts of secrets. A child thinking about punishment for poking into forbidden things might well fear separation, if only because isolation, rejection, or even outright threats of abandonment did serve as punishments in many families.

The symbolic associations between secret forces and victimization were described most fully by people who specialized in fear, that is, paranoiacs. It was common among paranoid schizophrenics to think about the end of the world, to feel a paralyzing loneliness as if they were lost in a frozen waste, to envision cosmic battles and revelations, and at the center of it all to see themselves, chosen for survival and rebirth to restore the life of the universe. Epileptic fits and hallucinogenic drugs could bring forth similar overwhelming images, while much the same visions came to apocalyptic mystics from medieval times onward, and sometimes to more ordinary poets and writers. In all these people, under extreme stress everyday consciousness peeled away to reveal a bedrock charged with mythic imagery. In their hopes for personal transmutation, victims and visionaries were attempting to invert the most common human fears. In place of mutilation they looked for health; in place of sterility, fecundity; in place of helplessness, magic power; and in place of loneliness, a new community of universal love.

The connections were reinforced by a literary tradition. Consider the prototypical mad scientist, Victor Frankenstein in Mary Shelley's immensely popular story. Frankenstein was determined to "penetrate into the recesses of nature" in order to create life, not in the ordinary way through marriage but all by himself, usurping divine powers. After he made his creature he abandoned it in disgust. The child-monster, a victim assailed by unbearable loneliness, sought murderous revenge. Scholars have suggested that the novel's theme of vengeance for paternal rejection had something to do with Shelley's own childhood problems with a cold, demanding father. Whatever the origin of her feelings, many people took the images to heart. The most famous of all horror movies, the 1931 *Frankenstein* by Universal Studios, diverged far from the original novel in plot, but this and many other versions of the tale followed Shelley closely in the underlying combination of uncanny birth, rejection, and victimization.[6]

The traditional mad scientist thus came to stand for a dangerous and uncaring authority, but not only that. In the end Dr. Moreau, Dr. Frankenstein, and their ilk died horrible, lonely deaths. That seemed appropriate punishment, for they had been prying into forbidden secrets—which was less the role of the parent than of the bad small boy. Who was really the authority, who the victim? This odd ambiguity points to the central knot in the puzzle of the nuclear energy legend, the least understood and most disquieting feature of all.

The key is not the dangerous scientist himself, but his creation. In most tales the audience's horror focused less on the scientist than on his monstrous creature or weird device. This creature, too, had a long tradition.

People of many cultures had imagined that the life-force could create living beings out of dead matter—another of the many forms of transmutation. Early Greek and Arab philosophers spoke of statues brought to life. Loosely associated with such beliefs were legends of the ecstatic rites of certain medieval Jews who reached not only for spiritual perfection, but also for mastery of the secret by which clay could be raised into a living creature: the golem.

Originally the danger was to the golem's makers, with their blasphemous ambition of displacing God as a creator, but gradually the danger extended more widely. Ideas of spiritual excess mingled with older, worldwide tales of witches who sent forth uncanny beasts. In the early nineteenth century the story was perfected. A rabbi of Prague, said the new legend, brought a clay servant to life, but it broke from obedience and went on a murderous rampage. The tale struck a nerve. In the early twentieth century, stories and movies about the golem flooded Europe.

Such a monster was easily associated with worries over science in general and atomic energy in particular. Reporters liked to say that radioactivity would be a "docile servant of mankind," but people began to wonder whether such servants would stay docile. The worry became a cliché, as when Raymond Fosdick asked whether technological civilization would become "a Frankenstein monster that will slay its own maker?"[7]

Even better at carrying such warnings was the golem's offspring, the mechanical robot. After all, it was technology that actually took away people's jobs, mangled them in wars, and insulted the countryside. Going beyond that, writers from the mid-nineteenth century on used "the Machine" as a compact symbol for threatening new social arrangements, such as political centralization, factory discipline, and bureaucracy. A more specific symbol was the first significant robot in film, a mechanical giant that went about in 1918 killing people with electrical rays. It turned out to be a fake with its inventor inside; the villainous creature was the evil scientist himself.[8]

Scholars who studied the many versions of *Frankenstein* similarly noticed that the writers conjoined scientist and monster. Mary Shelley's scientist himself cried that his creature was "my own spirit let loose." Not without reason did millions of people from the early nineteenth century on get confused over whether "Frankenstein" was the name of the scientist or the monster. They were two halves of one problem.

The division of a personality into two parts reflected a Western cultural tradition: a distinction between logical reason and emotional feelings. In the nineteenth and early twentieth centuries particularly, many people spoke as if human minds were split into halves, with a strict and rational side keeping precarious control over murky urges. The notion even pervaded politics, in claims that educated elites must hold in check the "dangerous classes" wont to break loose in crime and bestial rioting.

This duality was evident when fictional scientists tried to suppress their emotions and concentrate on research, only to find their evil urges set loose in their creatures. For example, Dr. Moreau said he found emotion meaningless, yet his lust for pain and blood could be guessed, and such lust was acted out in full when his band of creatures broke loose. Audiences brought up on this tradition, watching a movie scientist scoff at human feelings, would not be surprised when he himself turned murderous. That was the mad scientist's madness, whether it was himself or his monster that raged out of control.

But wait: wasn't the creature or robot a *victim?* In many versions of the story the Frankenstein monster, rejected by his own maker and all other humans and hounded to destruction, evokes our terror as much for his tragic plight as for his wild revenge. The word "robot" itself bears a heritage of victimization, for it is derived from a Czech word meaning compulsory labor. The word entered the world's languages through one of the century's most widely performed plays, Karel Čapek's *R.U.R.* The play featured an obsessed scientist who mastered the secret of life in order to create legions of impassive, sterile servants. Popular interest in the serf-golem was probably connected with the fears aroused in all classes as workers cried that factory life was turning them into mechanical slaves of production. By the 1930s social critics were warning that citizens would even be robbed of their thoughts and emotions: molded by corporations and bureaucracies, people were becoming frozen, rootless, isolated, controlled, or, in a word, robotic.

In *R.U.R.*, the creatures rebelled like suppressed proletarians and destroyed all humanity. After the apocalypse came a marriage of two robots who had become fully humanized—the birth of a new race. The story held personal meaning for Čapek, a shy man preoccupied with thoughts of illness, infertility, and death, raised by a stern father much like his fictional scientist. The blank-faced robot made an apt symbol for the frozen state of a victim, whether oppressed by society or by more personal problems, hoping for a rebirth.[9]

In sum, many of the tales show a bipolar structure. On one side stand the evil scientist, the corporate authority, the tyrant, the dangers of science and technology, not to mention the coldhearted parent; on the other side stand the golem, the assembly-line worker, the threatened citizen, the rejected child. Yet the monster and robot usually seemed to align as much with the attacker, hardly separable from the scientist who created them. If the creature embodied the mad scientist's murderous impulses, why did it also represent a victim?

A further complication is a third element in many stories: the hero who saves the world and goes on to marry the princess. Such fairy tales are structured like the old paintings of St. George spearing a dragon to rescue a maiden: hero, monster, victim. But the roles can overlap. The mad scientist often started out heroically taking up perilous labors in hopes of advancing humanity. And the victim could become a hero by surviving and moving ahead to found a new world. It is like a St. George painting seen in a dream in which each role keeps transforming into another.

The central meaning of a symbolic structure is often found right within its contradictions. As the anthropologist Claude Lévi-Strauss explained, symbolic structures may be powerful precisely because of their ambiguity. Balancing two or even three elements in a dynamic tension gives people a way to approach matters of the deepest social and psychological complexity, not through reasoned analysis but through intuition.[10]

The poet John Canaday has explained how symbolic structures work in the simple case of a literary metaphor. He points out that a phrase as straightforward as "Mary is a crab" can evoke a dozen different ideas about shells, claws, blind aggression, and so forth, all simultaneously in play. The linear, logical processing of ideas carried out by the most developed part of our brain (the prefrontal cortex) is accompanied by an older form of thinking. Our brains are wired to pursue several threads in parallel along different sets of neural pathways, sacrificing logical neatness in favor of speed and efficiency. Reading the phrase "Mary is a crab" activates an entire network of signals flashing among various parts of the brain, reaching deep into the places that store memories and even emotions. The result is a unified mental experience that cannot be reduced to a linear narrative. Thus a metaphor, and more generally any symbolic structure, is "particularly suited to expressing complex relationships . . . especially those involving contradictory or incompatible elements."[11] The confusion in nuclear tales among aggressor, victim, and hero is not so much a problem we can logically solve as a fact we may learn to accept as it stands. For each component represents an aspect of our own selves.

The reconciliation of contradictions through a symbolic tale was never done better than in a 1936 movie, *The Invisible Ray*. The star was Boris Karloff in a switch from his famous role as Frankenstein's monster. Now he was the scientific genius, the proprietor of a laboratory lit by bursts of lightning from a passing storm. Karloff's scientist built a cylindrical radium ray projector with the standard ambition of working miraculous cures. But he caught a sort of contagious radioactivity, glowed in the dark, and could kill with a touch of his hand. Meanwhile the scientist was so dedicated to his work that he ignored his young wife, and she left him, whereupon he set forth to murder. As advertisements for *The Invisible Ray* explained, he "sought to destroy the world . . . because his world of love crumbled." At the climax the radioactive scientist's own mother smashed the vial of a chemical he required to stay alive, and he burst into flames, consumed by radioactivity.

The movie studio's publicity releases claimed that all this showed scientific theory that might soon come true, but there was not a scrap of reality in the images. The images came, yet again, from the quest for "secrets we are not meant to probe" (as the mother had warned). Only now the punishment for probing was that the scientist himself became an uncanny creature, rejected by everyone he loved, dying in hellfire. The powerful adult who had imagined himself a hero ended as a punished child.[12]

The final goal of mastering secret powers, of transmutation, was to transform oneself. But the mad scientist's attempt stopped halfway, leaving him a monster, his own victim. And in truth, whoever set out to create something new in the world would run grave risks. Such an individual, whether trying to shape a new personal self or attempting a creative act in science, art, or politics, had to abandon patterns of belief that were familiar since childhood and were supported by society. Loneliness, disintegration, and chaos, as the alchemists and mystics said, had to precede rebirth. There was no guarantee that the pilgrim would not remain mired in those depths.

There was still worse to fear. The creature, however crushed, retained a fiery point of rage. From their early tantrums children knew what overwhelming emotions could erupt, up to an urge to run away forever from their parents or even to kill them. If such an outburst came after one had seized ultimate forces, what then? Might unbearable loss come as punishment not from an outside authority, but directly through one's own dire acts?

This was not a problem only for children. Many people, taught to distrust their emotions, feared that a loss of self-control would be disastrous. They became their own rigid oppressors and their own victims. The guilty fantasy of power and aggression would be thrust away and almost forgotten, a secret of its own, entangled with the secret life-forces and thoughts of universal devastation, all so hard for an adult to take seriously.

Sometimes a careful analyst could excavate the pattern whole. An Italian psychiatrist reported a chemist's nightmare in which the sleeper hurled miniature atomic bombs at a world globe until the globe exploded. The terror he felt from this dream reminded the chemist of the terror he had felt when, as a child, he imagined night attacks from a monster. Analysis identified this monster as the child himself, or rather as a ferociously angry part of the child, a part that wished to destroy everything. The cause of the childhood rage eventually came to light. The chemist's mother had died when he was a year old, and he had grown up feeling abandoned, furious at her and all the world, and terrified of his own fury.[13]

Čapek offered a more mature analysis in a 1924 novel. Here again a scientist was tempted to use atomic energy to make himself "uncontrolled master of the world." He succeeded only in accidentally blowing up a town. Most of the book was less about atomic energy than about the scientist's aggressive pursuit of women (at the time he wrote this, Čapek himself was frustrated in his longings for women). Atomic destruction, the novelist explained, was a force that came from within the scientist, expressing his willful passion. If it had not been in you, a wise man told the scientist at the end, it would not have been in your invention.[14]

Actual scientists were not pursuing solitary mastery of the forbidden life-force; they were grading student examinations and applying for money to buy instruments. They did sometimes ponder the old questions of life and mortality, of something greater than themselves. Scientific research was a way to identify with an eternal order, putting something of one's own into an imperishable structure of benefit to humanity. Scientists were driven less by aggression and the fear of death than by their opposites, love and the hope for life.

Hope, like fear, has its diseases. Science's difficulties have often been the diseases of hope. Some scientists became infatuated with their own discoveries, like Soddy with the redeeming potential of his atomic transmutation. Among the public, as among scientists, the image of science up to 1939 was warped less in the direction of fear than in the direction of unreasonable aspirations. Scientist villains in science fiction were outnumbered by scientist heroes by two or three to one. Meanwhile the ordinary mass fiction of magazines and books, when it showed scientists at all, portrayed them as clean-cut and useful characters. Nonfiction magazine and newspaper articles likewise had little but praise.

A popular movie, *Madame Curie*, captured the majority image of scientists as neatly as a butterfly on a pin. In a central scene the young Pierre and Marie Curie bent over a dish containing their first radium, faces bathed in its soft glow, enraptured by the powers for good that they had discovered. Along with their noble vision the movie showed their sacrifices, as Marie drove herself to exhaustion and risked cancer in hopes of conquering death. All this was factually accurate.[15]

Yet the film also suggested there was something odd about scientists. Neither Marie nor Pierre Curie could easily show human emotion. In the impressive closing scenes, when Marie learned that Pierre had died in an accident, for days she remained speechless and staring at nothing, unable to mourn. In historical fact, when the real Marie's friends took her to her husband's corpse she kissed his cold face passionately, then clung to his body, had to be dragged from the room, and wept like any suffering human. However, such behavior was not for the moviegoing public. Why not?

By the 1930s, the public's image of the real scientist had much in common with the stereotype of the evil one. To become a real scientist meant years of arduous study and long nights in the laboratory, requiring unusual self-abnegation. Some scientists also claimed that science could progress only if researchers saw the world in impersonal, emotionless terms. Few groups were as apt as scientists, then, to stand for suppression of human feelings. In fact many of the best scientists were warm and gregarious, much appreciated as colleagues and teachers, but this reality could not overcome the stereotype. To newspaper and magazine writers, the scientist was an odd character who ignored mundane concerns, risking his health and scorning riches (as scientists themselves claimed), an unworldly "wizard" who isolated himself in the pursuit of tremendous secrets. In short, every scientist was supposed to resemble a robot, if not the atom itself: prodigious energy locked up like a compressed spring.

The actual life of scientists was scarcely visible to the public. After all, a nation's physics laboratories in total cost less than one more battleship. Newspapers and radio networks of the 1930s poured forth millions of words about economic predictions and peace conferences. They paid little attention to the laboratories where the true history of the twentieth century was unfolding.

In the four decades after nuclear energy was discovered, scientists learned the main facts about the construction of atoms. Experiments revealed the subatomic protons and neutrons that constituted the nucleus, and atom smashers explored the peculiar force that held them together. It did not seem to be a master key to the universe. The real work of nuclear scientists became less a quest for overwhelming truths than a working out of puzzles about specific types of nuclei.

For example, the nuclei of the metals beryllium and protactinium did not seem to be radioactive in the way theorists predicted; during the 1930s much effort went into straightening out just what was happening. Another uncommon metal was uranium. The things that were known by 1938 about the behavior of the uranium nucleus did not quite fit with nuclear theory, so a few scientists began to take a closer look. This was routine science, which no newspaper would have dreamed of reporting.[16]

5

THE DESTROYER OF WORLDS

At Christmastime of 1938 the fantastic images of nuclear energy began to enter the world of physical reality. European nuclear scientists found that when they bombarded uranium with neutrons, sometimes a nucleus would split apart into two altogether different atoms. One of the scientists, Otto Robert Frisch, sought a new name for this new process. The uranium nucleus, quivering and elongating until it broke into two pieces, reminded him of the mysterious central transformation of birth, the division of a living cell. A biologist friend told him what that was called: fission.

Any physicist could calculate that when a uranium atom split in two it must release a large burst of energy, exactly the energy that had held the atom together until then. Now all the hoary clichés about atomic energy could march forth once more. As Kaempffert wrote in the *New York Times*, "Romancers have a legitimate excuse for returning to Wellsian Utopias where whole cities are illuminated by the energy in a little matter." Less optimistically, *Scientific American* warned, "The tabloids love to write of blowing up the world, and it's not such a sensational idea as one might think."[1]

There was exactly one scientist who had been thinking for years about nuclear energy as a realistic prospect. This was Leo Szilard, a bumptious physicist whose round face had a baby's look of mischievous innocence and secret wisdom. Since his childhood in Budapest Szilard had worried about the uncertainty of human life; he had been particularly struck by a classic Hungarian poem about civilization degenerating into savagery beneath a dying sun. He grew up with an ambition, as he put it, for "saving the world."[2]

When the news of fission broke, Szilard was in New York, and he set to work alongside a team of scientists headed by Enrico Fermi. Working neck and neck with a team in Paris, they determined that as a uranium nucleus fissioned, it spat out neutrons—which could in turn split more nuclei, continuing exponentially in a chain reaction. If this was controlled so that the reaction proceeded at a moderate pace, you would have a useful source of power: a nuclear reactor. But how to control it? A 1941 *Reader's Digest* article asked whether experiments on uranium might start a chain reaction that would unravel the whole planet. Other reporters and even a few physicists warned against playing the role of the "Sorcerer's Apprentice."[3]

If the things said about atomic energy had the flavor of science fiction, that was no accident. One member of the Paris team said the work reminded him of his beloved Jules Verne novels, and another said it recalled H. G. Wells's *The World Set Free*. That very book had been a main inspiration for Szilard in his fantasies of world renewal. The visions raised by such stories had much to do with the recruitment and funding of scientists that had sustained nuclear research over the decades until fission was discovered.

Nobody was as intrigued by the possibilities of uranium fission as the readers of *Astounding Science-Fiction*. The typical *Astounding* reader in 1939 was a college student majoring in science or engineering, eager for new ideas; the magazine was as close as you could find to a probe sunk into the back brain of American technology. *Astounding's* stories were the first to probe the full meaning of the discovery of fission. The most striking came from a new writer, Robert Heinlein. Heinlein had been trained to see ideas from the engineer's practical viewpoint, but in 1939 that training seemed useless, for he was ill with tuberculosis and not expected to survive. Many of his early stories were about loneliness and oppression, and some of his tales ended with a frankly paranoid vision of a world disintegrating into chaos. His first atomic energy story echoed that personal theme.

"Blowups Happen," published in September 1940, centered upon a future uranium power plant that the scientists called, with simple candor, "the bomb." Could any human being be trusted to operate such a plant, Heinlein asked, when one mistake might cause a catastrophe that would devastate thousands of square miles? The power company's board of directors refused to admit that any risk existed. They agreed to make changes only when worried scientists threatened to mount a damaging public relations campaign.[4]

If the story seemed to anticipate newspaper editorials of later decades, that was not because Heinlein had some uncanny gift of prophecy. Exploding atomic plants, and arrogant power companies too, had been banal ideas since the 1920s. Heinlein's story was not really about nuclear engineering but about what it might mean for vast energies to come within human grasp.

His next atomic energy tale was still more penetrating. Heinlein suggested one way in which radioactive substances could be used as a devastating weapon (not exploded, but spread to poison a territory). With that discovery, one of the characters exclaimed, "The whole world will be comparable to a room full of men, each armed with a loaded .45. They can't get out of the room and each is dependent on the good will of every other one to stay alive." His opinion of what could be done about atomic weapons was summed up in the story's title: "Solution Unsatisfactory."[5]

Through the war years, speculations about atomic weapons continued. For example, in late 1944 *Time* magazine reported that London was buzzing with rumors that Hitler's next deadly surprise, after his V-2 missile attacks, would be an atomic bomb. That same year a widely read Japanese magazine published a science-fiction story titled "The Obliteration of San Francisco," in which a Japanese uranium bomb destroyed the enemy city.[6] But Heinlein's "Solution Unsatisfactory" remained unique in the way it stripped down the consequences of such weapons to a basic logic. When the story appeared in May 1941 only a handful of people knew that, in our closed room, the guns were already near to hand.

At this point the story divides into two streams. One stream was the public history. The few facts that scientists explained about fission, before censorship shut them off, simply revived the talk common for decades. The other stream was the secret Manhattan Project, known in full by only a few hundred American and British officials and scientists. Here ideas ran ahead swiftly, going through all the main arguments that would later appear over everything from reactor safety to missile warfare. Yet at the end of the war, when the two streams rejoined, the fantastic and the scientific would merge with unreasonable ease.

In 1942, when Fermi's team was brought to the University of Chicago under physicist Arthur Holly Compton to build their first nuclear reactor, they took the responsibility most seriously. Indeed all the Manhattan Project workers were impressed by the thought that they were on the brink of unleashing cosmic forces. Some had recently read "Blowups Happen," and the rest could recall other warnings. General Leslie Groves, the career Army officer who had taken control of the Manhattan Project, was shaken when Compton told him the reactor would be started up in the middle of the city. Fermi took elaborate precautions, piling one safety measure on another as if he were bringing a perilous golem to life. The watchers who crowded into the room saw only a tedious inching out of a control rod while Fermi made calculations, until he announced that the chain reaction was self-sustaining. It should have been about as exciting as watching someone tune a radio, yet for those present it was unforgettable. Some, like Szilard, left the room troubled by thoughts of weapons, while others thought of the potential of a shining industrial future.

Now the Chicago scientists had to face the kilograms of radioactive isotopes generated within any working reactor. Compton had already set up a team to study radiation hazards, led by a feisty, no-nonsense physician, Robert Stone. The team first had to work out a unit of measurement, which they called the "rem." Experience in hospitals and the radium industry suggested that a person could regularly absorb a rem or so over a few days without suffering observable damage. Stone's team decided to allow Manhattan Project workers no more than one-tenth of a rem per day as an absolute ceiling.[7]

Plutonium was the biggest worry. Any reactor that used uranium as fuel would generate plutonium; that was the easiest way to get material for an atomic bomb. It was no coincidence that the newly discovered metal was named for the Lord of the Underworld. In its health effects the metal was roughly as dangerous as radium, and workers would be machining kilograms of the metal into bomb components.

The problem turned out to be manageable, because plutonium taken in through the skin or mouth is quickly excreted from the body. A worker who swallowed a speck of it would be foolish, but with ordinary luck he should suffer no harm. There was one glaring danger signal, however. A record of deaths among uranium miners had shown that if you inhaled even a tiny bit of radioactive dust, it could linger in your lungs and create cancer. When Stone had to decide what amount of plutonium it would be permissible for a worker to inhale, he gave a characteristically firm judgment: none.

To enforce their rules the physicians devised elaborate procedures that would set the tone for the future nuclear industry. Their precautions went far beyond any taken for products in the chemical industry (some of which were more toxic than plutonium), if only because the project leaders wanted to reassure workers who felt that nuclear energy was somehow uniquely dangerous. In the end, even workers who did accidentally inhale some plutonium, a group closely monitored over the decades, suffered no ill effects.

The Manhattan Project scientists sometimes took shortcuts, pressing their work forward with the haste of fear. These fears reflected a ghastly new fact that pressed on the Manhattan Project scientists, gradually reshaping their thinking: in modern war, nations would not hesitate to massacre civilians on the largest achievable scale.

The powers fighting the Second World War did exercise some restraint: contrary to expectations, they never used poison gas. Poison gas had acquired a taint of moral horror; besides, nations hesitated to set a precedent that might someday be turned back upon themselves. Anyway, they had more efficient means of slaughter.

There was nothing new about destroying cities. Possibly the worst single massacre in the Far East, not excepting the bombing of Hiroshima, had already set the precedent back in 1937 after the Japanese army captured the city of Nanking. Japanese soldiers had gone on an atrocious rampage, literally putting to the sword Chinese men, women, and children. The death toll from these old-fashioned methods was never established, but newspapers reported that some quarter of a million perished. A still worse slaughter, also done with traditional methods, befell Leningrad in the winter of 1941, when Germans besieged the city. More than a million Russians starved to death, the worst military massacre at one location in history.

Spurred by Russian demands for help, by 1943 waves of British bombers were unloading at random over German cities. Their first big success was a firestorm that engulfed Hamburg, turning the city's streets into a hell where shrieking mobs were reduced to charred heaps of corpses. By 1944 citizens of the Allied nations approved any kind of bombing. As news of enemy atrocities arrived, many decided that the Germans were naturally cruel and the Japanese worse, scarcely human. Their leaders, if they had moral qualms, groped for General Giulio Douhet's vision of a war made shorter as white flags sprang up over bombed cities. They only achieved something else Douhet had predicted: the ancient boundary between soldiers and civilians was erased.

In March 1945 the American general Curtis LeMay sent some 300 bombers over Tokyo dropping incendiary bombs. Fifteen square miles of the city went up in a firestorm, a far larger area than would burn at Hiroshima and Nagasaki together. The heat in Tokyo was so intense that small waterways boiled, bad luck for the people who dove in to escape the flames. The night's work caused roughly as many immediate deaths as would come at Hiroshima, plus far more burns and other injuries: a million casualties. LeMay repeated his tactics week after week, methodically burning down cities. By mid-1945 the U.S. Army Air Corps was destroying Japanese urban areas at an incremental cost of about $3 million per square mile—much cheaper than the first atomic bombs.

The nuclear scientists, however, were convinced that they were the ones who would revolutionize international relations. Since 1941 they had been insisting that atomic bombs were an ultimate weapon that could end any war in a matter of weeks. Political leaders up to Winston Churchill and Franklin Roosevelt agreed, impressed more by the traditional atomic imagery of science fiction than by any new facts. In the war and in the maneuvering against the Soviet Union that they expected would follow, these leaders came to see atomic bombs as a solution to all their problems. When would the bombs be ready?

Nobody felt the pressure more intensely than Robert Oppenheimer. Since childhood Oppenheimer had been troubled, and as a young man he had dismayed his friends and parents with bizarre actions, feeling so inadequate that he thought of suicide and murder. What helped him to overcome his problems was his hope for a future as a physicist dedicated to discovery and teaching. He won such wide respect that General Groves chose him to head the new laboratory at Los Alamos, New Mexico, where atomic bombs would be designed and assembled. Oppenheimer traveled up and down the United States to recruit physicists, warning them that Hitler must not be allowed to get atomic weapons first.

But there were other reasons, as Oppenheimer reflected later, that brought people to that secret city on top of a mountain. "Almost everyone realized that this was a great undertaking," a summons to a historic task. Once atomic bombs were made they would quickly end the war, and presumably spell the end of all great wars. Then the other long-promised benefits of atomic energy would be just around the corner: the world set free![8]

The task of exploding a mass of plutonium or uranium-235 was so tricky that nobody expected it could work efficiently. The best guess at Los

Alamos was that one atomic bomb might bring destruction equivalent to an ordinary air raid of perhaps 100 bombers. The first atomic bomb seemed to raise no new moral questions—except that it was, after all, an *atomic* bomb, a prodigy escaped from the pages of science fiction. In physical terms an atomic bombing might equal a night's work with incendiary bombs, but the psychological impact would be another matter.

Some leaders, especially at the summit of government, thought of the impact on the stubborn Soviet government. Others, notably Groves, considered the impact on the U.S. Congress, which would want to know where $2 billion had gone. But in the front of everyone's mind was news of the fierce defense of Saipan, where more than half the civilians (women and children included) killed themselves rather than surrender. Faced with this literally suicidal enemy, most leaders thought of how atomic bombs might impress the Japanese. That was why Oppenheimer and other project leaders, brought together to advise on the use of the first bomb, suggested dropping it on "a vital war plant . . . surrounded by workers' houses." They hoped the tremendous spectacle of flash and blast would bring the Japanese to their senses. The first atomic bombings would be an act of rhetoric, aimed less at the enemy's cities than at his mind.[9]

A bomb test in July would teach the scientists what they had actually achieved, but by then it would be too late to reconsider.

A different mood spread in Chicago. Compton's laboratory had no urgent tasks once the plutonium-production reactors were finished, and the scientists had time to think, perhaps even to read *The World Set Free* (someone had put a copy in the laboratory's library). A group of Chicago scientists with the irrepressible Szilard in the lead petitioned the U.S. government to set off the first bomb on a deserted island as a demonstration, hoping it would lead to an outbreak of white flags. They did not oppose killing Japanese; their concern was for the postwar world. There would be no safety from atomic bombs without revolutionary changes in international relations.

In a democracy there was only one way to proceed, as Compton had written on a memo in 1944: "public education." The Chicago scientists made ready to explain the perils of atomic warfare as soon as wartime secrecy was lifted.[10] And there were other things they wanted to say. Rough calculations suggested that nuclear reactors could produce electricity as cheaply as coal- or oil-burning plants. The Chicago scientists got excited about the prospects for a new field of work. Besides helping humanity toward a prosperous future, the rise of nuclear engineering would solve their problem of where to find jobs when the war ended. Provided, of course, the public could be taught the supreme importance of nuclear energy.

The wisest comment on these plans was offered by Karl Darrow, a senior physicist known for his caustic insights. "I take it that there are two main objects," he wrote to one of the Chicago scientists, "apart from inspiring interest in science for its own sake. One is, to please the public with the prospect of beneficial uses of atomic power, and the other is, to scare it out of its boots by threatening it with new weapons." He doubted that both goals could be reached at the same time.[11]

While the Chicago scientists thought over what to tell the public when secrecy was lifted, General Groves laid his own plans. He gave one journalist, and one only, access to the Manhattan Project. The man he chose was ideally qualified to exalt the work that Groves had accomplished, becoming the world's foremost prophet of atomic miracles.

Bill Laurence had been born in the fifteenth century. At any rate the Jewish hamlet in Russian Lithuania where he had grown up, with its wooden huts, muddy streets, and plodding horses, had nothing to do with the twentieth century. Yet even here Laurence came upon books about Mars, and swallowed whole the theory of an elder civilization. If only we could communicate with those wise and virtuous engineers, humanity could skip ahead over millennia of agony to reach its Golden Age! Laurence decided to go to America, where he might find the technology to communicate with Mars. He wanted, as he later recalled dryly, to learn "the secret of life—how to produce life synthetically. That was one of the little questions I was going to ask the Martians to tell me."[12]

Laurence's mother helped smuggle him out of Russia to Germany in an empty barrel, and he took passage to America. Abandoning his unpronounceable Lithuanian name for the patrician "William L. Laurence," he worked his way from penniless immigrant to a scholarship at Harvard. His body filled out, energetic and squat, topped by a flat, homely face with a thoughtful look around the eyes. In 1930 he joined the *New York Times* as a science reporter.

Laurence believed that science was the religion of the future. Like any religion, it needed missionaries, and that was what a journalist like himself should be. Science writers, he declared in an interview, "take the fire from the scientific Olympus, the laboratories and universities, and bring it down to the people." Atomic energy in particular had long fascinated him. General Groves knew his man when he asked Laurence to tell the Manhattan Project's story.

At Oak Ridge, Tennessee, they took Laurence to a mountaintop and showed him the valley below covered with a staggering secret city. One Oak Ridge plant for refining uranium-235 was the largest factory ever built, four stories high and stretching half a mile. As a senior administrator told Laurence, the Manhattan Project had outdone Jules Verne. Laurence warned his editor that he was onto a story bigger than anyone could have imagined, "a sort of Second Coming of Christ yarn." Finally Laurence reached the most secret of cities, Los Alamos. He called it "Atomland-on-Mars."[13] If the geniuses at Los Alamos could not tell him the secrets of life, they held knowledge that seemed almost as important.

Before dawn on July 16, 1945, Laurence joined the project leaders in a valley deep in the desert. They were awaiting a test of a plutonium device, a test Oppenheimer had given the code name "Trinity" in explicit reference to the terrifying mystery of divinity. When the fireball illuminated the valley, for a moment some scientists feared that they had set off a chain reaction that would consume the entire world. An Army engineer, General Thomas Farrell, cried, "Jesus Christ, the long-hairs have let it get away from them!"[14]

Surely the sight was awesome, but Farrell was bringing older images to the scene. General Groves said that Farrell caught the feelings of almost everyone when he wrote that the sound of the bomb, a long hard thunder echoing around the distant mountains, "warned of doomsday and made us feel that we puny things were blasphemous to dare tamper with the forces heretofore reserved to The Almighty."[15]

When Laurence asked Oppenheimer what he had felt at the moment of the explosion, the answer stunned the reporter. As Oppenheimer watched the cloud boil miles up over his head, multicolored and radiant, he had recalled a passage from an ancient Hindu scripture—a vision of the divine Krishna growing from earth toward heaven, expanding into a dazzling gigantic form with numberless flaming jaws that swallowed whole armies: "I am become Death, the destroyer of worlds."[16]

Death was paired as usual with hopes of rebirth. To Laurence, the rumbling echoes were "the first cry of a newborn world." He believed that he and Oppenheimer and probably many others there had shared a profound religious experience. He said later that witnessing the explosion was "like being present at the moment of creation when God said, 'Let there be light.'"[17]

The language of birth and apocalypse flowed outward from the Trinity test site. Coded messages described the success as a miraculous birth, with "Doctor [Groves] . . . confident that the 'Little Boy' is . . . husky." The new president, Harry Truman, remarked that the scientists might have discovered the "fire destruction" prophesied in the Bible. Churchill exclaimed, "This atomic bomb is the Second Coming in wrath."[18] All this before they received any detailed information about the explosion: it was enough to know that the thing had worked.

Laurence flew to Tinian Island in the Pacific. All that week fleets of ordinary bombers went back and forth from Tinian and other bases, smashing and burning Japan. There were few cities left to destroy. Hiroshima could have been leveled with some 200 bomber sorties, a routine night's work, but the Americans had reserved the city for another tool.[19]

6

THE NEWS FROM HIROSHIMA

It is an atomic bomb. It is a harnessing of the basic power of the universe." President Truman's announcement shocked the world, for few had expected that atomic bombs would come in their lifetime. The statement itself sounded like something William Laurence might have written for a 1930s Sunday supplement. In fact the president had received the statement from a committee, which had adapted a draft forwarded by General Groves, who had got it from . . . Laurence.[1]

The public, like Laurence, could understand the news only in terms of what they already had in their heads. In the first weeks there was little solid information about actual nuclear warfare. Most people filled it in with the myths that had come into being long before the first bomb burst.

The War Department gave the press a set of reports that Laurence had written weeks or months earlier. The reporter's language, pervaded by millennial awe, appeared in newspapers around the world. When other writers began to describe the emergence of nuclear energy, they either followed Laurence's lead or reached into the same stock of familiar images that he had used. A typical American newsreel exclaimed that Hiroshima had been annihilated by a "cosmic power . . . hell-fire . . . described by eyewitnesses as Doomsday itself!" This idea of apocalyptic power cropped up everywhere at once, like dormant seeds sprouting under a sudden rain.[2]

The idea was immediately connected with the issue of control over dread secrets. The bombs were a "revelation of the secrets of nature, long mercifully withheld from man," according to Churchill's official statement. Statements by Truman and others likewise spoke of incalculable forces, control, and secrets.

The leaders also hoped that the new power could be used to "serve" humanity, made into a helpful rather than a wrathful creation. Was the bomb a golem, then, a scientist's creature? Certainly the War Department's press release on the Trinity test suggested a scene right out of a 1930s movie:

"darkening heavens, pouring forth rain and lightning immediately up to the zero hour" had set the stage for the perilous experiment. (In reality the storm had passed hours before the test.)[3]

There were not many "atomic bombs" in all the talk during the next few years; everyone spoke of *the* atomic bomb, or just The Bomb, capitalized like a mythical demigod. Cartoonists drew brutal giants labeled "atomic power" or muscular genies released from bottles, looming over hapless scientists or citizens. Particularly explicit was H. V. Kaltenborn, a famous NBC radio pundit, in his broadcast on the day of Hiroshima. "For all we know," he concluded in hushed tones, "we have created a Frankenstein."[4]

Usually a large fraction of the public paid scant attention to world news, but a poll reported that 98 percent of all Americans had heard about atomic bombs. Surveys found ordinary people speaking much like the journalists. Instead of parading more examples of such talk, I will defer to an authority. Mr. Arbuthnot, the fictional creation of humorist Frank Sullivan, admitted in 1946 that he was the leading "cliché expert" on the atom. No question could stump him. For example, he was asked:

Q. Where do we stand, Mr. Arbuthnot?
A. At the threshold of a new era . . . Will civilization survive? Harness.
Q. Harness, Mr. Arbuthnot?
A. Harness and unleash. You had better learn to use those two words, my boy, if you expect to talk about the atom.

The cliché expert easily identified whose stone atomic energy was (the philosophers' stone); and whose dream (the alchemists' dream); and of course whose monster. He could have given exactly the same answers to questions about atomic energy a decade earlier.[5]

Mr. Arbuthnot, and the real-life cliché experts who staffed newspapers and radio networks around the world, insisted above all that atomic energy might open the way to a utopian civilization. Laurence told a huge radio audience that with the new energy, "We can air condition the jungles and make the arctic wastes livable . . . and we can lick disease." And of course ships would cross the ocean with only a little fuel—a lump the size of a pea, said Mr. Arbuthnot; "the pea is the accepted vegetable in these explanations." Nuclear energy could even solve the problem of war. In mid-1945 most people had expected many more bloody months of battle, and now two bombs seemed to have magically brought surrender and peace.[6]

Some fundamentalist Christians began to speak of true apocalypse. Atomic bombs, they announced, proved that the day foretold in biblical revelation was at hand. There had always been people at the margins of society preparing for Armageddon, but from 1945 on the most sober leaders, from presidents to popes, spoke in language that could evoke such thoughts. As sociologist Edward Shils remarked, atomic bombs made a bridge across which apocalyptic fantasies, marching from their refuge among fringe groups, invaded all of society.[7]

During the first few years, people did not fear anything specific or immediate. The public simply felt that the ground had fallen away from under them. One element in this was the realization, which struck many people right from the first news, that at some point in the foreseeable future no city on earth would be safe. A related element, harder to pin down in factual concerns, was best expressed in a famous editorial by Norman Cousins: "The fear of irrational death . . . has burst out of the subconscious and into the conscious, filling the mind with primordial apprehensions."[8]

There was one nation that had something original to say about atomic bombs: Japan. Imagine that you close your eyes and count to five, and when you open your eyes everything around you to the horizon is a smashed wasteland inhabited by wounded crowds. It mattered how the people of Hiroshima and Nagasaki felt, for some were determined to bear witness.

When the psychologist Robert Lifton interviewed Hiroshima survivors he found they tended to merge the bombing experience with childhood images of victimization and the end of the world—images of separation, helplessness, and annihilation. To have been in Hiroshima or Nagasaki was to know a terrifying "coming together of inner and outer experience," of death fantasies and reality. Decades later, studies of other communities that suffered disaster through human agency (chemical spills, for example) found the same effects. Many did not even know for sure whether or not they had been harmed. And their uncertainty never ended: they felt they were somehow damaged for life. Moreover, unknowns have a natural psychological connection with secrecy, a theme that immediately evokes distrust of authorities. Distrust redoubles when the authorities are in fact secretive (perhaps silent only from ignorance, but that is rarely believed). Worse, authorities have often been evasive or actively deceptive. The victims of a polluting disaster, one scholar concluded, suffer feelings of numbness and bleakness and a permanent change in worldview, distrusting the reliability of all human institutions and nature itself. Thus to the Japanese the bombings seemed less like a military action than a rupture of the very order of nature, an act of sacrilege.[9]

The feeling of uncanny attack that perturbed the survivors was redoubled by real radiation injury. Japanese began to tell one another that radioactivity would poison Hiroshima and Nagasaki for decades, that not even plants would grow there. They came to know the symptoms well: vomiting and diarrhea in the first days, in later weeks loss of hair and bleeding gums, sometimes ending in recovery but sometimes in death. Such injury was featured in the first news story out of Hiroshima, by a British correspondent who wrote that "people are still dying, mysteriously and horribly . . . from an unknown something which I can only describe as atomic plague."[10] Rumors spread that the bodies were so radioactive that they had to be buried using long poles, as if radioactivity were a ghastly contagion. Hundreds of thousands of Japanese who had been exposed to the bombs—a group that came to be known as the *hibakusha*, "explosion-affected people"— were suspected of carrying some uncanny malady.

As soon as it was announced that things called "atomic bombs" had been dropped, many people questioned the morality of the deed. Until this moment, only a few thoughtful people outside the bombed nations themselves had condemned the firebombing of helpless civilians. Now Hiroshima inspired more debate than the rest of the war's destruction put together. It was as if the entire moral problem of modern warfare could be concentrated into this one question.

The feeling that the bombs went beyond some human boundary strengthened as news arrived from Japan. The public was stunned by photographs and newsreels that swept across square miles of burnt-out wasteland. To be sure, sections of Tokyo and dozens of other Japanese cities looked little better, but it was atomic force that caught the imagination. The feelings of horror redoubled with the news of "atomic bomb disease." No matter that deaths by nuclear blast and fire had been similar to the deaths at Hamburg or Tokyo: here was something dreadfully new.

The linked ideas of poison and disease were viscerally disturbing, mobilizing primitive brain mechanisms. The basic emotion of disgust, and the gut feeling of nausea that comes with it, probably originated as organisms evolved to reject poisons and avoid disease. One of the main brain regions where disgust registers (the anterior insular cortex) is activated by bitter tastes and revolting smells, but also by disturbing pictures—like the photographs that were circulating of disfigured Hiroshima victims.[11]

Seeing an atrocity, it is natural to look for somebody to blame. When people say an injustice is morally "disgusting," it is no coincidence that they use the same word as they use for physical pollution. According to recent research, our emotional revulsion against an unfair act is a mechanism that has appropriated specific brain pathways: the ones that originally evolved to help us avoid poisons and disease. As human brains grew more sophisticated, the primitive mechanism of rejection and distancing was applied to other individuals—the unhealthy, or the untrustworthy—and eventually to despised groups of individuals, and possibly even ideologies. In short, our neurons have much the same reaction to a human parasite and an actual louse. It was at this fundamental level that many people began to see nuclear weapons as uniquely immoral.[12]

In this process one writer stood out. A year after the bombings the *New Yorker* devoted an entire issue to "Hiroshima" by the young journalist John Hersey. For days it seemed that everybody was talking about Hersey's report, and soon in the United States and elsewhere newspapers were running the entire text while radio networks read it aloud. It became a best-selling book. Within a few years many high schools were assigning it; the text would remain a favorite of teachers into the twenty-first century.

Hersey's approach was shaped by his tour as a war correspondent in the Pacific, where he had faced death time and again. Once he was ambushed with a jungle patrol; another time his airplane went down in the sea and he had to fight his way out of the sinking wreckage. Obsessed (as he admitted) with the human will to survive, Hersey was unusually qualified to probe Hiroshima's encounter with mortality. But he faded into the background like a good reporter, giving the *hibakusha* a voice. His work, along with many later reports, gave people a taste of the Japanese victims' experience with blast, fire, and radiation sickness. Like the famous shadows of humans scorched into concrete by the explosion, a set of horrifying images was burned into the public mind. Images from Hiroshima became emblems of nuclear energy, and at the same time emblems of modern warfare . . . and of death itself.

On the other hand, many people accepted a remark in Hersey's book that when vegetation sprang up lushly in Hiroshima, it did so not because of the rich ashes and suddenly open sunlight, but because the bomb had somehow "stimulated" the plants' "underground organs." Of all the mysteries of radiation, the most fascinating had always involved life and procreation. In the first significant postwar novel to invoke atomic bombs, Aldous Huxley's widely read *Ape and Essence* of 1948, the radiation left by a future nuclear war not only generated grotesque babies but also made women uncontrollably lascivious for one month a year.[13]

Some groups were not satisfied to let feelings of awe and horror develop by themselves, but worked to shape the world's image of nuclear energy for their own purposes. Most of these groups used scare tactics. However, laboratory studies have shown that if fear sometimes aids in persuasion, at other times the results may be the exact opposite of what is planned.

First to know of atomic bombs, the scientists of the Manhattan Project were the first to work out a public relations plan. As soon as the war ended the Chicago group launched their campaign, soon supported by scientific colleagues around the world. The Chicago scientists had some specific political steps to recommend, but first they meant to instruct the world in the dangers and opportunities created by their discoveries.

Leo Szilard saw the dreams of his youth coming true. Long convinced that scientists should join in a Wellsian band of virtuous technocrats to promote a better society, the bustling Hungarian took the lead in politicking. His first concern was that the U.S. government was planning to leave postwar atomic energy work in the hands of the tyrannical General Groves. When Szilard sounded an alarm, scores of scientists hopped on trains for Washington, formed a Federation of Atomic Scientists, and plunged into political battle.

The "atomic scientists" caught everyone's attention. Physicists were asked to attend countless private dinners, club gatherings, and government meetings, to write articles for all sorts of magazines, to speak everywhere from the "Quiz Kids" radio show to the White House. A sociologist who studied congressional hearings on nuclear energy concluded that even senators looked upon atomic scientists in much the way that primitive groups looked upon their shamans, as beings "in touch with a supernatural world of mysterious and awesome forces whose terrible power they alone could control."[14]

Some physicists were more modest. Theorists were coming to understand that the energy that bound nuclei was no supreme mystery, but on a level with any other property of matter. They began to suspect (although it would not be proved until around 1980) that the force revealed in radioactivity is simply one aspect of a force that also drives ordinary electrical and chemical processes. However, nobody explained to the public that a fission chain reaction was no more and no less mysterious than lighting a match. Atomic scientists were willing to be seen as masters of ineffable knowledge.

It was natural for scientists to be the main purveyors of information about nuclear energy. But information rarely comes all by itself; it is usually one ingredient in a mixture of ideas, feelings, and images. What emotional messages were scientists passing along, blended into their lectures on neutrons?

Time magazine suggested that the atomic scientists flung themselves into politics because they were "guilty men." Oppenheimer put it best, writing that atomic bombs "touched very deeply man's sentiments about the evil of having too much power." In 1947 he captured the idea in an instant cliché: the atomic scientists had "known sin." It was as if they had eaten the forbidden fruit of the Tree of Knowledge—a comparison that some physicists made explicitly.[15] Inquiries found that a majority of Manhattan Project scientists, like most Americans, denied feeling guilty; they thought the bombing of Japan had shortened the war and saved many lives. Yet guilt could be subtle. Psychologists have observed that people can feel secretly guilty when someone near to them has died. On seeing a photograph of Hiroshima, even people whom nobody could blame for the destruction might feel a gnawing anxiety not unlike guilt.

Everyone recognized that when the United States made atomic bombs it had opened up the possibility of being bombed in turn. Scientists thought more than anyone else about being paid back in their own coin. Some communicated their personal anxieties deliberately. "I write this to frighten you," the chemist Harold Urey told millions of readers of *Collier's* magazine in an article ghostwritten for him by the Federation of Atomic Scientists. "I'm a frightened man, myself," he continued. "All the scientists I know are frightened."[16]

Scientists spread fear not only because that represented their own feelings but also because fear could move a listener. Technical lectures drew yawns, but detailed descriptions of how radiation and blast could destroy a city brought audiences to the edges of their seats. The principle was explained by a committee of psychologists who made a study of atomic bomb anxiety. The psychologists warned that fear could lead to passive despair, panic, or aggression. But they said fear could also help move people in the right direction. "Our first objective," the committee reported, "must be to mobilize a healthy, action-goading fear for effective measures against the real danger—war."[17]

The atomic scientists and their allies had specific political goals. Already before Hiroshima was bombed the Chicago scientists had worked out their program. Now they drummed their reasoning into the public. Never again could a nation win security through a commanding military lead, for once each side had enough atomic bombs to wipe out the other's cities, more bombs would make little difference. And there could never be a defense, no matter how far technology advanced, when only a few small bombs would suffice to wreak unthinkable damage. Asked if there was a device that could detect an atomic bomb as it was smuggled into the country, Oppenheimer replied: "A screwdriver" (to open every crate).[18]

The only way to avoid doom would be for nations to abandon some of their sovereign rights; to be specific, they must put all nuclear energy projects under international control. It was plain logic; nobody since has offered a better permanent solution. Many hoped that the very existence of atomic bombs must force nations into cooperation. The philosopher Bertrand Russell explained to a radio audience that "no middle course is any longer possible" between annihilation and an age of peace under international government. The fictional Mr. Arbuthnot, with his unfailing knowledge of current clichés, said humanity was at a "crossroads." Once everyone was shown the two paths leading from the crossroads, "World Destruction" or "World Control" (as editorial cartoons portrayed it), surely they would choose rightly.[19]

It was as if nuclear energy were such a cosmic force that it would sweep away history, instantly replacing the web of international tensions with a millennial age of peace. The United States enunciated a detailed plan for international control of atomic energy. The chief developer of the scheme had been Oppenheimer. "Oppie's plan," as a sympathetic official put it, "would set scientists up as social policemen." Teams of physicists under an international authority would inspect every nuclear facility, imposing peace almost like Wells's virtuous technocrats.[20]

Scientists were joined by many prestigious groups that tried to promote the plan by spreading fear of the alternative. Through materials like the short film *One World—or None,* which began and ended with a picture of the globe consumed in flames, they taught that only international controls offered safety. Radio programs featured celebrities calling for world federation. Such shows could have a permanent impact. For example, a little girl who listened to a radio special called *The Fifth Horseman* recalled, decades later, that "it frightened me more than anything I can remember." Because of that show, the actress and antinuclear activist Jane Fonda continued, "I've always thought that we were tampering with things . . . We were playing God by creating such weapons, such material."[21]

The expert communicators could make people afraid, but they could not create a new world order. Indeed, the more powerful the bombs seemed, the more reluctant most Americans were to turn them over to an international agency. The negotiations broke down, partly because of this reluctance and still more because Soviet officials frankly scorned the plan.

From 1948 on the pivot of international Communist propaganda was a "peace movement" that proclaimed that using atomic bombs (what only the United States could do at that time) would be a crime against humanity. The bombs must be outlawed. Thus revulsion against nuclear weapons went on the agenda of the Communist propaganda network— the newspapers and magazines in many languages, the millions of party activists and tens of millions of fellow travelers around the world who in those years took their cue from Moscow.

As often happens with a propaganda campaign, few people were as strongly persuaded as those who waged it. The conviction that nuclear weapons were uniquely dreadful became seated in leftist thought. In the heat of the campaign, *Pravda* published remarks by Soviet scientists about "total destruction" and "mass extermination." Anti-American propaganda had joined the other political forces that promoted nuclear fear.[22]

In every language people were called upon to choose between the road to doom and the road to a golden age of peace. Few noticed that there were other ways to travel from the symbolic crossroads. Between the two highways shown in cartoons lay a rough countryside, leading toward a future in which nuclear energy would neither incinerate humanity nor redeem us.

———

While scientists, world federalists, and Communists sought to use nuclear fear in international politics, others wielded the same weapon in the domestic politics of their nations. The power of the atom was so staggering, they said, that radically new institutions must be built to handle it. At home as abroad, the only route to security would lie through the strictest controls.

The atomic scientists and many others said that controlling atomic energy required greater government power. Perhaps most insightful were the Alsop brothers, well-known political commentators. They wrote that henceforth the United States must be on a permanent war footing, ready for battle at an instant's notice. When saboteurs could smuggle in atomic bombs, "security police" with Geiger counters must be allowed to come snooping into any home. Worse, the constitutional rule that war must be declared by Congress would disappear, with authority over the very life of the nation placed in the hands of the president alone. Worse still, since Washington might be destroyed instantly, this ultimate authority would have to be at hand to unelected military officers. All this seemed unbearable; the Alsops and others felt that a nation with such practices would be no true democracy.[23]

The U.S. Congress created a Joint Committee on Atomic Energy with unprecedented privileges, sole overseer of a government monopoly that held sweeping powers: the Atomic Energy Commission (AEC). The AEC legislation jettisoned traditions as old as the nation. For example, basic scientific information, which had been freely available to all, automatically became secret government property if it had anything to do with nuclear energy. Control of nuclear energy was turning out in practice to mean control over secrets—and over anyone who might be near them.

The word "security" had originally conveyed a vague feeling of safety; General Groves had taught the Manhattan Project scientists a new meaning. To Groves, it meant hundreds of miles of fences with armed guards and special passes, censorship of private letters, and Army counterspies complete with hidden microphones, all far beyond anything seen in other wartime civilian work. Groves felt personally responsible for the great nuclear secrets, and he thought that such responsibility involved keeping everything under tight discipline—including his own feelings. This was the approach to life he had learned from his father, an Army chaplain and strict taskmaster. When the war ended, Groves warned Manhattan Project workers that "loose talk" still "jeopardizes the security of the nation and must be controlled." He had the ideas of control, secrecy, and safety all twisted up together.[24]

I was struck by the way secrecy and control were emphasized in the *New York Times*'s index of the articles it published in the autumn of 1945. Of all the articles about atomic energy, roughly two-thirds were mainly about international or other "control"; of these, nearly half were largely about "secrecy." A collection of mid-1947 digests of American network radio comments on atomic energy showed the same preoccupations in roughly the same proportions. Countless newspaper columns, radio shows, and entire novels were written about "the secret," as if it were a formula on a piece of paper in a safe somewhere.[25]

Secrecy became a mania. In 1949 when a press exposé announced that a bottle containing uranium was missing and perhaps stolen from an Atomic Energy Commission laboratory, the Joint Committee on Atomic Energy summoned top scientists and officials to explain the loss. In fact the bottle had not held enough uranium to drive a steamship across the Atlantic, nor scarcely a toy boat across a bathtub, but a desperate search overturned the laboratory until a bottle with uranium was produced. (Decades later, scientists smiled when asked if it was *the* missing bottle.)

In other ways secrecy was not funny. Congress passed a law that imposed the death penalty for revealing atomic secrets; in principle a scientist who told a colleague the results of pure research, done privately at home, might be legally executed. Spy stories, notably the treason of Klaus Fuchs and the trial and execution of Julius and Ethel Rosenberg, kept the pot boiling for years.

Even more frightening than secrets going out was the risk of bombs smuggled in. After the shocking news that the Soviet Union had tested an atomic bomb in 1949, it was obvious that the United States could easily be attacked, as an official statement warned, by "valise bombs." The government paid serious attention to the threat of Communist agents wielding infernal devices. Magazine and newspaper articles and a few Hollywood melodramas brought the concern before the public. This was not anxiety about individual terrorists, as would later emerge, but an extension of the Cold War fear of Soviet aggression. Millions of dollars were spent to deploy radiation detectors at airports and seaports, even though scientists doubted they could detect a weapon. All the more reason for intense efforts to search out spies and traitors![26]

No group was more closely watched than the atomic scientists. Physicists and mathematicians made up more than half of the people who were identified as Communists in congressional hearings. Hundreds of scientists were mercilessly pursued, often losing their jobs, some of them ending in exile or suicide. Americans came to accept something so undemocratic that they would have found it unthinkable a decade earlier: an elaborate peacetime system of guards and fences, locked safes, visitors making detailed inquiries about the personal lives of friends, and plain spying. This system was strongest in the Atomic Energy Commission, but it spread into many sections of government, industry, and even the universities. It remains fully in force today.[27]

The same passion for controlling secrets beset every other nation that planned a nuclear energy program. France, too, hounded leftist scientists from their jobs. The British government kept its work so secret that the scientists in charge did not know when the decision to build atomic bombs was made. In the Soviet Union, fission research pushed ahead under the command of none other than the secret police chief, Lavrenty Beria, with nuclear laboratories fenced off as separate islands within Beria's archipelago of prisons. Everywhere, control over atomic energy was in practice coming to mean control over scientists.

The anguish over spies—what Americans called McCarthyism—involved more than nuclear fear, of course. Everything that happened in these years was conditioned by the rise of Soviet-style regimes in Eastern Europe and Asia: the first battles of the Cold War. In the United States, accusations of treachery were useful, for example, to certain Republicans who sought to discredit New Deal Democrats. Moreover, possession of information was a touchstone of political power. For example, only a few top leaders in any nation were allowed to know how many nuclear weapons their nation owned. Others, left in the dark, could not offer rival policies with the same confidence. Symbolically, only the narrowest elite could share in the mystic power of the weapons themselves. In short, secrecy served as both an instrument and a talisman of political authority. Meanwhile, what politics drew from public anxiety about bombs it returned with interest, pouring new fuel on the smoldering fires of nuclear anxiety.

So much for the pressures of historical events and political structures in this period. I want to look at forces less often analyzed, but actually more obvious: things that not only officials but everyone from movie writers to common citizens were saying. Wasn't there something familiar about those scientists and spies with their incalculable secrets?

Several movies made a profit from the spy craze by showing traitors at work, and children's television shows such as "The Atom Squad" brought similar stories into the living room. Even Little Orphan Annie was kidnapped by comic-strip villains seeking to extort atomic information from her seniors. A new popular stereotype arose, the insidious Communist (or, in the Soviet bloc, capitalist) traitor who endangered people through stealing secrets. That meant atomic secrets more often than every other kind put together.

The stories became one more force promoting mythical imagery. For example, in each of three films in which spies tried to kidnap a man for his atomic secrets, the man in question had been in an accident that made him weirdly radioactive, like Boris Karloff when he played a glowing mad scientist. In yet another spy film the camera panned along a Los Alamos fence from a sign reading "Contaminated Area" to another reading "Restricted Area"—the peril from radioactivity merging with the peril from those who sought to know forbidden things.

Scientists had always been seen by many as queer, single-minded, powerful beings working outside normal society, and the stereotype was repeated in journalists' stories of atomic scientists. Whether dedicated and brilliant Manhattan Project workers or dedicated and brilliant spies or saboteurs, they seemed only too inclined to secretly inflict violent change upon society. People put their finger on the emerging pattern when they likened McCarthyism to a "witch hunt." Indeed conservatives used accusations of treason to discredit every upstart group that threatened prevailing social patterns, not only the newly powerful scientists but also Hollywood writers, State Department homosexuals, and so forth.

The underlying fears showed clearly in a British film about a childishly naive physicist. Dr. Willingdon cracked under the responsibility of working on atomic bombs. Stealing away a bomb in a satchel, he announced that he would blow himself up along with all London unless Britain renounced nuclear weapons. Overall the film resembled an ordinary police thriller; the mad scientist had stepped into the world of realism. What were atomic scientists really like? Mr. Arbuthnot the cliché expert had a ready answer: "Little boys playing with matches."[28]

A few observers remarked that mythical thinking was overpowering rational discussion, but even they barely glimpsed all the symbolism at work. Official secrecy was obviously connected with the desire to hold power above ignorant outsiders; but why then did so many citizens on the outside insist that officials should keep atomic secrets? There was, Stewart Alsop remarked, "a sort of Victorian reaction to the whole subject," as if atomic energy were best not discussed. A British member of Parliament complained that when he asked about bombs the government reacted as if he had "asked about something indecent"; another said that the press treated atomic energy in the cramped way it had once handled sex.[29]

The old innuendos had indeed broken out of the enclaves of science fiction and science journalism. For example, a U.S. Department of Defense film about the Bikini tests exclaimed, "Man has torn from nature one of her innermost secrets!" President Truman himself, in his final address to Congress, spoke of the decision to "probe . . . innermost secrets."[30]

Immediately on hearing about the bombing of Hiroshima, the poet Edith Sitwell sat down and wrote about Man as a tyrant "that conceived the death/Of his mother Earth, and tore/Her womb, to know the place where he was conceived." Nothing seemed more necessary than keeping such urges locked away, in others and in yourself.[31]

Atomic energy, a truly great and hidden power, neatly fitted this psychological pattern and its long tradition of imagery. Atomic bomb anxiety became a condensed way of thinking about more than the forces of science and technology in general, as had happened occasionally since the turn of the century. It meant more than an epitome of all the horrors of modern war, which had been a focus of concern since 1945. It also stood for the cruelest secrets of the heart: the drive to control others, or betray them; forbidden aggressive prying; and the urge to destroy, like Dr. Willingdon, even one's own city.

By 1949 the atomic scientists' attempts to impose international controls, and avoid domestic controls, had failed, and their movement was reduced to a small remnant. Citizens focused their apprehensions about bombs on domestic traitors and foreign enemies. After all, it was reasonable to take the fearful hostility provoked by nuclear weapons and displace it onto those who might help to build or use the bombs. The warning of the psychologists' committee was confirmed: propaganda that attempts to frighten people toward peace may instead rouse aggression.

What role did nuclear fear play in creating the Cold War, that tide of fear and hatred between and within rival blocs? Historians of the period agree that, at the least, the threat of the new bombs intensified feelings of suspicion and hostility on both sides. Within the United States, some historians have argued that "the Bomb altered our subsequent history down to its deepest constitutional roots" by redefining government as a "National Security State, with an apparatus of secrecy and executive control."[32] Nuclear fear permitted, indeed demanded, a monopoly of information. The supreme power over life or death—the very survival of the nation—was put in the hands of the president alone, if not nameless military officers in a bunker. That responsibility fostered a progressive expansion of presidential powers, decade after decade, far beyond anything imagined in the original Constitution. It was just as the Alsop brothers had feared, or worse.

When groups of the left and right spread nuclear imagery, they did not recognize how powerful a force they were deploying. Talk about secrets, control, and security was less likely to reassure citizens than to remind them of their most intimate problems. Nuclear fear was like a Chinese finger-trap: the harder people tried to pull out of it, the tighter it gripped them.

7

NATIONAL DEFENSES

Much of the talk about security in the 1950s was about "national security" in a specific sense: safety through military means. Surely the bombs, as physical weapons, could be countered with physical defenses? Generals, admirals, and other authorities up to President Truman insisted that every weapon must eventually meet its match. Attempts to stave off the bombs mobilized millions of citizens and billions of dollars. The effects on nuclear imagery were correspondingly powerful, but only left citizens feeling more insecure than ever.

Although huge sums of money were spent on well-publicized fleets of interceptor airplanes and far-flung radar systems, polls in the United States found a majority admitting that if war came, their own city at least could be blasted. The bomber would always get through. Could bomb shelters protect us? Mr. Arbuthnot and his fellow cliché experts suggested that atomic bombs might "drive cities underground," a fantasy of hiding from atomic bombs in the most literal way. Others began to discuss a more plausible defense: "civil defense."[1]

Governments were under pressure to do something, if only to calm the public's anxieties. President Truman created the Federal Civil Defense Administration (FCDA), which soon had a staff of more than a thousand people, while local agencies added thousands more. This effort was far too small to do much in the event of an atomic attack, but the government was unwilling to pay the tremendous cost of a full-scale program. Civil defense officials decided the only way to do their job was to find and train millions of volunteers through a massive public relations campaign. They knew they would have to adjust their campaign carefully if it was to ease fears rather than stimulate them.

Experts felt that the handful of bombs the Soviet Union owned in the early 1950s might cause more damage by inducing panic than by direct destruction. Most people assumed that a bombing would mean shrieking mobs and cities disintegrating into chaos; by 1953 the American press was using the word "panic" fourteen times more often than in 1948. In fact, wild mobs are scarcely ever seen in disasters. At Hiroshima, for example, most people had reacted with stunned inactivity, meek random actions, or attempts to help one another. But people had always expected that atomic bombs would signal the apocalypse, which traditionally included collapse of the social order.

The first civil defense instructions were accordingly aimed at forestalling panic if war came—not to mention soothing public qualms in the meantime. The basic message was "Keep Calm!" The FCDA's booklet *Survival under Atomic Attack*, printed in tens of millions, insisted that most citizens could survive if they simply learned a few facts (which the booklet conveniently provided).

But to prepare citizens for actual war, a program would have to instruct them in what atomic bombs could do. The civil defense agencies accordingly taught their millions of volunteers some raw facts about blast, fire, and radiation injury. Beyond simple training, officials hoped that exposing citizens to selected images of atomic war would "inoculate" them, getting them accustomed to horrible sights so they would not run around screaming when the bombs came. Civil defense materials therefore showed carefully adjusted scenes of destruction. No actual corpses or gore appeared, but the images of wreckage and of first-aid practice on mock wounded told a story that was, deliberately, about as frightening as the public could bear.

Some private individuals, deeply worried about the future, joined the cause. The most effective was a famous novelist, Philip Wylie. His stunning civil defense novel *Tomorrow!* portrayed a near-future Russian attack on America with unforgettable scenes. Descriptions of the look and smell of burnt flesh, or a woman flayed by a burst of shattered glass from a window, no doubt instructed some readers about civil defense. But overall, Wylie's novel, along with a good many realistic radio shows, magazine stories, and so forth, written by others in the same cause, did less to instruct the public than to frighten them. As Wylie later admitted to a group of fellow writers, "We have taught the people to be afraid—because most of us are afraid."[2]

Besides training citizens and inoculating them against panic, civil defense advocates had a third reason to spread images of atomic war. While congressmen often said civil defense was important, they never voted sums of money to match the enormous task. Worse, not enough volunteers stepped forward. Civil defense leaders constantly worried about what they called a "criminally stupid" lack of support. If only the public could be made to understand the terrible dangers, wouldn't they be more supportive?[3]

In the end, civil defense agencies spread images of nuclear disaster more efficiently than even the atomic scientists had done. Other agencies joined in. For example, the U.S. Air Force helped radio networks to produce dramatizations of Russian attacks on American cities. These were frightening demonstrations of what enemy bombers could do . . . and thus of the urgency of supporting a powerful Air Force. Within the White House itself, beginning in 1953 James M. Lambie Jr., an aide to President Eisenhower, got free advertisements for civil defense placed everywhere from newspapers to bus placards. Like many of Eisenhower's supporters, Lambie wanted to stir Americans to do their utmost in the Cold War. He worried that unless he made citizens recognize their peril, they would demand "dangerous reductions in our expenditures for armaments."[4]

The American press and other institutions cooperated wholeheartedly. Newspapers and magazines printed countless civil defense articles. Every radio listener from 1953 on was periodically jolted when the emergency warning network interrupted a broadcast at random ("This is a test . . ."). Towns set up air-raid sirens and tested them at intervals. Spurred by Lambie's advertisements, hundreds of thousands volunteered for the Ground Observer Corps to watch the skies for enemy bombers sneaking in. Most impressive of all was a series of "Operation Alert" exercises held from 1954 on. As fictional Russian bombers approached, citizens in scores of cities obeyed the howl of sirens and sought shelter, leaving the streets deserted. Afterward, photographs of empty streets offered an eerie vision of a city without people. The press reported with ghoulish precision how many millions of Americans "died" in each mock attack. The drills were supposed to show the public that a nuclear war could be survived, but many people reached the opposite conclusion.

Eisenhower's cabinet spent many hours discussing the details of these "war games." On only one occasion did the cabinet discuss how they might affect public attitudes. In 1956 Secretary of Defense Charles Wilson suggested that the upcoming Operation Alert would "scare a lot of people without purpose." A fiscal conservative, Wilson warned that such drills might provoke people to demand inflation of the military budget beyond all reason. He thought the drills would "strengthen and confirm the views of what might be called the fear lobby here in America—the people who might be said to have a vested interest in massive preparations for war." Wilson was overruled, and the exercise went forward.[5]

Another impressive exercise in civil defense instruction took place in the Nevada desert in 1953 when the government brought hundreds of reporters to witness an awesome atomic explosion. Three-quarters of the nation heard about the bomb test or saw it on television. Of special interest were houses constructed nearby, with mannequins in the living rooms and kitchens. Wylie had helped promote the idea, hoping that a grisly demonstration would help awaken the public. After the test, television and magazines showed the mannequins lying twisted amid broken glass and collapsed lumber, as macabre a scene as Wylie could have wished.

Year after year the Nevada tests continued. A quarter of a million men experienced the tests much closer in. The Army hoped to learn whether troops would panic in the face of atomic bombs, and claimed to be heartened by the results. But many young soldiers who felt the blast shake their bones knew terror and nightmare images of apocalypse.

Still more susceptible to civil defense messages during the 1950s was an even larger audience: children in nearly every school in the United States. Thirty years later I asked audiences of American adults if they remembered what to do if a teacher sounded the alarm, and about a third would put their hands up. The procedure recommended by the FCDA for a sudden attack was "Duck and Cover," and in tens of thousands of schools the children (I among them) practiced ducking under their desks and covering their heads.

Older children shared some of the training materials designed for the several million adults who took civil defense courses. The tone was set by confident, calm voices uttering reassurances, but in an undertone the materials whispered a different message. For example, the *Survival under Atomic Attack* booklet noted cheerfully that "lingering radioactivity . . . is no more to be feared than typhoid fever." A film of the same name (which sold more prints than any film before it) showed a window bursting and plaster raining down on a table set for a family dinner. Many children understood that no matter how nicely they obeyed instructions, their chances of surviving an attack might not be high.[6]

The children seemed to take it all in stride, and few adults asked what was really going on in those young heads. Not until long after, in the late

1970s, did some people begin painfully to describe their childhood thoughts. A woman remembered how as a fourth grader she was sometimes frightened at night if she heard an airplane drone overhead, and would sit up begging, "Please don't let them drop the bomb on me!" A college student said that as a little girl she had had nightmares of running all alone for shelter from bombs; this nightmare of separation, of desperately seeking home or a hiding place, was particularly common. Other adults would still occasionally shrink at the sound of a siren or a passing airplane, wondering for an instant, as in their childhood, if the moment had finally come.[7]

Psychologists interviewing young adults in the 1970s dug up recollections of terror. As children these Americans had shifted back and forth between fear of atomic bombs and more general anxieties, dreams and fantasies of dying or being separated from their families. Nuclear weapons had become almost inseparable in the minds of many young people from overwhelming death itself.

Ever since the news from Hiroshima, many adults had similar feelings. Yet once the initial shock was over, that is, by the end of 1946, most people had decided the problem was not immediate but in a nebulous future, and stopped talking about it. After a first great burst of attention the media gave less and less space to nuclear weapons. After all, once a magazine has carried several stories about a particular topic it must turn to something newer. More surprisingly, most people said they trusted in scientists to devise a defense, or trusted their government to keep the peace, or simply trusted God. Typical was the Texas rancher who admitted that nothing could protect him against atomic death in some future war, but who added, "Most of my friends are more interested in this year's calf crop."[8]

The unconcern struck contemporary observers as peculiar. People who confessed that they understood they were in deadly peril declared with the next breath that they weren't worried. Asked questions about foreign affairs, most people did not even bring up atomic bombs at all unless asked. The whole question was set aside as incomprehensible, something only scientists could deal with. "Their planes could get through," a worker admitted tensely in 1950. "I hate to think of what would happen."[9] Indeed, most people hated to think about the danger—so they didn't. Like death itself, atomic energy seemed so overwhelming that most people shied away from the subject. A 1954 survey found nearly all the respondents admitting that civil defense was a sensible precaution, yet very few of them had taken the least concrete action to safeguard their own families.

Much later, laboratory studies demonstrated that statements designed to inspire fear may only push people to ignore or minimize a threat. But that tendency was already understood by sociologists pondering survey results in the 1950s, and likewise by a psychologist who speculated that everyone was so afraid of apparently "supernatural" atomic forces that "we hide our heads in the sand."[10]

Around the world people were coming to realize that civil defense could never guarantee their safety, any more than international controls could, or spy hunts, or soldiers with antiaircraft cannon. But such insecurity was intolerable. A solution had to be found.

By the early 1950s many people had concluded that there was only one thing powerful enough to defend against nuclear weapons: more nuclear weapons. The result was a race to command the most apocalyptic devices. But this began to look like something apart from previous arms races, in which common sense had called for tanks and battleships to bar the approach of enemy tanks and battleships. Atomic bombs could not stave off the enemy's bombs in that way. How, then, were they to be used?

Farsighted people suggested that the bombs might be less important as physical weapons than as mental influences—a "deterrent." Foreshadowed by some thinkers since the 1920s and publicized by atomic scientists in 1946, the idea was precisely stated by an American academic strategist, Bernard Brodie. "Thus far the chief purpose of our military establishment has been to win wars," he said. "From now on its chief purpose must be to prevent them." During the 1950s the U.S. government adopted the policy of threatening "massive retaliation" for any incursion. Without ever being used, the bombs would quench the enemy's aggressive urges.[11]

Rather more common during the early years was a more traditional idea of using nuclear bombs as ordinary military tools. The Joint Chiefs of Staff in 1946 foresaw using atomic bombs chiefly as a psychological weapon, "with a view to forcing capitulation through terror and disintegration of the national morale." After all, the terror of atomic attack seemed to have worked on Japan.[12] But there would be no pause to watch for white flags: a spasm of destruction using all available means became official strategy. Such a strategy was barely enough for the commander of the American bomber forces, General Curtis LeMay; he reportedly told weapons de-signers that he wished they would build him a bomb that could destroy all of Russia in one blast.

American strategy, this mixture of deterrence, munitions, and fantasies of obliteration, was questioned by only a few penetrating thinkers. The clearest-headed was a British physicist and expert military analyst, P. M. S. Blackett. He argued in a 1948 book that no sustained military advantage, indeed no sane human purpose, could be achieved if generals actually took to dropping atomic bombs.

President Truman said privately that Blackett was making a serious mis-take in analyzing atomic bombs as instruments of war. After all, the presi-dent exclaimed in a wild but significant exaggeration, the United States was building enough bombs to "blow a hole clean through the earth." Atomic bombs were "not a military weapon" at all, Truman and many others admitted. Their real use was as a threat during disputes, that is, as a diplo-matic and political weapon.[13]

Yet Blackett had a point. Truman found that hints about bombs did nothing to turn the Soviets from their obdurate rejection of American plans for the postwar world. By 1950 a small number of leaders were com-ing to a surprising conclusion: nuclear weapons might be useless in every way.

The Korean War brought the first clear test of whether atomic energy could be an irresistible force in world politics. In late 1950, as Chinese troops routed Americans, talk about using atomic bombs grew loud. Most American citizens were terrified by the threat of a nuclear war. In Europe the war scare was even stronger, alienating America's closest allies. Tru-man had to work hard to soothe everyone, promising he was far from ready to drop a bomb. In fact the Joint Chiefs of Staff saw no military advantage in an exchange of bomber raids.

If the bombs could not be used as weapons, they were no better as threats. Their existence had not prevented the Korean War from starting, nor kept it from dragging on for years. In other cases, too, such as the fight-ing in French Indochina, the Joint Chiefs felt the bombs would make poor military tools, while veiled diplomatic threats only brought cries of defiance from the enemy, threw sand into the gears of relations with allies, and fright-ened Americans themselves. Yet leaders and the public alike could scarcely believe that nuclear weapons were useless. If atomic bombs would not protect us, didn't we need something even more powerful?

Edward Teller had an answer for that. At first sight the physicist was only a stubby man with big brooding eyebrows, yet his intelligence and diffident charm made him the most persuasive of all the atomic scientists. Teller's background had encouraged an interest in cosmic, personal, and social catastrophe. There was his enthusiastic reading of science fiction. There was his mother, who was abnormally fearful for her children's lives. Above all, there was his experience as a Jew of Budapest, a doomed community. If Teller was concerned with insecurity it was because he felt, more personally than most, how horribly precarious life can be.

Faced with the forces of death, many people yearn to join with something beyond the mere individual self. Teller had joined other atomic scientists in preaching the need for a new international political order. When that failed he sought to defend his nation through mastery of the greatest power within human ken: hydrogen fusion. Earlier, at Los Alamos, Teller had abandoned work on fission in order to pursue vaster explosions. There was only one other problem that had really excited Teller, he recalled later: whether the first atomic bomb test could trigger a chain reaction that would wrap the earth in fire. That was the kind of brain teaser he found "really delightful."[14]

In 1949, after the first Soviet bomb test redoubled Americans' fears, Teller and others secretly urged the United States to leap ahead by developing a hydrogen fusion bomb. They were certain that if the Soviets got such a weapon first, the result would be catastrophe. Oppenheimer and others disagreed. Ordinary fission bombs could be made big enough to destroy any military target; if fusion weapons could be built at all, they would be too powerful to use against anything short of an entire metropolitan region. Fermi and other physicists added that such a massacre could never be justified "on any ethical ground which gives a human being a certain individuality and dignity." A hydrogen bomb, they said, was "necessarily an evil thing considered in any light."[15]

Although the technicalities were top secret, the public quickly learned that weapons far more powerful than fission bombs might be developed. The news media filled with vigorous debate over whether fusion weapons should be developed. The most thoughtful debaters, both in public and in secret, agreed that a hydrogen bomb would not be so much a thing built of metal as an uncanny force—a "Frankenstein," as one scientist called it.[16] The Joint Chiefs of Staff themselves said they wanted hydrogen bombs not for actual warfare but as a psychological weapon to keep the Soviets mentally on the defensive. Truman, too, protested that he only wanted to have the weapon in his pocket when he bargained with Stalin. A more immediate psychological use was a domestic one, for the administration was under acute pressure to do something, anything, to answer the menace of Soviet atomic bombs. With the backing of a majority of the public, the president decided to press at full speed toward hydrogen bombs.

Congress heaped money into the lap of the Atomic Energy Commission, and immense construction works got under way on a scale beyond even the Manhattan Project. Reactors were built to make plutonium by the ton, while new factories covering dozens of acres began to separate more uranium-235, which is required for efficient hydrogen bombs. Within a few years the American nuclear program was using about one-tenth of the electricity produced in the United States, more electricity than was used in all of Britain.

The Russians, too, believed that they had to possess thousands of nuclear weapons of every variety, and zealously pushed ahead immense construction projects. During the next decade they were joined by Britain, then China, and finally France, all building their own reactors and isotope separation factories and eventually testing hydrogen bombs. Other nations stood aside. Yet by the 1960s Germany and Japan, despite their lack of bombs, would seem no weaker in the international arena than Britain, France, or China. It would not be in world politics, but in the development of civilian nuclear energy, that the gargantuan factories would one day exert their pressure.

8

ATOMS FOR PEACE

The island that held the device was over the horizon from the ships where observers waited, yet the explosion turned their tropical night to dazzling noon. A fireball heaved itself up from the sea, growing and growing, much larger than anyone had expected; sailors thirty miles away felt the heat sear their skin as if a furnace door had been opened; some scientists thought that this time they had finally gone too far, that they had set off the last experiment. Then the familiar cloud began to mushroom upward, but enormously larger than any atomic cloud ever seen. November 1, 1952: a test of the first fusion device.

The newspaper stories were melodramatic, and although they carried little information there was enough to remind people of talk that had been circulating for several years. Back in 1950, for example, Drew Pearson had told his radio audience that 2,000 hydrogen bombs could "blow up the world." From the outset everyone saw fusion as something that went far beyond fission bombs into the realm of apocalypse.[1]

Most leaders met the appalling news of the fusion test with intensified effort along the lines established for fission, from arms control negotiations to fleets of warplanes. But some people felt that only a tremendous change of mind could meet the new reality. Two main movements arose. One was a campaign against radioactive dust, which I will come to later; the other, which started first, was a crusade called Atoms for Peace.

It began in the White House. The ground was prepared by the Eisenhower administration's conviction that in holding off Communism, public relations and "psychological warfare" were as important as bombers. To wage this propaganda war Eisenhower had brought in C. D. Jackson, a mature, balding man who looked like an energetic vice president of a major corporation; and so he was, for he had come to the White House on leave from Time Incorporated. Jackson searched for ways to "go on the moral and ideological offensive against the Communists." An opportunity came when Eisenhower asked him to draft a speech about hydrogen bombs.[2]

Eisenhower had been disturbed about the new weapon ever since he learned the results of the first test. Most impressive was a top-secret film, called *Operation IVY* after the test's code name. It was screened in the White House on June 1, 1953, for the cabinet, the Joint Chiefs of Staff, and others like Jackson; the few dozen people on that level were the entire audience for whom the Air Force and Atomic Energy Commission had made the film. Nobody who saw it was likely to forget its picture of an entire atoll vanished into a crater, or the fireball with a dwarfed New York City skyline printed across it in black silhouette. Worse news arrived in August: the Soviets, too, had tested a fusion device. Americans would have to be told something about all this.

After reading Jackson's drafts that told of the awful damage Soviet bombs could inflict, Eisenhower reportedly complained, "We don't want to scare the country to death." Jackson went back and redrafted the speech to emphasize America's enormous power to retaliate. That sounded frightening too. As Jackson admitted, the proposed speech boiled down to "bang-bang, no hope, no way out at the end."[3]

A better approach was suggested by a phrase that Jackson's first drafts had mentioned only in passing: the "peaceful uses" of the atom. For Jackson, bringing this aspect to the fore would be just good public relations. But for the deeply worried Eisenhower it offered grander hopes: a way to set a precedent for cooperation with the Soviets. That December Eisenhower delivered a much-revised speech on hydrogen bombs before the General Assembly of the United Nations. His eyes shining with emotion, the president offered to support an International Atomic Energy Agency to develop—and perhaps ultimately control—the new power by turning it to benevolent uses.

The speech had an unexpectedly strong impact on the world public. The final result would be to promote a nuclear power industry far more rapidly than ordinary business practices would have advised. This happened partly because of the universal demand for an answer to hydrogen bombs, and still more because plans for civilian benefits had been prepared in advance by certain elite groups.

Since 1945 cliché experts had never ceased to tell the world it stood at a crossroads, and many journalists had said less about the road to doomsday than about the road to an atomic White City. The best known of these was William Laurence ("Atomic Bill" as some called him), now more of a missionary than ever. He explained that atomic energy could turn deserts and jungles into "new lands flowing with milk and honey," making "the dream of the earth as a Promised Land come true in time for many of us already born to see and enjoy it." He admitted that fission reminded him of his adolescent daydreams about Martian technocrats, and even of a new Tree of Knowledge that would return us to Eden. Many other journalists, and scientists as well, spoke as if ancient millennial dreams were almost established fact.[4]

The titles of articles in the *Readers' Guide to Periodical Literature* suggest what Americans were reading in the 1950s. Many of the articles were about military uses, and many of these naturally had titles that might provoke anxiety. But writings on civilian uses were at least as numerous as those on weapons. When neutral titles are left aside, the emotion-laden titles on civilian atomic energy were almost entirely positive.

Nobody was as well qualified to understand the forces of optimism as David Lilienthal. His enemies sometimes made the mistake of seeing Lilienthal as a simple man, for his face, whose drooping features might have seemed melancholy, carried a quirky grin that gave him a gentle and elfin look. Yet he had risen like a rocket through government service, and by the age of forty he was running the Tennessee Valley Authority (TVA), a New Deal program that fought backwoods poverty with monumental dams and power plants. Here he had faced all the contradictions of his progressive beliefs, the desire for both compassionate government plans and efficient private competition. In the Tennessee Valley he went a long way toward combining the White City of technology with the green hills of Arcadia. If anyone deserved to be called chief of the Martian engineers, it was Lilienthal. President Truman was making a natural choice when he picked him in 1946 to become the first chairman of the Atomic Energy Commission.

Briefings from Oppenheimer and other scientists were a "soul-stirring experience," Lilienthal told his private diary. He felt as if he had been allowed "behind the scenes in the most awful and inspiring drama since some primitive man looked for the very first time upon fire." He noticed that experts facing the apocalyptic potential of nuclear bombs kept looking for some grand Answer. His experience had taught him that such a thing rarely existed: faced with a problem, people normally had to stumble through a maze of partial measures. But as he later wryly admitted, he, too, "became emotionally committed to the search for an Answer." Anything that could yield such a dreadful weapon simply *had* to have marvelous peacetime uses. At the atomic crossroads, all that was necessary was to set people on the correct path. Lilienthal insisted that "my theme of Atoms for Peace is just what the country needs."[5]

Another force added its weight to the balance: money. Business magazines regularly printed information and speculation on the future of atomic energy. Most businessmen were cautious, with no inclination to invest much money yet. But a few enterprising men resolved not to be left behind when the atomic revolution arrived. Demand for electricity was doubling every decade; sooner or later the most economical fuel might be uranium. When that day came they were determined that the uranium would be in the hands of forward-looking capitalists like themselves, rather than controlled by a socialist "atomic TVA."[6]

In some other nations, leading groups were still more strongly impelled toward nuclear power. In Europe and Japan the most easily mined coal was gone, the best hydroelectric sites were already in use, and almost every barrel of oil was imported. The lesson of fuel vulnerability had been driven home by the Second World War, when lack of gasoline had forced most citizens to get about on bicycle or foot, and when the German and Japanese armed forces had been virtually immobilized by late 1944 for lack of fuel. The war was followed by painful coal shortages. Atomic scientists of every nation promised to solve the problem with nuclear power.

The U.S. government began to worry that other nations would seize the lead in this grand new industry. The British were visibly leaping ahead, while the Soviets voiced total enthusiasm and seemed bound to join the race. In March 1953 the National Security Council decided that "economically competitive nuclear power" must become "a goal of national importance." A healthy American industry could capture the world market. Moreover, the threat of Communism in impoverished countries could be countered by donating atomic energy plants to sit in jungle and desert, bringing economic blessings (and prestige for the donors). Most importantly, a civilian industry would make a solid foundation for America's military programs.[7]

All these discussions were held among elites in the absence of any well-formed public opinion about peaceful uses. A 1946 survey concluded that "to the general public atomic energy means the atomic bomb." A 1950 survey noted that, despite the extravagant promises in the media, "involvement with the atomic energy process is restricted to the upper socio-economic and relatively well-educated groups in the population." As of 1953, the drive toward a civilian nuclear industry was confined to some nuclear scientists, their followers in journalism and government, and a small minority of industrialists. But those groups could be persuasive, fired as they were not by dry facts but by a vision of saving the world and leading it to atomic Utopia.[8]

Then came Eisenhower's "Atoms for Peace" speech. Jackson exploited it with a blizzard of press releases ("Era of Atomic Power Is on the Way" and the like) that got into newspapers around the world, followed by countless magazine articles, pamphlets, radio broadcasts, traveling exhibits, and films prepared by the U.S. government. It was a psychological warfare triumph beyond Jackson's dreams. A secret report to his office in November 1955 boasted that the campaign had successfully "detracted popular attention away from the image of a United States bent on nuclear holocaust," diverting the public eye to "technological progress and international cooperation."[9]

The campaign taught people everywhere to believe what until then had seemed convincing only to a few elites: an atomic Eden could be reached within their lifetimes. Most other governments rushed to show that they, too, could use atomic energy for something besides destruction. The volume of magazine and book publication on civilian uses of atomic energy doubled or tripled between the five years before Eisenhower's speech and the five years after, more than at any time before or since. Even the newsreels (a major source of simplistic information before television reached everyone) found ways to dramatize civilian uses.

The Atoms for Peace crusade broke through to a new level of credibility at an international conference convened in Geneva in 1955 amid a frenzy of publicity. More than money was at stake, as might be seen in the working nuclear reactor that was the centerpiece of the American exhibit. Magazines around the world ran photographs of this deep cylinder of crystal-clear water with its rods of uranium, where visitors could see the water glow with a ravishing blue light—like "the light of Aladdin's magic lamp" according to a German industrialist. Another onlooker told how one young scientist spoke with such intensity "that he seemed the priest of a mystical religion, to which his listeners could not but become converts."[10]

Industrial leaders of many nations went home eager to build or buy re-actors. The plans took on unexpected urgency in 1956, when war over the Suez Canal plunged Europe into its second postwar fuel crisis. As cars lined up for rationed gasoline and economic ruin threatened, everyone saw the weakness of an energy system built on the shifting sands of foreign oil. Officials in Europe and elsewhere drafted plans to build dozens of nuclear electric plants as soon as possible.

To spread the enthusiasm the United States offered to give reactors to almost anyone. The Americans hoped to score propaganda points against the Communists, and meanwhile make future exports easier by embed-ding their technology in foreign nations. But there was a more specific way in which the program was meant to enhance security: the United States would help a nation to build reactors only if these were subject to inspection, for example, by an International Atomic Energy Agency. The aim was to keep plutonium produced in the reactors from being diverted to make bombs. Above all, Atoms for Peace was promoted as the key to controlling the proliferation of nuclear weapons.

By the end of 1957 the United States had signed bilateral agreements with forty-nine countries from Cuba to Thailand, and American firms had sold foreigners twenty-three small research reactors more or less like the Aladdin's lamp of Geneva. The campaign set down reactors in many na-tions that lacked the skills to exploit them, nations that had far more need for fertilizer or high school teachers. Lilienthal later called the program an "absurdity" driven only by "the desire to prove somehow that atoms were for peace."[11]

What had begun as psychological warfare against foreign enemies would have its greatest impact at home. Noting the success of Atoms for Peace exhibits overseas, Eisenhower told the Atomic Energy Commission that he "would favor additional exhibits being prepared and displayed to a large number of our own people."[12] The AEC scarcely needed encouragement; the commission and its laboratories had been deep into public relations from the start. After 1953 the AEC stepped up its efforts, for example, ar-ranging many interviews for journalists and distributing copies of hun-dreds of speeches, nine-tenths of them dealing with peaceful uses. It offered film and television producers quantities of footage and made free loans of its own movies, such as *Atoms for Peace*, which was originally assembled by the United States Information Agency for distribution abroad.

A main force behind the public relations work was Lewis Strauss (pronounced "straws"), a Wall Street financier who was now chairman of the AEC. He seemed master of any situation with his courtly Virginia accent and a grin resembling that of his good friend Eisenhower. In his childhood Strauss had avidly read Jules Verne and dreamed of learning the secrets of nature, and in high school he had been excited by Millikan's textbook with its speculation about whether man could "gain control of this tremendous store of sub-atomic energy." In maturity the devoutly religious Strauss thought mainly of using that energy against atheistic Communism; he did much to promote the development of hydrogen bombs. Strauss was not at first enthusiastic about Atoms for Peace, although he told Eisenhower that at least it "might have value for propaganda." But gradually he became wholly converted to civilian nuclear energy. "Beneficent use of power which the Almighty has placed within the invisible nucleus," he declared, "will prevail over the forces of destruction and evil."[13]

The press was his most important ally in this struggle, Strauss told a meeting of the National Association of Science Writers in 1954. He was concerned, he said, that the AEC's funds might be cut back unless the public approved of its programs. Therefore he asked the science writers to emphasize the bright side of the AEC's work. Our children, he said, will enjoy "electrical energy too cheap to meter," an "age of peace," and so forth. Laurence and others relayed such claims to the public verbatim.[14]

The science journalists had their own reasons to laud Atoms for Peace. Besides honestly reporting what experts said, they made their living from tales of future amazements. After the bombing of Hiroshima the number of science writers mounted rapidly, and the new generation, like their elders, were either trained in science or otherwise inclined to view it with respect. As one of them explained, the first atomic bomb explosions drew him into science writing because "I realized then that science was going to rule us." Such journalists were the last to doubt that atomic energy was going to do wondrous things.[15]

Alongside nuclear authorities and science writers, some businessmen had a personal stake in Atoms for Peace. A vice president of General Electric (GE), noting the ambitious British and Soviet plans to export reactors, warned that "already the contest is on." Courageous entrepreneurs must march forth on behalf of the nation, democracy, and private enterprise, not to mention their company profits.[16]

Few were as enterprising as the president of the Detroit Edison Company, Walker Cisler. An engineer by training, Cisler had risen to the top of this respected electric utility because of his ability to make things work. With a strong jaw and silver hair framing his impassive face, he looked a model of industrial leadership. Nuclear energy caught Cisler's attention in 1947 when he joined a committee formed by the AEC to make connections with private industry. He set to work to persuade Detroit Edison and other companies to found an atomic age within a capitalist framework. Shortly after Eisenhower's Atoms for Peace speech, Cisler founded the Atomic Industrial Forum (AIF) to coordinate lobbying and public relations.

By 1956 the AIF had some 400 companies as members. It became a main agent for spreading nuclear visions within the business community. Its specialty was meetings that brought representatives of corporations together in a hotel to sit through lectures or join discussion groups. Typical was a 1955 meeting with the topic "Atomic Energy, a Realistic Appraisal." The chairman opened the proceedings by joking, "Perhaps it is a little early in the morning to work up a keynote mood of wondering and of passionate awe. Nevertheless . . ." Nevertheless, such was the mood that speaker after speaker evoked.[17] An international network was growing, a community of people who had decided to stake their companies' money and their own careers on nuclear energy.

These people worked to tell not only their fellow businessmen but the whole world about Atoms for Peace. The Atomic Industrial Forum did so much public relations work that White House staff praised it as a valuable "cold-war weapon" (even though the AIF chiefly addressed the domestic rather than the international public).[18] Individual companies did still more. Foremost was General Electric, which ran the AEC's plutonium-production reactors and by 1957 was employing more than 14,000 people in its nuclear divisions. GE public relations experts lavishly praised nuclear energy through advertisements in American magazines, and promoted it to the tens of millions of television viewers of *General Electric Theater*. Similar efforts by other American and foreign firms helped spread similar messages around the world.

The optimistic pronouncements of national nuclear agencies, scientists, science writers, and industrial companies confirmed other leadership groups in attitudes that already attracted them. For example, in 1955 the World

Council of Churches called for energetic development of atomic energy for peace. Meanwhile the American Federation of Labor and individual unions formed study committees that endorsed the swift development of nuclear power as a source of jobs and prosperity.

Most enthusiastic of all were American students. Since 1945 their teachers had believed, naturally enough, that the primary response to bombs must be "education." Some American teachers also said they hoped to counter the anxiety that civil defense drills aroused in their pupils. From state education organizations to local advisers of high school Atomic Energy Clubs, everyone wanted students to face the future not only with knowledge but with good cheer. In support of the teachers, the AEC developed teaching materials and sent traveling exhibits to high schools. General Electric helped, too, distributing millions of copies of its comic book *Inside the Atom*, and reaching about two million students a year with its 1952 animated color film *A Is for Atom*.

Perhaps most effective of all was Walt Disney's *Our Friend the Atom*, shown on television and in schools beginning in 1957. The great storyteller naturally introduced the subject as something "like a fairy tale," indeed the tale of a genie released from a bottle. The cartoon genie began as a menacing giant much like the bomb-monster of editorial cartoons. But scientists turned the golem into an obedient servant who wielded the "magic power" of radioactivity, symbolized as glittering pixie-dust.[19]

An example of the results of all this work was an essay written by a second-grader: "Good Atoms. Everything is made of atoms. When we learn more about how valuable these atoms are, people will be very happy . . . the business man will have machines and better things to sell. Every body will be happier."[20]

9

GOOD AND BAD ATOMS

Scholars studying fairy tales find that the stories can be classified, hundreds at a time, into one or another traditional pattern. In the same way, we can find much in common among the various Atoms for Peace productions: *Our Friend the Atom*, the American newsreel series *The Atom and You*, the corresponding newsreel by Actualités Français, the U.S. Information Agency's *Atoms for Peace* film, the long Russian documentary by the same name, the scores of books in various languages, the exhibits, and so on. As a scholar might explicate the structure of a set of folktales, so I wish to unpack the pattern of themes within Atoms for Peace productions—themes that would become basic to public attitudes toward the nuclear industry, toward science and technology in general, and eventually toward our entire civilization.

At the center of the structure was the polarity of weapons versus peaceful uses, the atomic genie that could be either menace or servant. Facing this imagery, some people would focus on wonderful benefits while others would see mainly death. Atoms for Peace productions were based on thrusting positive images to the fore. The producers typically gave a glimpse of an atomic explosion near the beginning or a brief, banal warning against catastrophe near the end—but that mention of bombs was scarcely ever omitted. Since it was precisely because of the bombs that so much attention focused on peaceful uses, in the end the Atoms for Peace publicity gave greater prominence than ever to the whole ancient tangle of transmutation imagery, the bright and the somber together.

Public anxiety about nuclear weapons was so strong that Atoms for Peace productions could not hope to outweigh it except with the most wondrous visions. The core vision, offered by Soddy at the start of the century, was rephrased by Laurence in 1946: atomic energy was "a philosopher's stone that not only could transmute the elements and create wealth" but could also provide an "elixir of life" and "mastery over time and death."[1]

Nuclear healing was the most prominent theme of all. No production about Atoms for Peace seemed complete without a patient lying calmly on a table and gazing up at a gleaming white radiation mechanism. Medical radiation was so popular that U.S. officials had to issue a warning against "atomic" potions peddled by quacks. During the 1950s X-rays were often used to kill unwanted body hair; thousands of fluoroscopes in shoe stores across the United States and Europe showed people the bones in their children's feet; some hospitals routinely X-rayed infants simply to please parents with an inside view of their offspring. The accumulated radiation sometimes reached hazardous levels, yet the public continued to trust in rays.

Medical hopes were not all unfounded, for artificially radioactive isotopes could reduce tumors and aid diagnosis. Soon the isotopes were successfully treating hundreds of thousands of patients every year. By the late 1950s the number of lives taken by radioactivity, including those in Hiroshima and Nagasaki, was surpassed by the number of lives saved by radioactivity. Still more important in the long run, isotopes became an invaluable tool for research. Much of the tremendous progress in biology and medicine from the 1950s on owed a debt to radioactivity.

Popularizers exaggerated even that, promising a future with everything up to immortality. And, as a narrator explained in a CBS radio program, "when you get deeper and deeper into the secrets of life, you find them so fascinating you sometimes forget that the atom can kill." (This 1947 program was meant to bring just such a change of mind, and testing showed that it did leave listeners less fearful.)[2]

Atomic rays could also be applied more directly to promote life through increasing the supply of food. Fantastic growth was the property most often stressed, as in a film that showed a little boy gaping at oversized peanuts "created by radioactivity."[3] Radiation could cause a variety of mutations in seeds, and photographs of jumbo vegetables grown from irradiated seed became commonplace. The old idea of a life-force within rays was so appealing that biology, medicine, and agriculture together took up between a quarter and a half of typical Atoms for Peace productions.

In terms of hard cash, the life sciences actually took up only a tiny fraction of the budgets of the various national nuclear establishments. The U.S. Atomic Energy Commission spent less than a tenth of its budget on *all* civilian uses in the mid-1950s. The nuclear weapons industry dealt with uranium-235 and plutonium by the ton, while Atoms for Peace imagery relied upon a stock of isotopes that could have been stored in a drawer.

Atoms for Peace publicists also spoke of industry, but still with particular attention to research with isotopes: "Experiments with Radioactive Piston Rings Hold Promise of Benefits to Motorists," and the like. Substantial money was going into reactors, but early Atoms for Peace productions often gave nuclear electric plants less than half of the total space devoted to industrial applications. The emphasis would change rapidly, for by 1960 electrical power was recognized as the center of gravity of civilian atomic energy. But on its first look the public beheld not prosaic electric generating stations so much as wonders akin to magical elixirs and talismans.[4]

Popular thought was already in the realm of fantasy. After 1945 even more than before, storytellers made use of atomic rockets, atomic rays, and the "atom-powered two-way wrist radio" that Dick Tracy wore from 1946 on. Using uranium simply as a replacement for coal to generate electricity did not seem exciting enough. Scientists, journalists, and officials claimed that big reactors were also ideal for propelling ships, rockets, and even railway locomotives. Children played with "atom-powered" toy cars; a Ford Company officer said real atomic cars were on the way. David Sarnoff, chairman of RCA, declared that by 1980 every home would have its own atomic power plant.

The height of organized fantasy was a plan for an atomic aircraft to go farther and faster than mortal man had ever flown. Here the duality underlying Atoms for Peace was unmistakable, since civilian transport would surely come only after the military version—a bomber that could stay aloft for months on end. The United States threw more than a billion dollars into the effort until President Kennedy put the program out of its misery in 1961.

The only nuclear propulsion schemes that actually worked involved ships, for in truth a kilogram of uranium-235 could get the proverbial steamship across the Atlantic. In 1959 the Soviet Union launched a nuclear-powered icebreaker with great fanfare, and later came an equally well-publicized American cargo vessel plus more modest German and Japanese ships. Here, as so often with new technologies, a good idea proved to be not quite good enough for commercial success. It was in another area, where cost meant little, that atomic ships would become important.

The world's first nuclear-propelled vehicle, the *Nautilus* (named after Jules Verne's fictional submarine), hit the water in 1955. The product of a U.S. Navy crash program, it was finished perhaps five years before a normal technological program would have produced such a thing. In retrospect it made little military or political difference that the United States had such a submarine so early, but at the time it impressed everyone.

There are many ways to build a reactor, and during the 1950s engineers experimented with dozens of designs, each as different from the next as a steam-powered locomotive differs from a diesel truck. For example, a group initiated by Edward Teller took safety as their prime criterion and designed reactors for which a major accident would be physically impossible. The prime criterion for the *Nautilus* reactor, on the other hand, was that it must fit inside a submarine. Engineers solved that problem by using highly concentrated uranium-235 as fuel. They had no problem getting this exotic substance, thanks to the enormous plants the AEC was building for its hydrogen bomb program.

The submarine reactor was elegant and compact, but it was not optimized either to make accidents impossible or to turn a profit. However, it was the only developed reactor at hand when the AEC decided to build a civilian power reactor. They found a partner in Philip Fleger, chairman of a Pittsburgh utility. Fleger later said he was attracted by the prospects for nuclear energy in the hands of private industry rather than government, calls to help the nation to a Cold War triumph, and a shrewd understanding that the first company to master nuclear power would gain matchless publicity (in particular, Pittsburgh was suffocating in smog, and citizens would welcome a supermodern source of clean electricity). In short, when Fleger teamed up with the AEC to build a reactor, nuclear imagery meant more to him than short-term profits.

The reactor's design by the Westinghouse Company was simply a scaled-up version of the submarine engines it was building for the Navy. Constructed at Shippingport, Pennsylvania, the plant came on-line in 1957 to universal acclaim. More would follow, most of them likewise based on technology derived from military requirements.

Over the next few years many of the glittering trappings of Atoms for Peace faded into the background while nuclear reactors stood more and more in full public view. These reactors were basically prosaic devices, sources of hot steam. Yet from their birth they had been seen as part of a grand mythical structure; that was a main reason they were built. In that mythical structure they were fated to remain.

An ethnologist studying Atoms for Peace publicity might be reminded of the many folktales in which there are a virtuous brother and an evil one—a bipolar structure like the classic confrontation of hero and monster that I mentioned in Chapter 4. Claude Lévi-Strauss pointed out that in such tales the main point may be the structure itself, the fact that the world is divided, rather than just who is virtuous and who is evil. Sometimes, in related stories gathered from different tribes, the brothers changed sides while the structure persisted. Often myths and symbols made an impact precisely because they contained opposite meanings simultaneously. Gathering contradictions into a package that made intuitive sense, they met the human need to get some kind of hold on life's paradoxes. In just this fashion the Atoms for Peace tales built their message on the division between what a *Time* magazine article called "Good & Bad Atoms." It was the structure already seen in the editorial cartoons that showed one bright and one dark path leading from a crossroads: the point was that extremes of power now lay within human choice.[5]

As in sets of folktales, the sides in nuclear tales could switch. For example, while some saw nuclear bombs as wholly evil, the U.S. Air Force saw them as a deterrent to Soviet attack and thus the very mainstay of peace. Similarly, the more some people spoke of the wonders of peaceful reactors, the more others thought of danger. If a civilian nuclear energy plant could turn a desert into a garden, how easily could it turn a garden into a desert!

When polls showed that a majority felt optimistic about civilian nuclear energy, they also showed that a minority did not. The rising nuclear industry realized that, as a pollster told an Atomic Industrial Forum meeting, a "hard core of about one-fourth" were "dominated in their thinking by fear." The expert had asked Americans what came to their minds when they heard the word "atom," and two-thirds immediately spoke of bombs and destruction.[6]

The nuclear industry urgently wanted to dissociate civilian products from bombs. Publicists encouraged everyone to speak not of "atomic" but of "nuclear" power; aside from being scientifically more accurate, they hoped the new usage would disentangle reactors from "atomic" bombs. Vocabulary began to shift. However, the replacement word was immediately applied to bombs as well as reactors, and the word "nuclear" in turn soon took on a frightening tinge.

Atoms for Peace publicity itself cast inadvertent shadows. For example, countless exhibits, films, and photographs showed concrete shields with thick glass windows, workers hidden in white protective suits, and robotic "slave" hands for manipulating radioactive substances from a safe distance. The images were supposed to show how carefully experts protected everyone, but they carried a more archaic message. A French author put it precisely in his caption to the typical picture of a face peering intently through glass at robot hands opening a bottle of isotopes: didn't that symbolize "an ever more mechanized Humanity, avid to tear from Nature her eternal secrets, and at the same time fearful of the unknown that can escape from a simple bottle"?[7]

Many such hints about the mad scientist and his golem could be found in the publicity. Everyone from advertising men to presidents talked about making atomic energy an "obedient and tireless servant"—a formidable creature indeed.[8] It had to be controlled, of course: with civilian energy as with weapons, the word "control" was on every tongue. Some of the most widely seen productions, such as Disney's *Our Friend the Atom*, even got their facts wrong, speaking as if an out-of-control reactor could suffer an explosion as enormous as the Hiroshima bomb.

And there were so many images of . . . victimization? When Atoms for Peace productions featured a patient prone on a hospital table, surrounded by gleaming machinery and waiting to be penetrated by rays, the picture looked strikingly like the mad scientist's victim in comic books. As if to drive home such ideas, the AEC's stock of free film was especially generous with footage of helpless white rats that scientists inspected and injected.

A 1950s poll found that a significant fraction of Americans associated civilian nuclear energy with things "dangerous to touch or be near." The category of perilous things might include human beings. When two workers in Texas were accidentally contaminated with a little radioactive material in 1957, neighbors shied away from their children, as if fearing a weird plague. It became a widespread joke that anyone who took in radioactivity would glow in the dark, like Boris Karloff playing the mad scientist who killed with a touch.[9]

Nuclear industry workers were beginning to carry what social scientists call a stigma. Originally a "stigma" was a brand burned into the faces of slaves or criminals, to signal that they were untrustworthy. But we can trace it much farther back, to the signs of disease that evoke feelings of disgust and rejection—deep instincts that impel animals to shun those of their kind that might be contagious. The Japanese *hibakusha* were similarly becoming stigmatized. Young people from Hiroshima began to have trouble getting married, for who knew whether the irradiation they had suffered might produce defective offspring?

It was not only industrial workers and *hibakusha* who could be tainted by their association with atomic energy. When Atoms for Peace enthusiasm turned a spotlight on the leadership of the Atomic Energy Commission, citizens did not like everything they saw. Often enough they saw arrogance and incompetence. The atomic officials were no more arrogant or incompetent than officials in any other large organization, but they were subjected to a more intense and anxious scrutiny. In charge of unthinkable power, they could not be forgiven their occasional mistakes.

The first public complaints were about excessive secrecy. At the outset the AEC had been trusted to act as it chose behind an impenetrable wall. Some officials came to feel that only they knew enough to make decisions about the cosmic secrets for which they were responsible. They adopted what critics called a condescending "father knows best" attitude. Even the Joint Committee on Atomic Energy, which had its own sense of superiority and special knowledge, complained bitterly that the AEC was barring them from important facts.

It was inevitable that people would suspect whoever was a keeper of secrets. And the AEC, constrained by strict laws, could never give its critics all the information they sought. But some of the AEC's political decisions reinforced the suspicion. One notable mistake was secret machinations by Strauss and others that ended in 1954 with the highly publicized removal of Oppenheimer's access to secret information. The impact on public opinion was as great as if the physicist had been accused of treason. When it was revealed that Teller had raised doubts about his former boss's commitment to building hydrogen bombs, liberals began to call Oppenheimer a martyr in the cause of peace. The AEC had begun as the great hope of atomic scientists and their liberal allies, but its security policies had increasingly alienated them. The Oppenheimer affair was the last straw. The AEC lost the trust of many forever.

Trust matters. In recent years specialists in fields ranging from economics to international relations have developed the thesis that a stockpile of trust is necessary to maintain workable relationships in any enterprise. From buying milk to negotiating terms of an alliance, people expect their counterparts to be trustworthy at least most of the time. In modern civilization trust became even more central as people had to rely on distant organizations for vital services. It was especially problematic in areas relating to technology, where each of us depends helplessly on experts in many fields.

Trust is difficult to create. We must become convinced that an individual or organization is not only competent but will exercise that competence in our interest. And trust is easy to lose. It is enough to see someone using his or her authority not in the common service but for selfish gain or personal politics.[10] That was the effect of the Oppenheimer controversy.

The Oppenheimer stories sounded like tales of a hero in cosmic battle against a villain, and such tales can be read either way; many believed Oppenheimer was indeed a traitor. Others projected the division between good and evil onto Oppenheimer's own soul. For example, a 1964 play that reached large audiences portrayed the physicist as a flawed martyr who confessed that in building bombs he had been "doing the work of the Devil."[11] Whether it was Teller or Oppenheimer or both who were called morally sick, the message was the same: the world's foremost nuclear authorities now carried a stigma, a mark of distrust.

Other mistakes followed, above all in a fight between government-owned and private electric utilities. The fate of the nation's entire economy was thought to be at stake in this struggle literally for power. Strauss marched into the minefield bearing the flag of free enterprise. He was backed up by Eisenhower, who was determined to reverse what he called the "creeping socialism" of enterprises like the Tennessee Valley Authority. A passionate battle broke out in 1954 over public versus private control of civilian reactors. It climaxed in a record-breaking thirteen days of Senate filibuster as the Republicans forced through legislation that encouraged privately owned nuclear power plants.

The AEC had originally seemed a haven of idealism above party squabbles. Now nuclear energy, in the eyes of every citizen who followed the news, had lost its political virginity. The self-righteous AEC chairman had provoked widespread distrust even within the nuclear community itself. It was with this burden that the AEC would face its most severe test.

10

THE NEW BLASPHEMY

The fireball heaved itself up from the sea, growing larger and larger; then it froze. At the base of the hemisphere of flame appeared tiny skyscrapers in silhouette. "The fireball alone," said the narrator, "would engulf about one-quarter of the Island of Manhattan." On April 2, 1954, the public was watching the *Operation IVY* film. American television stations played it repeatedly all day, and it was soon riveting audiences (including me) around the world.[1]

Since 1945 most people had shrugged aside warnings of nuclear Armageddon as fantasies of some fairly remote future. That future had edged closer in late 1952 with rumors about the first fusion test. Many people hounded Eisenhower for solid information, but it was the *Operation IVY* film that forced his hand. For one thing, civil defense officials wanted to use it in their campaign to spur citizens to action. For another, its information was leaking into the press. Eisenhower reluctantly agreed to release a censored version.

The film's message was clear . . . and terrifying. The Hiroshima fission bomb had caused no more devastation than the conventional air raids on Hamburg, Tokyo, and many other cities. A single hydrogen bomb was a thousand times mightier. Churchill told Parliament in 1955 that fission weapons "did not carry us outside the scope of human control," but with the coming of hydrogen bombs, "the entire foundation of human affairs was revolutionized."[2]

Many people only now gave serious attention to nuclear energy. In the mid-1950s the number of publications on the subject increased sharply around the world. Part of the increase resulted from the Atoms for Peace literature; magazine articles dealing with civilian uses of the atom increased, as a proportion of articles on any subject, by more than half between 1948 and 1956. But still more striking was the rise in articles on nuclear weapons—a fourfold jump. The number of novels and feature films centered on nuclear energy, nearly all of them about weapons, likewise doubled and redoubled.

However, over the next decade the public would not give its main attention to the fact that a hydrogen bomb could reduce several square miles to incandescent gas. What dominated attention was a fact barely hinted at in the *Operation IVY* film. It turned out that a hydrogen bomb could kill people not only nearby, but hundreds of miles away. This fact opened the way for a dismaying idea that had long been buried within nuclear imagery to come into the open: the idea that releasing nuclear energy was a blasphemous violation of the entire planet.

It all began with dust. Already in planning the Trinity bomb test the Los Alamos scientists had worried about the dust that the explosion would hurl into the air. Passing through the swarm of neutrons and fission fragments in the fireball, the dust would become dangerously radioactive, then drift downwind to eventually fall out on the desert. It became part of the civil defense program to warn children that after an atomic attack, they should not eat a sandwich or drink a glass of milk until authorities had checked for radioactive "fallout."

When the AEC ran a series of bomb tests in Nevada, a few citizens began to wonder about risks. Tests in 1953 produced so much fallout that citizens of nearby towns were ordered to stay indoors while the clouds passed; thousands of sheep, weakened by bad weather and then exposed to fallout, died. Songwriter Tom Lehrer suggested that visitors to the Wild West should bring along lead underwear.

The AEC was acutely concerned about the growing anxiety, for the agency was determined to let nothing impede its tests. Despite the doubts of some of its own scientists, the AEC mounted a public relations campaign insisting that there was no chance whatsoever of harm. To keep local citizens from worrying, some overzealous officials failed to warn them of precautions they should have taken. The press repeated the reassurances, not only because most reporters believed the AEC experts, but also because nobody wanted to interfere with tests that were deemed essential for national security. Pointing to the cloud of radioactive dust that drifted off from the televised 1953 Nevada explosion, network anchorman Walter Cronkite assured the audience, "It's not dangerous."[3]

Attitudes began changing after BRAVO, an American fusion test on March 1, 1954. The explosion was more than twice as powerful as expected, and the fallout went far beyond what had been predicted. Gray dust drifted downwind onto a Japanese fishing vessel, the *Lucky Dragon*, coating the crew. When they got back to Japan two weeks later they were showing the classic signs of radiation sickness.

As the *Asahi Shimbun* reported, everyone in Japan was "made to feel acutely once again the horrors of an atomic bomb."[4] Adding to Japanese dismay was Strauss's refusal to admit that the AEC had made any mistake. When the crew member who was most weakened by radiation died, the AEC insisted he would have lived if he had received perfect medical care. Respected journalists accused the AEC of feeding the public pap. These were the same months when the AEC was losing the trust of many Americans because of the Oppenheimer affair, but for the rest of the world it was chiefly evasions about fallout that poisoned the commission's reputation.

Meanwhile fine radioactive dust, catapulted into the stratosphere by the explosion, drifted on the high winds. Independent scientists who had already been worrying about the effects of radioactivity now warned that a test might cause birth defects among people hundreds or even thousands of miles away. Few images are as likely to stir deep emotions as damaged babies.

What could fallout really do? Month after month the American government issued no more than scraps of ambiguous information mixed with reassurances. The fact that hydrogen bomb fallout could dangerously pollute entire territories was treated as a military secret, for it undercut the prevailing strategy for nuclear war and civil defense. The government scarcely knew what to say to itself, let alone to the public. In any case the AEC, the world's main source of atomic information, was largely discredited. Lacking reliable facts, distrust and anxiety could only grow.

Open fear erupted among the Japanese when they learned that a few tuna fish caught in the Pacific had excess radioactivity. Thanks to the exquisite sensitivity of Geiger counters, Japanese scientists could track faint radioactivity from the BRAVO explosion for thousands of miles through the wandering currents of the Pacific. The isotopes could become concentrated as they worked their way up through the food chain, so that a fish could be much more radioactive than the surrounding water. A person who ate many such fish might have an increased risk of contracting cancer.

Millions of Japanese stopped buying fish. In terms of diet and emotions, it was as if Americans did not dare to eat beef. Japanese government agents with Geiger counters selected fish for destruction by the ton; parents refused to let their children swim in the ocean. They were reacting to a prodigious reality hitherto unknown: a single human act could physically affect the environment not just in its own locality but far across the planet.

In 1956 presidential candidate Estes Kefauver declared that hydrogen bombs could "right now blow the earth off its axis by 16 degrees"; in 1957 Soviet Premier Nikita Khrushchev reportedly boasted of owning a bomb that could "melt the Arctic icecap and send oceans spilling all over the world"; a British screenwriter drafted a movie about nuclear tests that set the planet adrift toward the sun.[5] More plausibly, many worried that a bomb test could cause earthquakes thousands of miles distant, and even the AEC took that idea seriously enough to seek reassurance from professional geologists. The most persistent rumors were about weather. In 1951 the AEC began receiving letters from all over the world saying that bomb tests were upsetting weather patterns, and beginning in 1953 the world press often discussed such complaints. Farmers blamed the AEC for heat waves and cold spells, cloudbursts and droughts.

The irrepressible rumors, viewed as a group, made symbolic sense. To say that nuclear bombs were polluting fish or causing birth defects, to say that they could disrupt the weather or set the planet's axis askew, in each case was to say that nuclear energy violated the order of nature. This idea was bound up with one of the strongest of primitive themes: contamination.

In most human cultures the violation of nature, and indeed all forbidden acts or things, are identified with contamination. The anthropology theorist Mary Douglas explained that whatever is "out of place," whatever goes against the presumed natural order, is called polluting; dirt seems right in a vegetable garden but disgusting in vegetable soup. Violating the correct order of things, for example, by breaking a taboo, was seen as dangerous; the transgressor might be accused, for example, of bringing down sickness on himself or even on his whole tribe (which could be true, if the taboo happened to guard against transmission of disease). This fit a familiar pattern: the bad small boy who defies rules is liable to make a filthy, damaging mess, defiling himself and perhaps others too.[6]

Anthropologists noted that the fear associated with taboos would fasten especially on a person who was "out of place" in the accepted social structure, such as a person who broke caste rules or made unreasonable demands. As I noted in Chapter 6, recent research indicates that the same brain pathways that react in disgust to dirt and rottenness are activated by people we find "rotten," untrustworthy. In many communities individuals who were thought to be acting outside accepted boundaries were stigmatized as witches. Some tribes were convinced that if anything happened that seemed outside the natural order, for example, an untimely death, a witch must be at work. It was a witch who prevented the conception of healthy babies, or brought an unseasonable storm—perverting nature itself. The damage was often explicitly connected with contamination: witches were said to attack the inner purity of others by projecting a deadly substance into their bodies.

After 1945 scientists and nuclear officials made particularly apt targets for suspicions about the disruption of childbirth, the weather, and so on. It was scientifically true, after all, that radioactivity could damage people. Equally important was the long tradition of accusing scientists of transgressing the natural order, which was akin to saying they were contaminating the world.

The great original sin, many felt, was the bombing of Hiroshima. Thanks partly to Communist propaganda but also to spontaneous revulsion, millions around the world came to believe the bombing had no connection with normal wartime actions. Hadn't the Japanese already been on the brink of surrender? Wasn't it obvious that Truman had obliterated Hiroshima for no reason but to overawe the Soviets, or for some other indecent purpose? A controversy began over the decision to destroy the city with an atomic bomb instead of with incendiaries.

It was by no means clear whether the war couldn't have been ended just as soon without dropping the bombs—especially if the huge scientific and industrial effort had been dedicated to other weapons instead. (Much the same was said of the conventional bombing of cities in Japan and Germany, but that question attracted little public interest.) The debate spread widely in the 1960s and continued to agitate the public for decades, inspiring a shelf of books and countless newspaper items and letters to the editor. As late as 1995, vehement objections to the text of a historical exhibit about the Hiroshima bombing forced the director of the Smithsonian Air and Space Museum to resign his post.[7]

The arguments were remarkable for their passionate tone and limited scope. Few Americans and Japanese who had personally experienced any of the countless atrocities of 1941–1945 saw the bombing of Hiroshima and Nagasaki as exceptionally guilty; the arguments usually said little about the Second World War as a whole. The real debate was over whether or not American authorities were moral monsters for using weapons that many now saw as immeasurably loathsome. The truly interesting question was not "Should atomic bombs have been used in 1945?" but rather "May hydrogen bombs legitimately be used now?" A horror of hydrogen bombs was at the root of the blame—and of a rising tendency to see authorities as people who would not scruple to contaminate the world.

In sum, the powerful theme of uncanny pollution was in a good position to link up with nuclear energy. And it happened that certain facts strengthened the association. Most important was the fact that radiation can cause genetic defects. This resonated with old and widespread ideas about contamination. Traditionally, defective babies were a punishment for pollution in the broadest sense, violations such as eating forbidden food or breaking a sexual taboo. Incest in particular, according to many groups both primitive and civilized, would almost inevitably bring unnatural progeny.

The association between nuclear radiation and pollution was strengthened by the fact that radiation can cause cancer. By the 1950s the word "cancer" had come to stand for any kind of insidious and dreadful corruption. Demagogues labeled Communists, prostitutes, bureaucrats, or any other despised group as a social "cancer," while people afflicted with real tumors came to feel shamefully invaded and defiled. Thus the symbolism of cancer coincidentally reinforced the idea that nuclear weapons meant odious violation of the proper order of things.

Nobody set out deliberately to make any of these associations; they came to many people at once. As early as 1950, liberal newspaper and radio commentators had exclaimed that the proposed hydrogen bombs, wrongfully exploiting the "inner secrets" of creation, would be "a menace to the order of nature." On receiving news of the BRAVO test, the conservative publisher William Randolph Hearst told millions of readers that such explosions "could cause dangerous changes in the orderly processes of natural law." Even Pope Pius XII, in Easter Sunday messages heard over the radio by hundreds of millions on every continent, warned that bomb tests brought "pollution" of the mysterious processes of nature.[8]

No group felt polluted so strongly as the survivors of the Hiroshima and Nagasaki bombings. Like people anywhere who have helplessly watched family and neighbors perish, the *hibakusha* had tended to feel personally in the wrong, unworthy of surviving when others did not. Simply by being present when their bustling city changed instantly into a charnel house, they felt they were contaminated.

Those who tried to forget the experience were forcibly reminded in the early 1950s, when dozens of survivors came down with symptoms reminiscent of radiation sickness: bleeding under the skin, weakness, death. This was leukemia, which arises after a few years in a small fraction of irradiated individuals. By the mid-1950s, when one or another survivor of the atomic bombings died, perhaps of some ordinary illness, newspapers around the world might announce the cause of death as a new and uncanny affliction, "A-bomb disease." Robert Lifton, who interviewed many *hibakusha*, reported that each of them revealed a "sense of impaired body substance."[9]

Much later, in the 1980s, psychologists began to understand that many of the *hibakusha* also suffered from the newly recognized syndrome of post-traumatic stress disorder (PTSD). Their depression, anxiety, nightmares, and even physical symptoms like fatigue were identical with those experienced by rape victims, combat veterans, and others subjected to an unbearably terrifying experience. Whether they recognized it or not, they were "bomb affected" not only in body but in mind.

Into the mid-1950s most *hibakusha*, preoccupied with their sense of shameful contamination, were not inclined to speak out. Besides, the American occupation authorities discouraged talk of atomic bomb horrors. In other countries, too, the media continued to follow the advice of authorities, saying little about pollution from bomb fallout. It was not in newspapers that attitudes about contamination first emerged strongly, but in another form of discourse—a far more popular form, where myth became ferociously visible.

What was it that bomb tests brought from the deeps of the Pacific in 1954, radioactive and threatening, to terrorize millions in Japan—that is, aside from tuna fish? The answer was Godzilla, the 400-foot prehistoric reptile that stomped Tokyo flat in the first Japanese movie to achieve international financial success. The world had seen dozens of earlier films about cities wiped out in a natural disaster or a war, or about prehistoric beasts, or about monstrous results of scientific experiments, but *Godzilla* combined all these themes. It was not the bombs themselves that the movie's director said he had in mind in crafting "a warning to mankind," but the radioactive contamination emerging from the ocean waters, and nuclear energy in general.[10] The most disturbing elements of modern imagery were coalescing into solid form and connecting up with bomb tests.

The connection became entirely clear in the popular 1954 movie *Them!* Killer ants the size of buses crawled out of the desert near the Trinity test site, "a fantastic mutation," as the film's scientist explained, "probably caused by lingering radiation from the first atomic bomb." Moviegoers found this pseudoscience plausible, and it put them into a cold sweat.

A cold sweat: the sort of visceral reaction that can place a permanent imprint in the brain. In watching movies, especially, we can be flooded with emotions (and since millions of people may share the experience, cultural historians like myself take a special interest in movies). It is worth pausing to look into this experience more deeply, for it differs from the mental processes I have been discussing up to this point.

The associations formed through the first half of the twentieth century between atomic energy and other things—healing elixirs, birth defects, utopian cities, death rays, and so forth—were largely forged by repetition, from science-fiction tales piled atop newspaper articles. It was the classic Pavlovian way in which a dog learns to associate food with the sound of a can opener: if A happens in conjunction with B enough times, then the pair becomes linked. This model of how we think appeals to biology without the need actually to understand the neurological mechanisms it assumes.

In recent decades psychology has made great progress in understanding brain processes. Research has focused attention on the way in which an image may be associated not only with other images or ideas, but also with emotions. A monster, for example, can evoke fear, and perhaps other feelings such as astonishment or disgust. This association with an emotion is a connection to the body itself. Fear means widened eyes, a wrenching in the stomach, sweating, and even trembling. At some deep level, to remember the monster is also to remember the visceral feeling of fear.

A single powerful experience can be enough to sear a permanent neural trace. The event at Hiroshima was singular, yet its impact on the survivors was stronger than many things that are repeated a hundred times. Specifically, a memory of any instant of terror is imprinted permanently in a primitive brain organ called the amygdala. This is the mind's "guard dog," ever on the alert for something that we should react to instantly by attacking or fleeing.[11] The imprint typically classifies an experience as "good" or "bad" (for example, delicious or frightening). As a matter of promoting survival, not only the most immediate stimulus but anything near it is likely to be associated with the memory: the wounded soldier remembers the smell of the street where a bomb exploded.

The memory of an emotion can influence even the most refined logical thinking. Our evolution (to appeal to another area of study that has made important advances in understanding) shaped our ancestors' brains so that they could survive on the African plains, where decisions had to be made with limited information and less time. The result is a set of short-cuts. These ways of thinking usually function with astonishing efficiency, but they can be tricked into error. The emotional punch associated with fire would be a help in running from a conflagration in the grasslands, but it can push people to their deaths as they crowd at a door to escape a burning room, too preoccupied to look for other ways out. Overall, however, emotion not only benefits thinking but is essential to it. Unfortunate people whose damaged brains do not connect the emotional memory organs to the cognitive organs prove to be incapable of making commonsense decisions, for they feel no attraction or aversion to any choice over other choices.[12]

The brain's record of emotions ("affects" in psychologists' lingo) becomes doubly important when people think about a risk, for example, bombs or radioactivity. That naturally draws on the most primitive survival mechanisms. As one expert put it, in judging risks "people consult or refer to an 'affective pool' containing all the positive and negative images associated with the object or activity being judged." The process is automatic and very rapid, preceding any conscious, logical calculations we may try to employ.[13]

All these mechanisms are evoked not only by direct experiences but by vicarious ones. Recent experiments show that our brains use "mirror neurons" to represent actions that we observe: if you see a picture of someone's face being slapped, neurons are active in the same brain regions as if your own face were being slapped. Thus in audiences watching a movie, everyone's brain is activated in similar patterns. Indeed the same parts of the brain that show activity in someone's actual experience will become active in a person who is only told about the experience. Human communication may depend on this alignment of neural patterns between the person telling a story and the person hearing it.[14] Although the science is still being worked out, it appears that such vicarious experiences can act much like real experiences to cause permanent changes in the brain's emotional/thinking circuits.

Moreover, there is evidence that the regions of the brain that organize memories also organize imagined experiences. I do not suppose that an imaginary trauma can leave as powerful an impression as a real trauma. But imaginary events do leave permanent neurological traces that will affect our thinking, using the same brain regions and pathways as memories of actual events.[15] This phenomenon suggests how the media, and especially movies, can permanently influence our attitudes. After all, as one expert remarked, there were no movies "in the ancestral environment . . . whatever you saw with your own eyes was true." Even when we read about something, especially in gripping fiction, our brains' default is to accept everything presented as "fact": unless something prompts us to pause and analyze our perception, we take its truth for granted.[16]

Physiology aside, many experimenters have shown that even a brief exposure to an image will act as an "anchor" to thinking, biasing the observer's judgment on subsequent decisions. One reason advanced for this bias is that our brains must work with what is "available" to them; we find it hard to believe something unless we can easily picture it. Thus people may choose to travel several hundred miles by car rather than by airplane even though statistics show air travel is far safer, for it is news stories of airplane disasters that have impressed their minds with the most dramatic images and associated emotions.

Memories can influence thought processes well before they reach consciousness. For it is an uncomfortable fact that many of our decisions are not made consciously, even though we may feel they are.[17] In the survival situations that our ancestors faced, it would have been impossible to think everything through in detail under time pressure. But the price of mental efficiency is that we form prejudices quickly and shed them with difficulty. First encounters are thus crucial even when they do not impact our emotional systems . . . and doubly so when they do.

The demands of survival in our evolutionary past have given special power to memories of terrifying events. Physiological embedding is especially strong and immediate for the emotion of fear, much more so than for other emotions such as shame, lust, or even happiness. The one other emotion that has such a strong effect is disgust. Like fear, disgust is a bodily state that in a few seconds can imprint the brain for a lifetime. For example, a study of birds that tried just once to eat a Monarch butterfly, and were sickened by its poison, found that those birds would never approach another Monarch.

Now consider again the gigantic mutant ants of *Them!* They evoked both fear and disgust—which is to say, horror. There is much anecdotal evidence that these particular images had a lasting impact on many who saw them. A link was forged between the idea of atomic bomb tests and the memory of a gut reaction to something horrid.

———

At the conclusion of *Them!*, after the Army exterminated the creatures, an official worried that if such monstrosities followed the first test, what would come from all the bombs exploded since? The answer soon appeared in theaters. The financial success of *Them!* inspired a crowd of movies about oversized creatures engendered by radiation, including a giant radio-active squid, giant leeches, giant scorpions, a tarantula the size of a house, and a pair of 25-foot crabs. Never was so much faith shown in the idea that radiation could promote growth. As if to verify the Atoms for Peace promises, one low-budget thriller described a radioactive substance invented to enlarge vegetables; insects ate the produce, and an army of gigantic grass-hoppers advanced on Chicago.

These were no detours of the popular imagination but well-traveled highways. Millions of people would pay to see the cheapest production if it had a radioactive creature in it. Meanwhile comic books also impressed millions of youngsters, and adults too, with tales of monsters as well as of contamination, world destruction, and perilous rays, all usually associated with atomic energy. For example, after 1946 the immensely popular Super-man found himself susceptible to weird "kryptonite" rays, and his authors revealed that both he and kryptonite had originated on a planet blown up in an atomic apocalypse.

Critics quickly declared that the science-fiction creatures were personi-fications of nuclear bombs. It might be better to say that in the popular imagination the creatures filled a vacant space where the public declined to look at real bombs—and radioactive contamination too. The films scarcely referred to nuclear war itself, except indirectly. Not one of the countless futuristic movies of the 1950s dared to show an audience the realistic results of atomic bombing. At most a film might mention nuclear wars on some faraway planet, or tucked out of sight as a past event. Scriptwriters preferred symbolic war. For example, an American film treated *The Deadly Mantis* exactly as if it were an approaching Russian bomber, tracked by radar and Air Force interceptors. Almost every monster film showed troops mobilizing, mass evacuation, and collapsing cities, as actually experienced by millions in 1944–1945, as well as perplexed officers with scientist advisers hovering about, just as in the real Cold War. Not all these scientists were dangerous; equally important in the 1950s were virtuous scientists, dedicated and responsible men who devised clever means to destroy the monsters.

In a famous essay Susan Sontag suggested that audiences, not daring to think about nuclear war, projected their fear onto a monster in order to be able to defeat it vicariously—trusting that dedicated scientists, heroic military officers, and other authorities would take care of us in the end.[18] However, the movies did more than simply deal with people's fear of being bombed. The creatures introduced a new and peculiarly horrible imagery of their own. Unlike the Frankenstein monsters, Wolfmen, and Draculas of an earlier generation, most 1950s creatures were altogether inhuman, devoid of intelligence and feelings. Now the world was menaced by things as impersonal as X: *The Unknown*, no living animal at all but a sort of glittering sludge that oozed about in blind search of isotopes.

Did filmmakers see nuclear weapons as altogether remote from human emotion—or did the weapons have a human meaning too overpowering to face? Consider the sludge in X: *The Unknown*: it was inspired by a disgusting mass in an immediately preceding film from the same studio, and there the blob had indeed once been a man, an astronaut who degenerated after probing into forbidden realms of radiation. Many other films began with people scientifically victimized, turned into inhuman creatures rejected by all the world. The prototypical nuclear monster, the Japanese Godzilla, had this dual character: as well as a raging menace, he was a tragic victim of the hydrogen bomb test that drove him from the ocean depths.[19] In the end, he perished pitifully (in later revivals he even became a hero, battling rival monsters).

In sum, the new radioactive horrors were descendants of the alchemist's chaotic mass and the clay golem. They apparently held the same frightening meaning: violation of forbidden secrets punished by loss of true life and feeling, a takeover by the most inhuman and bestial parts of oneself, a wrong turning in the attempt at rebirth.

Most of the films combined victimization with references to failed re-births and forbidden secrets. *The Thing, The Beast from 20,000 Fathoms*, and *The Deadly Mantis* came to doomed life when they were melted out of a literally frozen state. Meanwhile giant ants or blobs emerged from dark, secret holes in the earth; in the Japanese movie *Rodan* scientists penetrated a tunnel to find a monster emerging from its egg, what the actors called a forbidden sight. If moviegoers did not already associate radio-activity with such primal problems when they went into the theater, they would do so by the time they came out.

In an old sense of the word a "monster" was a malformed offspring. If bomb tests violated the natural order, then perhaps it was only to be expected that monstrosities would emerge from the womb of the contaminated earth. Movies did not dare touch the question of defective babies directly, but since the 1930s science-fiction magazines, less prudish, had run stories about children deformed by radiation. By 1952 the idea was so hackneyed that the editor of a major pulp magazine begged authors to stop, for his desk was buried under tales of atomic-bomb children with too many fingers or heads—the monster as victim. Nobody suggested openly that the movie creatures were likewise a symbolic progeny of the breaking of taboos, but the idea was clear to critics: the 1950s monsters symbolized a retribution for "man's technological tampering with nature."[20]

Many mad-scientist stories could be interpreted as showing victimization by coldhearted authority in general. The 1950s movies, aimed chiefly at teenage audiences, might show adolescents struggling less against the monsters they discovered than against unmindful parents. "Dad, you've got to *listen* to me or everybody will die!"—a plea that was also a threat. Critics noticed that the monsters themselves, raging and lonely, seemed like rejected children. An especially direct movie featured an officer irradiated at a bomb test who grew into precisely that: a gigantic, outcast, violent baby.

As a critic remarked, one could no longer believe in werewolves and vampires, but "I am still prone to terror at authentic-looking young atomic scientists talking what may well be scientific nonsense." When the Geiger counter began to clatter and the Professor whispered, "It's radioactive!" audiences learned to expect something slimy and vicious to crawl out. These stories not only exploited, but thoroughly reinforced, the association between nuclear radiation and frightening monsters, dragging in all the implications of disgusting pollution and tragic victimization.[21]

If *The Beast from 20,000 Fathoms* represented nuclear energy, then it was easy to see what was represented by the hero, a white-gowned scientist who destroyed the monster with a new isotope—good atoms mastering bad ones. But the neat plot endings were less impressive than the central image. Overshadowing the clever scientists loomed their bastard offspring, unkillable despite all, 400 feet high and growing.

11

DEATH DUST

If bombs were bad, then why not stop testing them? This idea, like almost every idea involving nuclear energy, was raised first in the scientific community. Physicists argued that an international moratorium on testing nuclear weapons might forestall the development of usable hydrogen bombs. It would also serve as a first step toward mutual trust. In short, a test ban would raise both a practical and a symbolic obstacle against the spread of weapons.[1]

There was another argument for halting bomb tests. A few experts warned that fallout from the tests might cause leukemia or other proven forms of harm, not only near the explosions but around the earth. Could the tests cause "A-bomb disease" everywhere?

Citizens who followed the news with care began to understand that a bomb brought three sorts of radiation hazards. First were the direct rays from the fireball; this was what brought radiation sickness to thousands at Hiroshima and Nagasaki. Second was the fallout of heavy radioactive ash that an explosion could scatter hundreds of miles downwind; in a hydrogen bomb war this would be a major killer. Last was the tenuous dust that a hydrogen bomb lofted into the stratosphere to drift about the globe for years, settling almost undetectably into people's air and food. It was this last type of radiation—the faintest, most widespread, and least understood—that became the focus of world debate.

The first important stirring of mass dissent came in Japan. The tens of thousands of people in Hiroshima and Nagasaki who shared an experience of victimization also shared a deep and understandable distrust of all authorities. With so much in common they began to come together, finding hope in mutual emotional support and practical action. Some groups began to demand special medical aid or financial compensation from the American or Japanese governments; others launched campaigns for international peace. Both the people seeking a stipend and those seeking universal amity had reason to emphasize the special horror of nuclear energy. Besides explaining their sense of personal contamination, shame, and outrage, their unique experience could give the *hibakusha* the right—more than the other millions of war wounded—to make demands on the world.

The *Lucky Dragon* incident and the tuna contamination scare unlocked anti-American feelings that since the war had lain hidden within most Japanese. Now all Japanese could see themselves as victimized by a nuclear experiment. When American scientists came to study the *Lucky Dragon* crew, Japanese newspapers cried, "We are not guinea pigs!" Meanwhile the *Operation IVY* film and other hydrogen bomb news convinced many Japanese that they would perish if war came.[2] In an initiative that was loosely connected with the bomb survivors, Japanese housewives concerned about fallout began to circulate petitions against further hydrogen bomb tests, and soon had tens of millions of signatures. A mass political movement was taking shape.

A big rally in Hiroshima on the 1955 anniversary of the bombing caught the eye of the world press. It marked the opening of a worldwide crusade against bomb tests and the authorities who sponsored them. As a response to hydrogen bombs, the movement would become more powerful even than Atoms for Peace in its effect on public imagery.

The first risk of global fallout that scientists talked about was the most primal of threats: genetic damage. Scattered early warnings converged at the 1955 Atoms for Peace conference when biologists attracted attention by insisting that even the tiniest amounts of radiation could injure human genes. The message was reinforced by a study that the National Academy of Sciences of the United States published in June 1956. The academy's experts denied that the weak radioactivity from bomb tests could cause cancer, and pointed out that in any case the fallout was slight compared with all the other radiation that bathed the public, whether from cosmic rays or medical X-rays. However, the group's Genetics Committee said that any amount of radiation, no matter how small, would cause some genetic damage. Newspapers displayed that conclusion on the front page.[3]

Many scientists disagreed, declaring that the extremely dilute global fallout could do no harm. AEC officials issued a barrage of reassurances, and the American press tended to go along. During the mid-1950s the mass circulation press printed a few soothing articles insisting that there was "not a word of truth in scare stories" about deformed babies. Mostly the press just ignored fallout amid the flurry of Atoms for Peace articles. Only a handful of scientists and their followers dissented, writing mostly in small intellectual journals like *The Nation* and *Saturday Review.*[4]

Under the calm surface, however, strong forces—political forces—were aligning with each side of the bomb test issue. When Strauss clashed with Democrats over private ownership of nuclear power and other matters, it became clear that the AEC chairman scarcely distinguished between factual statements and special pleading for Republican policies. Even magazines that supported bomb tests began to question whether Strauss was telling the truth about hazards. One reporter remarked in 1955 that the AEC's critics might be right about fallout—"Death Dust I like to call it."[5] In his choice of phrasing he was already choosing sides. The sides began to sort themselves out along traditional political lines; it was characteristic that the leftist *Nation* worried about bombs and fallout while the rightist *U.S. News* did not. Somehow the issue touched upon the same fundamental beliefs that determined political choices.

This curious correlation between political ideology and attitudes toward risks has been studied by social scientists over the half-century since. To crudely summarize a sophisticated body of work: people can be placed along a spectrum according to certain fundamental beliefs. Toward the right are people who see society as necessarily and properly a field of struggle among rugged individualists, requiring a hierarchical authority to keep the worst among us in check. These people trust traditional authorities ranging from generals to priests and parents. They tend to accept risks as inevitable. Toward the left are people who think society should aspire to be a field of cooperation among equals. They tend to distrust all hierarchical authority, preferring bureaucratic regulation. And they seek to minimize risks, especially environmental ones that seem to be imposed unfairly. In short, the right/left division between Republicans and Democrats over fallout was no accident, but a reflection of the deepest personal convictions.[6]

In April 1956 the Democratic Party's candidate for president, Adlai Stevenson, brought the issue into the limelight by calling for a moratorium on tests of large fusion devices. In harsh language he warned against radioactive isotopes—"the most dreadful poison in the world."[7] Polls showed that Americans supported bomb tests by a heavy majority, but Stevenson kept returning to the issue, for he was impressed by the outcries of the Hiroshima victims, and convinced that hydrogen bombs were of central importance. Some prominent scientists backed his warnings, and Eisenhower was deluged with mail opposing bomb tests. The letters typically said that the scientists' warnings raised a moral problem. The letter writers not only were concerned for their own children, but condemned the immorality of AEC "Frankensteins" (as some put it) who endangered innocents around the world.[8]

A pressing concern about children had already been prominent among the Japanese bomb test protesters. The image of the victimized child, perhaps unconsciously connected to the deep symbolism of mad-scientist-monster stories, proved tremendously effective in mobilizing protest— especially for women. Women were exceptionally active among the Japanese protesters, and made up two-thirds of those who wrote Eisenhower to oppose the tests.

The most effective protester of all was a biochemist, Linus Pauling. His power came partly from his stature as an outstanding scientist, but still more from his personality. With his passionate convictions, well-crafted sentences, and rubbery features that stretched into an infectious smile, he was an irresistible persuader. Constitutionally disinclined to trust officials and at the same time deeply concerned about nuclear war, Pauling decided that fallout was causing many deaths already.

Pauling's skill as a publicist became apparent in his calculations, which boasted a precision that conventional scientists would never dare to claim. In a scholarly article that immediately became famous he predicted that each year of continued testing could mean another 55,000 defective births and 100,000 stillbirths. Others took up this imagery of large numbers, speaking with all the authority of statistics about thus and so many thousands or millions of deformed babies. The numbers suggested a dreadful picture, already popularized by numerous science-fiction stories: a world infected with grotesque mutants.

Although the consensus of scientific opinion did not go nearly so far, it was turning against fallout. Influential journals such as *Science* and *Scientific American* gradually became more willing to call bomb tests a risk to health. Even the AEC's own scientists now admitted that there was at least a chance of cancer from fallout, and that in theory some level of genetic harm was certain.

Many experts nevertheless insisted that the harm from fallout was too slight to worry about. Most widely heard was Edward Teller. He explained over and over that global fallout involved very low levels of radioactivity, and that harmful effects from such faint traces had never been detected. For all anyone knew, he said, a little extra radiation might even be good for people. Natural radioactivity was all around us; some locales had water supplies that gave residents more radiation than they got from bomb tests, yet nobody was demanding a change in water supplies. Teller could not understand how any rational non-Communist would object to bomb tests unless from ignorance of the facts.

Such soothing and condescending statements failed to calm public worries. If anything they made citizens suspect that the AEC was dishonestly concealing the truth—after all, the AEC insisted on secrecy even where it did not seem necessary. Eisenhower got more and more letters opposing bomb tests, not only from pacifist and religious groups but from labor union locals, businessmen, and mayors. Writers joined the campaign, producing dozens of stories specifically opposed to bomb tests. By the late 1950s a majority of informed Americans (and still more in other nations) believed that fallout held serious dangers. In 1958 Lilienthal wrote privately to another former AEC leader, "I don't remember any instance in which a major public body has lost public confidence in its integrity to the extent that has happened to our old friend the AEC."[9]

By the end of the 1950s anxiety over fallout had become a powerful force around the world, with the full support of many governments. India's prime minister cried that fallout "might put an end to human life as we see it." Scandinavian officials issued alarming emergency warnings whenever fallout from a test drifted near. Most anxious of all was Japan, where the government responded to Soviet tests with loudspeaker trucks in the streets telling citizens to wash their fruits and vegetables. Never before in history had there been such worldwide concern about a scientific issue.[10]

Fallout was perfectly situated for this role, even aside from its associations with nuclear energy and contamination. Studies of how people respond to risks in general have disentangled a number of factors that bring unusual concern. A risk is especially likely to be feared if it is largely unknown—not only invisible but something new, outside normal experience, mysterious. A risk also seems magnified if it has elements evoking dread: something bearing a stigma of horrid associations, something with a potential for unbounded catastrophe. Adding to this are social elements:

a risk is more dreaded if it is unjust because (unlike, for example, a powerful medicine) it can harm individuals who did not volunteer for it and can get no benefits, or if it can harm future generations, or if it is uncontrollable or under the control of organizations that seem untrustworthy. Every one of these elements applied to fallout.[11]

In addition, news reports of a controversy about a risk, no matter how evenhanded, will tend to make people more wary of whatever is under discussion. One reason for this may be that the reports make images of harm available—once a fearful concept is stored in the brain it will have a strong anchoring effect on unconscious attitudes. Beyond this, the media are by nature biased toward stories of exotic dangers; there is little news value in everyday reassurances. For example, when two articles on the connection between radiation exposure and cancer were published in the same issue of the *Journal of the American Medical Association*, one reporting an increase in leukemia and one reporting no increased risk, the one reporting a danger got far more news coverage.[12]

The most widely publicized and convincing warnings came from the U.S. government itself. After the BRAVO test civil defense authorities realized that it was hopeless to try to save anyone within a city struck by a hydrogen bomb. But citizens who lived elsewhere might be protected from the dangerous fallout that would drift 100 miles or so downwind. Beginning in the mid-1950s civil defense officials bombarded Americans with pamphlets, magazine articles, and films instructing them how to shield their families from radioactive dust. Such local fallout in a possible future war had little to do with the far weaker global fallout from current bomb tests. But many people failed to notice the distinction and came to fear all sorts of "death dust."

Local fallout became a source of gnawing anxiety in some regions after 1954. Soldiers who had cowered in trenches in Nevada as a test explosion erupted almost over their heads quietly began to seek financial compensation for all sorts of maladies. Over the years some of them began to feel the same worries about lingering "A-bomb disease" as the Hiroshima survivors, and the same resentment (some of them were probably suffering from what we would now call PTSD). Communities that lived within a few hundred miles downwind of the tests also began, like the *hibakusha*, to demand attention for their exposure to radiation. For example, in 1958 the respected newsman Edward R. Murrow gave a Nevada critic a national television audience for bitter attacks on the AEC. A little boy had played in radioactive dirt, the critic said, and later died of leukemia. Was he not a fallout victim? The AEC stoutly denied that there had been enough fallout from bomb tests to cause any damage. Increasingly the "downwinders," typically patriotic ranchers with respect for authorities, began to declare that AEC spokesmen were lying to them. This suspicion later proved correct.[13]

Fallout was well suited to induce a lingering anxiety, as psychologists since Freud have defined the emotion: something that rests upon helplessness and uncertainty, on the feeling that a threat cannot be escaped or perhaps even comprehended before it is too late. The biggest uncertainty came when you knew that your family had been exposed to radiation but you did not know what the effects might be. Even if you fell ill with leukemia, no physician could tell whether your particular case had come from fallout, or from natural radiation such as cosmic rays, or from some other cause altogether. The mother whose boy had played in radioactive dirt and a few years later died could never prove that the AEC had killed him; the AEC could never prove that it had not.

Americans exposed to bomb tests began, like people in Japan, to call themselves "guinea pigs." The words recalled the irradiated animals in Atoms for Peace films, ignorant of what was happening to them and helpless to prevent it—the model of a scientist's victim. It did not help when AEC spokesmen said that bomb tests were merely "scientific experiments." And it was not only downwinders, but everyone in the world, who was now exposed to a trace of extra radioactivity from the tests that continued year after year. Contamination brought on by overweening scientific authorities was no longer a vague phrase or a horror movie trope; it was real dust drifting into every home.

How does nuclear radiation in fact affect living creatures? That question was burdened by facts so unhappy that few cared to discuss them. One fact is that roughly one out of every ten children conceived will have a congenital defect. Another is that, among healthy people, roughly one out of five will die from cancer. If radiation raised the rate of birth defects or the rate of cancer in some group by a small fraction, say from 10.0 percent to 10.1 percent, the change would be invisible among the normal fluctuations in the rate. Thus Teller was correct: harm from global fallout was invisible, compared with ordinary ills that the public rarely discussed. Yet if such a slight percentage increase did occur, among 100 million deaths in the world it could cause a million more. Thus Pauling, too, was correct: nobody had proved that fractional increases were *not* adding to the total burden of human suffering.

There was never any doubt that radiation in bursts of several dozen rems or more causes injury. By the mid-1960s it had been proved that a few dozen fetuses exposed to heavy radiation from the Hiroshima bomb had been born mentally deficient as a result. More important, on top of the tens of thousands of cancer and leukemia deaths that were normally expected among the Hiroshima and Nagasaki survivors, by 1980 there were a few hundred additional deaths among people who had been exposed to some dozens of rems. Similar risks threatened those who had ingested a large amount of radioactive dust (although the most common harm was treatable thyroid abnormalities). Genetic defects were far harder to detect; for humans this problem was never found even at Hiroshima. But nobody doubted that this was another real risk. The result would not be outlandish two-headed mutants but a few more of the usual defects such as muscular dystrophy or an unusual allergy.

What about smaller doses of radiation? An average person was exposed to nearly two-tenths of a rem per year. Roughly half of this came from cosmic rays, from traces of radioactivity in ordinary rocks, and from other natural sources; the other half of a person's normal irradiation came chiefly from X-rays and other medical procedures. In the early 1960s, global fallout added about a hundredth of a rem per year to the normal two-tenths. The question was whether that addition made any difference.

According to one hypothesis, there was a threshold. Five hundred rems given in one burst will kill most people, but cancer patients are routinely exposed to several times that amount in total, spread out over a long period; the patient is weakened but the cancer cells may die. Many experts believed that if the dose was spread far thinner, for example, to less than a tenth of a rem per year, it would cause no harm at all. According to a different hypothesis, which became scientifically respectable only in the 1950s, there was no threshold for cancer or birth defects; all that mattered was the total quantity of radiation no matter how it was divided up. Under this no-threshold hypothesis, whether 100 people get one rem apiece or 100 million people get one-millionth of a rem apiece, in either case the radiation would produce the same total number of cancers. That was how Pauling calculated that the many rems' worth of fallout from tests, though thinly spread around the world, would harm a great many people.

Scientific evidence gave no way to choose between these two hypotheses in the 1950s, and a half-century later it still gives no way. Hundreds of studies suggest that radiation will indeed cause some types of damage even at rather low levels. Hundreds of other studies found very minute amounts of additional radiation causing no harm—or actually provoking the body to strengthen its immunity to disease and stimulating mechanisms that heal genetic defects. Thus Teller's remark that a slight addition of radiation might be beneficial to the world's health, which his critics took as an obvious lie, was just as valid in terms of scientific knowledge as Pauling's claim that fallout might possibly kill millions.

The natural medical response to ignorance was to play it safe. Panels of experts began to choose the no-threshold hypothesis, and demanded that exposure to radiation be kept as low as could reasonably be achieved. No doubt global fallout was insignificant compared with natural radioactivity; nevertheless it just possibly might be causing a worldwide tragedy. By the late 1950s even AEC staff members were admitting that bomb tests might bring random deaths or genetic damage—yet few of them said the tests should cease. Meanwhile, knowledgeable critics understood that the risk from test fallout was far less than the damage done by the widespread overuse of X-rays—yet few spoke out against misused X-rays. On both sides, most people were less interested in radioactive dust than in the bombs themselves.

Convinced that the future of the world lay in the balance, both sides reached for the most emotion-stirring arguments. In a prime example of the power of imagery, the question of whether to ban nuclear weapons became almost indistinguishable from the question of whether or not radioactivity was an uncanny horror. On one side, a formidable governmental public relations apparatus replied to the critics by proclaiming ever more loudly the trustworthiness of government plans and the safety of radioactivity itself. Within the testing nations (the United States and the Soviet Union, joined in 1952 by Britain and in 1960 by France) the highest levels of government labored to convince people that a little fallout radiation was tolerable. For they urgently wanted to test new warheads. Halting tests would mean halting a vital project: the creation of bombs good enough to deter nuclear war.

Teller more than anyone else devoted himself to this project, which he thought necessary for the survival of freedom and of life itself. As he struggled to brush off the personal attacks of his critics, he rose to a sort of optimism. Surely the AEC weapons laboratories could create truly useful explosives? In June 1957 Teller and other physicists visited Eisenhower to say that tests must continue so they could perfect a device free of all but a minimum of fallout. Eisenhower, impressed, promptly announced that the government was working to devise bombs as "clean" as possible. (By implication, of course, normal bombs were disgustingly *unclean*.)

A fallout-free device might grant civilian benefits too. Taking their motto from the Bible's finest passage about a future Golden Age, Teller and his AEC colleagues proposed to beat nuclear swords into plowshares. Project Plowshare would "make deserts bloom" (yet again) as well as gouge out harbors and carve a sea-level canal across the Isthmus of Panama. "We will change the earth's surface to suit us," Teller exclaimed. For a few years nuclear explosives became part of the worldwide Atoms for Peace enthusiasm.[14]

Not everyone was attracted by the idea of such "control over nature," as Teller called it. The AEC's critics pointed out that excavating a canal with nuclear explosives could scarcely fail to produce radioactive fallout. But the most striking fictional picture of a Plowshare program was a 1965 movie in which an overweening scientist fired an atomic bomb down a shaft into the earth, hoping to release "limitless, clean" energy. He only broke open the disastrous *Crack in the World* of the film's title.

There must have been some particular meaning in that image of dangerous probing down a deep shaft, for the same image showed up in many other tales, from the original *Last Man* book to the popular movie *Beneath the Planet of the Apes*. In fact the secret tunnel is one of the most common images I have found connected with nuclear energy in fictional books and movies. There is no way to say whether many people made the further association between this image and infantile fantasies of forbidden excursions toward the mother's womb. Certainly, in more general terms many people feared that scientists pursuing peaceful benefits would end up, as a classic science-fiction story put it, "digging too deep and blowing up the whole shebang."[15] Whether for proposed Plowshare experiments or actual bomb tests, thoughts of radioactive pollution quickly veered toward thoughts of punishment by vast explosions.

It became common for critics to deliberately associate ideas about test fallout with images of war. An example was a five-minute 1961 film called *Falling Out*, a collage of images of rain falling, charred bodies lying in a bombed street, a red telephone, and charts of ever-larger bomb tests, all without commentary. Such indiscriminate association of images dismayed scientists whether or not they favored tests. As one biologist put it, there were many ways in which the United States affected people's lives around the world, for better or for worse, more than with traces of radioactivity. He thought arguments for and against the tests were a diversion from the real issue, which was nuclear war itself.[16]

That is what psychologists would call "displacement," a hostility that shrinks away from what is too threatening, directing itself onto some other target instead. A familiar example is the man who suffers silently through a bad day with his boss, then goes home and scolds his dog. A combination of fear and hostility can turn up far from the original source, as in children who cannot admit their feelings about angry parents or death itself, and instead become fearful and hostile toward ghosts or nightmare monsters.[17]

All the classic elements of displacement were present in the fallout debate. There was a threat that some people (as they admitted) found too great to face directly, namely the threat of war; there was a connected hostility; and there was a convenient nearby target. Displacement en masse can never be proved for certain, but I suspect that many of the feelings expressed toward fallout from bombs were originally provoked by the bombs themselves.

The displacement of feelings made a sort of sense. If a problem seems insoluble we can go after a piece of it, not so much displacing activity as concentrating it at one point. Pauling said frankly that although his agitation against bomb tests stressed fallout, it was the weapons themselves that most worried him and his fellow critics. Other leading opponents, while pressing their fight against fallout, agreed that their primary goal was to reduce the likelihood of future war. They believed a halt to testing could be a first step in reducing the tensions of the arms race; besides, the fallout issue drew attention to the entire problem of the bombs.

The opponents of testing found they could draw attention above all by talking about milk. The U.S. Public Health Service had begun to monitor fallout in foods, and samples of milk happened to be particularly easy to gather over a wide area. By the spring of 1959 the laboratories noticed a sharp rise in amounts of the isotope strontium-90, which tends to settle in bones, and especially the rapidly growing bones of children. Within a few years the concentration of strontium-90 in American children had doubled.

Elite magazines like *Consumer Reports* and *The New Yorker* carried frightening warnings. Even *Playboy* published an editorial attacking "The Contaminators," exclaiming that all children "may die before their time . . . or after having spawned grotesque mutations." Editorial cartoonists and groups favoring disarmament propagated a new image, a milk bottle labeled with a skull and crossbones. As dairy companies worried about sales, authorities insisted there was no hazard; in January 1962 President Kennedy publicly drank a glass of milk and announced he was serving the delicious beverage at every White House meal. Radioactive milk was in fact rarely the worst hazard from bomb tests; the radioactive content of vegetables or wheat, for example, was often much higher. The authors of a newspaper advertisement against bomb tests explained frankly why they chose this issue: "Milk is the most sacred of all foods."[18]

Thoughts of poison in a sacred food called up old ideas going back to accusations of witchcraft. Poison, violation of nature, obscenity, pollution— increasingly, people were applying such words to fallout. The imagery was epitomized by a 1961 editorial cartoon published after the Soviet Union suddenly resumed bomb tests, showing Khrushchev flying over the world on a witch's broom trailing noxious black clouds. All the feelings that nuclear bombs were filthy corruptions of nature had spread into feelings about radioactive isotopes as well. The underlying metaphor was best expressed by a five-year-old who warned a little friend that he shouldn't eat snow "because there is a piece of the bomb in it."[19]

By 1962 the AEC had lost its most important battle. Most people had come to see "death dust" as a condensed representation of nuclear weapons or even as a displaced substitute for the weapons. By extension *all* radioactive isotopes, whatever their origin, were openly and permanently associated with horrible contamination. Any link to the ancient idea of life-giving radiation now seemed ludicrous.

Of course a radioactive atom is only a particle of matter, neither a magical friend nor a monster of corruption. Radioactivity is a physical process going on in every pebble, as normal as the rusting of iron. But by association with diseased *hibakusha*, malformed babies, Hollywood monsters, bomb "experiments," and more, radioactivity had come to seem irredeemably disgusting. This new attitude, resembling a primitive taboo, like fallout itself was settling invisibly into every home.

12

THE IMAGINATION OF SURVIVAL

In 1961 President Kennedy delivered an address to the United Nations with his customary eloquence, declaring, "The weapons of war must be abolished before they abolish us."[1] It was already a tired cliché that "If we don't end war, war will end us"—a line from a 1936 movie inspired by H. G. Wells's *The World Set Free*. But with the coming of hydrogen bombs, citizens began to suspect that the end of humanity was not a science-fiction story, but an imminent possibility.

Starting around 1955 a few presumed experts began to say that a war with hydrogen bombs really could mean the end of everything. The idea was spread around the world by a Communist appeal asserting that "the use of atomic weapons will result in a war of extermination of the entire human race." Over half a billion signatures were said to have been gathered on the related petition to abolish the bombs.[2] But the greatest pressure came from events. Now and then an American president hinted that he was thinking of using nuclear weapons, for example, when Eisenhower insisted the United States would defend islands in the Formosa Straits against a prospective Communist Chinese invasion in 1955 and would defend West Berlin against a threat of a Soviet takeover in 1959. The American military in fact doubted that nuclear weapons could be used to good effect if a conflict broke out in those places, and the impact of Eisenhower's veiled threats on diplomacy seemed slight. The chief impact was on the world public: each event triggered an avalanche of outcries foreboding universal doom.

There was no single moment when most people began to feel they were living on the edge of universal death. Even in the 1960s a substantial minority around the world felt that although a war might kill them personally, their nation would survive handily, while an equally substantial minority had feared since the early 1950s that all civilization would soon perish. For many Americans the turning point came in October 1957, when the first

Soviet Sputnik orbited overhead. Unstoppable missiles had seemed like a fantasy of a remote future; now they could drop on the United States next year. The American press erupted in almost hysterical alarm. An urgent question weighed on any thought of the future: What would the world look like if the bombs fell?

In the imagination, nuclear war often meant no future but an empty one. That was the image carried, for example, by the statement that the world's armaments included "enough explosive force to equal ten tons of TNT for every man, woman, and child on the face of this earth," as President Lyndon Johnson said on television in 1964. The image assigned each of us a sort of personal bomb. That future was a void.[3]

Still more purely symbolic was the "clock of doom," reportedly suggested by Teller, featured on the cover of every issue of the *Bulletin of the Atomic Scientists*. The clock first appeared in 1947 set at seven minutes to midnight, and the editors put it forward to a hair-raising two minutes in 1953 when hydrogen bombs arrived. The symbolism was plain: nuclear midnight meant the end of time.

The idea was developed to greatest effect by an aeronautical engineer, Nevil S. Norway. With his weary eyes and long nose, he had the look of a bloodhound sniffing after something unpleasant, and indeed his autobiography was much concerned with death and disaster. In 1955, when he started writing a new book, a series of heart attacks had sunk him in thoughts of mortality—it would in fact be his last book. But the engineer never said much about his morbid preoccupations; he did not even say why he adopted as a pseudonym his mother's maiden name. It was as Nevil Shute that he published *On the Beach*.

The novel imagined that in 1963 a Third World War wiped out all life in the Northern Hemisphere. Australians, untouched by bombs, tried to carry on normal lives while awaiting inevitable extermination as deadly fallout crept south on the winds. The book was serialized in more than forty newspapers in 1957, and by the 1980s the paperback edition had sold over four million copies, the highest sales of any novel centered on nuclear energy.[4]

The veteran film director Stanley Kramer saw a chance not only to market a good story but to caution the world. In 1959 audiences from New York to Tokyo left the theater stunned or weeping; I found that even a quarter-century later people remembered *On the Beach* with strong emotion. The audiences were drawn into Kramer's busy world of decent, ordinary people, then watched that world gradually die away. At the end there were only miles of empty streets. It was as if the pictures of cities with deserted streets in civil defense drills were to become a permanent reality.

U.S. government experts insisted that the story's premise was scientifically ridiculous, for all the weapons in existence could not possibly spread enough radioactivity to sterilize the planet. That argument missed the point. Szilard and others had speculated about 1,000-megaton infernal machines encased in cobalt metal to generate tons of fallout, enough to sterilize the globe—a "Doomsday Machine." Nuclear strategists pointed out that you could build such a device and set it to go off automatically if your homeland was attacked; that would be the very epitome of deterrence. To be sure, no such devices existed. But as we have seen (Chapter 10), when people vicariously experience something by seeing it in a movie, or even just by hearing it in a story, they may viscerally accept it as factual, almost as if they were witnessing it in person.

Teller rightly remarked that "The cobalt bomb is not the invention of an evil warmonger. It is the product of the imagination of high-minded people who want to use this specter to frighten us into a heaven of peace."[5] If doomsday bombs had not been built, what mattered psychologically was that such things *could* be built. Szilard's point was that as the number of bomb-tipped missiles mounted into the thousands, nations were already moving toward the equivalent of a Doomsday Machine. For the first time in history it was not humanity's limited capabilities that prevented us from committing racial suicide, but only our good sense.

This made another criticism of *On the Beach* the crucial one. Would people really act as the film showed them, exploding cobalt bombs and then living on almost normally, finally taking suicide pills? Teller, for once agreeing with *Pravda*, cited a Russian comment: "What manner of men will accept the end without resistance?"[6] The implicit reply was that nearly everyone was already passively accepting a gradual approach of the nuclear midnight. Was not the world public like the fictional Australians who shut their eyes to imminent death?

If that was Shute's and Kramer's message, it eluded most people. *On the Beach* said far less about how to prevent nuclear war than about a more personal problem that Shute, a dying man, felt keenly: acceptance of separation and of death. Audiences came away from *On the Beach* and similar works not with questions about military policy but with a sense of tragic fate.

Much later, some observers remarked that people could not think about the end of the world in the same way they thought about ordinary events or even traditional wars. An entirely different, "apocalyptic," mode of thinking came into play, a realm of nightmares, fantasies, and religious visions. The death of one child would upset you; the end of everything could be contemplated only abstractly as if from a great distance.[7]

People looked for more than blankness in the space following a future nuclear war. But that was a space outside authentic experience, a space visible only to the imagination. People who peered into the fog ended up projecting images already within their heads.

"Who survives?" asked a newspaper reporter in 1947. At best a few remnants, he answered, "too weary to rebuild, too weak even to dream, dully scrabbling like their Neanderthal foresires in the primordial ooze."[8] The first widely seen movie to show a world after atomic bombs, the 1950 *Rocketship X-M*, showed "crazed, despairing wretches" shambling through a radioactive desert. A ridiculously cheap production, the movie made enough money to found a genre. In 1952 *Captive Women* featured barbaric tribes of survivors, more or less warped by radiation, battling one another in the ruins of New York; in the next dozen years *World without End, Teenage Cave Man, The Time Machine*, and *The Time Travelers*, and beginning in 1968 the six *Planet of the Apes* movies and two derivative television shows, won an enormous audience with variations on the same theme.

These savages had a heritage. People who nodded wisely at editorial cartoons of cavemen shambling through the ruins after a nuclear war had forgotten that long before *On the Beach* and civil defense exercises, the nineteenth century had appreciated the image of an Empty City. Romantic painters had depicted the proud landmarks of London or Berlin in some future age, fallen like the broken temples of Rome, overtaken by vines in a resurgent wilderness. The downfall of cities had a certain appeal. Farmers squeezed by urban creditors and clerks ground down by mass society listened to populist politicians and intellectuals who argued that the world would be better off without its cities, those garbage heaps of grasping bankers and criminals. Traditionalists everywhere longed for a world without the cities where "the good habits and simple customs of the rural communities are gradually being destroyed" (as explained in a 1913 report by the military commander of the Hiroshima district).[9]

The image of a savage clad in skins gazing at the wreckage of Wall Street appealed directly to longings for an Arcadian idyll. By 1941 the theme was so common that a reader complained to *Astounding Science-Fiction*, "Why must we always have Rousseau's Noble Savage, with his biceps, his stone ax and his mate, crawling around the ruins of mighty Nyawk or Chikgo?"[10] After 1945 authors only needed to add atomic bombs to transform this pulp genre into an image familiar to everyone.

The symbolic meaning became clear when the survivors were pared down to a few individuals, as happened, for example, in two important American films: *Five*, in 1951 the first serious film on the aftermath of nuclear war; and *The World, the Flesh, and the Devil* in 1959. Each put a handful of survivors into the empty streets of cities where everyone else was gone, cities that somehow lacked landscapes reduced to rubble and rotting corpses. By the end the survivors seemed happy enough. "I hated New York," the young hero of *Five* confided to his girl; "I'm glad it's dead." Now they could enjoy one another without inhibitions—as in the private fantasies that some teenagers had about sex play in bomb shelters. When villains showed up, they could be hunted and shot like beasts. Movie audiences or science-fiction readers could picture themselves indulging in gunplay and sexual adventures that would have been forbidden in the old society.

Even the thought of being a Last Man had its appeal. Survivors in after-the-bombs movies helped themselves to abandoned cars, food, and empty houses. A more subtle liberty was noted by an adolescent girl who told her therapist that she liked to imagine herself alone on the earth, free from her anxieties about relations with other people.[11] Such narcissistic dreams of escape haunted many a nuclear war story. Best of all, when a Last Man roamed through the hackneyed science-fiction magazine stories and television shows, audiences would hardly be surprised when he met a Last Woman. After-the-bombs stories usually ended with survivors setting out amid sunlight and twittering birds to begin the world anew. Unfettered barbarism led to renewal, perhaps another turn of the Wheel of Time as civilization prepared to rise once more. Nuclear weapons brought nothing new to the theme. The author-director of *Five* had conceived its plot in 1939. He reached still farther back for its ending, quoting from the Revelation of St. John (21:1), "I saw a new Heaven and a new earth."

The whole cycle might be shortened, according to civil defense training films of the early 1950s. Millions of Americans watched actors portray families that took shelter from bombs, then emerged a little later with their clothes only slightly rumpled. The next step, the father might explain, would be to "wait for orders from the authorities and relax."[12]

Only slightly more realistic was Pat Frank's 1959 novel *Alas, Babylon*, which next to *On the Beach* was probably the best-selling book about a world after the bombs. Frank depicted a community in rural Florida, far from the explosions. It suffered little worse than attacks by a few marauders who were hunted down by vigilantes in an exciting adventure. At the conclusion the town had become admirably self-sufficient. This was not a realistic postwar society of squalid camps filled with sick refugees, but a haven for sturdy individualists with fishing poles, determined (as an advertisement on the paperback put it) "to build a new and better world on the ruins of the old."[13]

Survival of the fittest, Frank suggested, would kill off the weak and the evil: a world not just restored but improved. Other American storytellers agreed that nuclear war would at least rid the world of Communism. Everywhere it was hinted that society might be purified by its passage through flames and chaos, like lead transmuted into gold.

Social transmutation was traditionally associated with spiritual renewal of the individual. In *Alas, Babylon* and other stories, disaster did indeed leave protagonists more strong and loving. Nuclear war novels and films also showed a remarkable affinity for straight religious feeling, inserting quotes from the Bible with an obsessive frequency not found in most other genres. And the only after-the-bombs novel of the period to win some respect from literary critics was frankly centered on religious salvation.

The work of Walter Miller, an ordinary science-fiction writer and devout Christian, *A Canticle for Leibowitz* sold over a million copies in the two decades following its publication in 1959.[14] The tale followed an order of monks who preserved scientific knowledge through a new Dark Age following nuclear war, although they suspected that overweening humanity would misuse the knowledge and rebuild a wicked secular civilization. Their only solution was the traditional one, to look for redemption by divine grace—that is, the ancient passage through suffering to a better life.

Decades later Michael Ortiz Hill explored in depth the combination of the struggle through the "dark night of the soul" with nuclear fears, by studying dreams with nuclear thematic content reported by nearly a hundred people.[15] Hill found frequent accounts of a mythic struggle with a personalized Bomb, symbolized in countless ways. But surprisingly, the dreamer rarely fought by matching the bomb-monster's magical force with the corresponding power of a hero or heroine. Sometimes the dreams showed only a victim. More often the dreamer won a kind of victory simply by living on: the Hero as Survivor. This is the simplest form of passage through the dark valley, without being transformed or reborn but finding enough victory in just getting through. In other dreams the hero did better, perhaps managing to prevent nuclear war. Typically this was done not through force, but through craft and subterfuge. The bestial monster's heroic counterpart was its opposite even in the manner of fighting, suggesting the immemorial Trickster, found in many folk cultures. Perhaps the modern nuclear apparition was simply too overwhelming to be speared like a dragon.

Often, however, the survivor-hero was a guileless innocent. Some cultural observers remarked that children in particular tended to show up frequently in nuclear productions. They were prominent not only in fiction but in wrenching stories about victims of Hiroshima and in the fallout protesters' warnings about deformed babies. "Without a question," wrote Hill, "children are the largest discernible group of people that inhabit the dreamscape of Apocalypse. Children appear in over a quarter of the dreams I've collected." The innocent child may represent simply the war-monster's victim. But it is often a survivor, or even a heroic protagonist: in both dreams and fiction a child may help prevent the war from ever starting. This is just what we would expect in a complex of transmutational imagery centered on rebirth, which necessarily calls up the image of a child.[16]

The most profound symbolic use of the innocent child survivor was in the resolution of A Canticle for Leibowitz. In the book's final section civilization was indeed rebuilt, only to fall again in nuclear war. As the last abbot died in his blasted monastery he witnessed a saving miracle: a pure birth, possibly the Second Coming itself. It was as if nuclear war, like individual martyrdom in pain and guilt, was a gateway to spiritual redemption.

No idea could have been more dangerous. Perhaps an individual personality does need to go through chaos before it can hope for full rebirth. However, to suppose that an entire civilization can be redeemed by passing through catastrophe—that is a dangerous fallacy. Wars do not leave most people more loving and creative.

If the wretched survivor became a new person, if escape from victimization equaled rebirth for the individual or all humanity, there could be a chilling deduction. Shouldn't people *want* the bombs to drop? The idea was seldom discussed, for it led into murky questions that few people cared to face.

Deliberate destruction, already central to the old mad-scientist stories, reappeared in the postwar movies about anguished monsters that stamped around smashing everything. The idea could also be uncovered in the clinic, for some angry paranoiacs felt that the way out of their misery was not merely to escape the attacks of their oppressors, but to abandon self-control and bring down wholesale destruction. The victim who became a destroyer, and then a survivor, would seize the magical secret of death, life, and rebirth itself. Did these turbulent fantasies, structured like the traditional mad-scientist tale, appear in stories of nuclear war? The answer is yes, only with a crucial omission.

Every after-the-bombs book and film incorporated standard items of victimization: mutilation, incapacity, lack of healthy progeny, separation from loved ones (perhaps from all human society as a Last Man), and simple death. Most of them worked through this to a transmutational happy ending. The tale that deployed the most complete set of themes was a widely read English science-fiction novel of 1955, aptly titled *Re-Birth*. John Wyndham described a primitive after-the-bombs village bordered by a forbidden zone, a radioactive "Badlands" infested with degenerate mutants. The villagers failed to notice that a few of their children could exchange thoughts by telepathy. When the protagonist's father discovered this new mutation in his son, he condemned him and his friends to death as "deviants." Just in time a group of adult mutants showed up, a new and superior species that slew the adult villagers and carried the mutant children to a new White City where thoughts mingled in communal friendship and wisdom. These themes were not the idiosyncrasies of one writer; each of them—radioactive Forbidden Zones, degenerate creatures, murderous fathers, telepathic prodigies, and harmonious new communities—appeared again and again in one or another book or film connected with nuclear war.

Most widespread were the horrible creatures. Entire tribes of loathsome mutants, like the rubble-strewn Badlands they inhabited, put the bombs' blasphemous pollution of the earth into visible form. Supposedly descended from irradiated humans, they were in fact descended from mad scientists' monsters. Other after-the-bombs productions, like the archetypal 1981 movie

The Road Warrior and sequels, used humans themselves, in the form of criminal gangs, as the bestial enemies of civilization. Almost as common was the trope of the secret. In many an after-the-bombs tale survivors ventured into Forbidden Zones or even crawled down tunnels into the earth in search of the mysteries of the old civilization. The search might be opposed by an oppressive tribal leader, the one who had made the zone forbidden and otherwise played the dangerous parent.

At this point the argument runs into a difficulty. Explorers in after-the-bombs tales usually found not dire punishment but revelation and renewal. Something essential was missing. The real violation, the blasphemous act that had created the victims and monsters in the first place, was invisible—somewhere back at the time the bombs fell. If after-the-bombs stories resembled mad-scientist stories, then where was the mad scientist himself?

Virtually no author of the 1950s offered a character who was directly responsible for any of the political or technical acts that caused nuclear catastrophe: the Man Who Started the War did not show his face. It became a convention to avoid naming even the nation that triggered the fictional war. When writers did name someone, Americans of the 1950s took it for granted that Communists would be to blame, while Communists emphatically asserted the reverse. If anybody was responsible, it was nobody the writer knew. In this central omission, the first wave of nuclear-war stories seemed very distant from the mad-scientist tradition. Even radioactive-monster films omitted him: Godzilla and his ilk were not created by any identifiable villain. Like the mutants that scuttled through radioactive ruins and like the primitive human survivors themselves, they were victims of distant and nameless acts.

During the 1950s mad scientists continued to flourish in their own movies and stories, remote from nuclear war. A revealing example was the most highly praised science-fiction movie of the decade, *The Forbidden Planet* (1956), in which the traditional island stronghold of the sorcerer-scientist became a faraway planet. There Dr. Morbius meddled with an underground nuclear energy device that could make wishes materialize. Morbius inadvertently created a murderous monster—what he finally had to admit was his own "evil self," a condensation of his "subconscious hate and lust for destruction." Morbius's "monster from the Id" might have been an apt metaphor for the true problem of nuclear war—but neither the script-writers nor film critics seem to have noticed that.

It was not just the normal human lust for destruction that people some-how failed to connect with nuclear weapons; there was a related idea even less often brought up. Morbius exposed it in the end when he got rid of his monster by blowing it up along with the entire Forbidden Planet and him-self too. He was simply following his prototype, Jules Verne's Captain Nemo of 20,000 *Leagues under the Sea*, who (in the highly successful Walt Dis-ney film of 1954) ended his quest for a better world in an atomic blast. The inventor destroying his enemies along with his own island, like the mad scientists who tried to blow up their own cities or the entire planet in other stories, stood for a most unsettling idea: mass destruction combined with suicide.

Suicide was a serious matter. Every year hundreds of thousands did per-ish that way, far more than were murdered. And suicide-murder combina-tions were only too familiar. Indeed, many psychologists believed that the combination was a fundamental one. Some would-be suicides cherished a hatred of inner evil, a self-loathing, that might easily be projected outward to inspire attacks on others. There were even a few authentic cases of dis-turbed individuals who worked on scientific plans for the destruction of all humanity.[17] Many suicides held fantasies of death as the doorway to a peaceful afterlife, or said that at least their death would make the rest of the world better. Psychopaths who longed to destroy not only the wrongness within themselves, but everything around them, imagined destruction as a route to control over death, to magical survival, and to a better world.

Nevertheless the theme of deliberate global suicide scarcely appeared in after-the-bombs stories. *On the Beach* and *A Canticle for Leibowitz* did show crowds lining up to get poison pills, but audiences and critics did not seem to notice any connection with the forces that could lead to war. From the 1950s on the word "suicidal" became a favorite polemical description of nuclear war, but as an abstraction whose real meaning few explored. Only a very few psychologists suggested that war might in fact start from a sort of "Samson complex," the deliberate desire to destroy one's enemies and oneself, to wipe the slate clean.[18]

Had the motives that had razed a thousand cities since ancient times— the desire to survive one's enemies and to build a better world upon their ashes—evaporated with the arrival of hydrogen bombs? There is a better way to explain the remarkable omission of a major traditional character, the self-destructive mad scientist, from a collection of tales in which he should have been perfectly at home. Perhaps mass destruction had become so plausible and dreadful that people did not dare to connect it with the desire to bring it about. If so, then at least some of us did resemble the chemist I mentioned in Chapter 4, who dreamed at night of vengefully destroying the world with atomic bombs while in daytime he hid his anger even from himself.

To investigate this I will take a few authors and try to see what web of associations lay within each mind. When that mind originated influential tales, it can show something about what underlay the images impinging on the rest of us. I have already mentioned that the author of the original *Last Man* book was a lonely man who killed himself soon after it was completed; the book itself included the idea of a relentlessly probing researcher working in a laboratory deep within the earth. Evidently an embryonic form of the mad-scientist theme shared space with world doom and suicidal urges in the writer's thoughts. Biographers of the author of the next major *Last Man* book, Mary Shelley, found she was troubled by conflicts with a coldhearted father and by the loss of her mother at birth, problems that had already helped her to create Dr. Frankenstein. The world-catastrophe novel and the self-destructive mad scientist seem to have had a common foundation in her anger and loneliness. For other authors who developed the mad-scientist stereotype, from Nathaniel Hawthorne through Jules Verne to Karel Čapek, biographers have in each case suggested a connection with unhappy relationships with the author's father. But to combine the mad scientist with nuclear doom required further personal complexities.

A particularly important case was Philip Wylie. After his 1954 atomic-bomb-survival novel, *Tomorrow!*, he grew more pessimistic with the advent of hydrogen bombs; during 1963 he showed millions of *Saturday Evening Post* readers a nuclear war of the 1970s that left America empty of life. His scenes of mobs trampling one another were so gruesome that the magazine refused to print the details, but Wylie restored them when he published the tale as a book with the strange title *Triumph*.[19] A particularly familiar symbolism had appeared in Wylie's first nuclear doom story, which was also the first such fantasy to reach a huge general audience when *Collier's* published it in January 1946: Scientists sank a deep shaft in the earth to conduct atomic experiments, and accidentally set off a chain reaction that incinerated the planet.[20]

As a novelist Wylie saw himself as a crusader struggling to purge the world of hypocrisy, but he was best known for his vituperative attacks on the modern "Mom," a far from maternal creature. Wylie's biographer suggested some motives. Not only was the novelist's father a hypocrite, but his mother had died when he was five years old, leaving him with "an abiding sense of irreparable loss" and perhaps also "a smoldering rage." As he wrote nuclear war stories Wylie was ill and despondent, despising himself, ultimately killing himself with pills and alcohol. Annihilation of everything hateful, perhaps followed by survival and a new birth of virtue, was something he often wrote about but never achieved in himself.[21]

For these key writers, and others I will mention later, the end of all things on the one hand, and dangerous scientists on the other, while not necessarily linked in any single story, were bound together by a common origin in problems near the core of the writer's personality. All these writers guessed the secret of the mad scientist—his homicidal or even patricidal or matricidal lust for destruction, often mingled with suicidal thoughts and a desire to seize the forces of rebirth. When these writers, and the public who read them, failed to discuss such things in the same breath as nuclear war, the reason was not that the ideas were far from every mind. They were only too close.

In one popular medium in which the images were more impersonal and thus easier to face, the Man Who Started the War did become visible, if only as a nameless symbol. Drawing on an old pictorial cliché, editorial cartoonists drew the globe as a round bomb with a fuse, and sketched in alongside a generic fat politician or pompous general playing with matches. The planet could be recognized as a bomb because generations of cartoonists had pictured such a spherical device in the hands of wild-eyed anarchists out to blow up oppressive society and perhaps themselves too. The threat of nuclear weapons—the threat of global doom—the threat of dangerous authorities—the threat of childish meddling with forbidden secrets—the threat of an explosion of rage against modern society and against oneself: all merged into one compact symbol.

Most adults realized that real nuclear war would have little to do with the empty streets of *On the Beach*, or the mutant monsters of *Re-birth*, or the heroic survivors of *Alas, Babylon*. People who tried to think seriously about war pushed colorful fantasies and ingenious theories into an out-of-focus background, leaving actual experience in center stage. This experience was limited. The public had become familiar with the basic bomb effects described by atomic scientists, some tales of Japanese victims, and a few dozen photographs of Hiroshima wreckage. Many people were like a woman who said, "What's imprinted on my brain is those photographs that I have seen, so if I ever did think of it I'd see a city destroyed and blackened and burned." In the terms of neuroscience, the woman's memory of the images was permanently linked with memories of her reaction to them, presumably the emotion of horror, along with the abstract idea of nuclear war. Every literate person saw these photographs, and they became the world public's fundamental image of a world after the bombs.[22]

It was a dangerously incomplete image. Although some exceedingly gruesome photographs had been taken of the wounded at Hiroshima, the U.S. government released very few of these through the 1950s and 1960s. The commonly seen Hiroshima pictures mostly showed vast landscapes of rubble without people, an Empty City if not a Forbidden Zone. This impersonal image was reinforced by newsreels of empty houses flattened in bomb tests and newspaper drawings of skyscrapers snapped in two. The destruction was usually viewed from an Olympian distance, as in the frequently published maps that showed with concentric circles how many square miles of a city would be pulverized. The best examples were paintings done by the master illustrator Chesley Bonestell for magazines in the 1950s, views gazing down from a great height upon a city lit by a nuclear fireball. Earlier illustrations of cataclysm going back to medieval times had put human victims in the foreground, but Bonestell's paintings might have depicted (what was his usual subject) a distant astronomical event. Nonfiction writers offered little more. Usually they confined themselves to brief, hackneyed phrases about the dreadfulness of war and some abstract statistics about how many millions would be killed.

When pollsters in the 1950s asked Americans what the world might look like after a nuclear attack, most people could scarcely come up with an image. Some talked vaguely about blasted landscapes as in the Second World War or worse. A small minority offered hopeful images of survival and recovery, while another minority spoke of utter doom. About a third gave emotional responses such as, "Oh my God it would be terrible, I can't imagine what it would be like."[23]

Although a few civil defense films of the early 1950s had offered more personal scenes, these were quickly outmoded by hydrogen bombs. The same obsolescence overtook early attempts at realistic fiction, such as Wylie's *Tomorrow!* After the IVY fusion test, fiction and movies were dominated by fantasies of mutant tribes and so forth. Three decades passed without any technically accurate and widely seen portrayal of hydrogen-bomb war.

There was almost an exception: *The War Game*, a television film produced for the British Broadcasting Corporation (BBC) and released in 1966. The carefully researched scenes, such as a group of maimed children sunk in apathy and a bucket full of wedding rings stripped from the dead, seemed too grim to show the public. *The War Game* was the first BBC film that the corporation refused to broadcast. The film played only to relatively small audiences at colleges and the like.

In the early 1960s several polls studied how people reacted to the idea of nuclear war. I will summarize the results in terms of the way eight representative Americans felt. One of the eight was certain that nuclear war would mean the end of life on Earth. A second representative citizen expected the end of civilization, or at any rate unbearably brutal conditions for the survivors. At the other extreme, one of the eight was confident that the United States could come through without much damage (further surveys showed that this person knew little about nuclear weapons). The remaining five, in the middle, felt that the United States had a chance of eventually rebuilding its society, but they expected great destruction first. The best-informed half of the public thought that their own towns would be annihilated or at best harmed by fallout, and felt their personal chances for survival were not good. (In Western Europe, the Soviet Union, and Japan the range of opinions seems to have been similar, perhaps on average a bit more pessimistic.) The most typical image was of the entire world as one boundless Hiroshima, but as the woman who spoke of the Hiroshima photographs "imprinted on my brain" explained, the image was "not elaborated."[24]

A hydrogen-bomb war would in fact have reduced nations to scattered islands of ramshackle industry in a sea of dejected, disease-ridden refugees. But aside from a few conscientious civil defense officials and *The War Game,* hardly anyone put the facts in such concrete terms, let alone tried to make them vivid. From the corners of their eyes the public continued to glimpse lurid visions of a sterilized planet, uncanny mutants, or a new Adam and Eve with backpacks.

Western Europeans and Japanese had more to draw on. Unlike the Americans, they knew from direct and recent experience what it meant for their cities to be burned to ash and smashed into rubble. It was in Western Europe and Japan, above all, that visceral fears of war became most closely connected with a determination to take practical action against the forces tending toward destruction.

13

THE POLITICS OF SURVIVAL

A ll the early active responses to the coming of nuclear weapons, from the atomic scientists' movement to civil defense exercises, were driven by small elites—mostly people working for a government. Aside from Communist propaganda organs and a few tiny pacifist groups, organized public opposition to nuclear weapons was unknown outside Japan. But in 1957 the world's newspapers began to show a few photographs of protest demonstrations elsewhere. By 1961 such photographs were appearing frequently, showing larger and larger crowds. As the news about hydrogen bombs sank in and governments pursued Atoms for Peace as an answer, ordinary citizens were throwing themselves into a very different crusade.

The breakthrough came during November 1957 in response to the shock of Sputnik. The advent of ballistic missiles spurred liberal intellectuals and pacifists to appeal to the public, and the response went beyond their hopes. When the writer J. B. Priestley published an article in *The New Statesman* calling on Britain to disarm, he got more than a thousand spontaneous letters of support; when a group led by Norman Cousins and others published an advertisement in the *New York Times*, they drew 2,500 responses. In various countries people set out to found a movement, but they did not guess what they were starting. The following Easter showed them.[1]

Some little-known pacifists had announced a protest walk from London to a nuclear bomb production plant fifty miles distant. They hoped for something like the demonstrations of a few dozen people that had been held here and there over the preceding decade; but when the column left London on Good Friday, 1958, it was two miles long. Cold wind and driving rain met the marchers during four days on the highway, but at the end 10,000 people stood shivering in silent vigil outside the barbed wire fence at Aldermaston.

Liberal and pacifist leaders meanwhile formed a Campaign for Nuclear Disarmament (CND) in Britain, a National Committee for a Sane Nuclear

Policy (SANE) in the United States, and similar groups in other countries. Like the Japan Council against Atomic and Hydrogen Bombs, founded in 1955, these organizations served as coordinating centers for swarms of small groups with little funds or staff. The movement lived on the donated services of hundreds of talented people, from artists to songwriters; on the long hours that thousands of volunteers gave to distributing petitions and arranging meetings; on the attendance of tens of thousands of protestors at demonstrations like the Aldermaston march, which became an annual event imitated around the world; and on the silent sympathy of millions.

The movement's message was embodied in a set of familiar images, bundled together and thrust in the public's face. A British pamphlet explained how bombs would kill people everywhere "in a lingering and horribly painful manner . . . just as they are still dying in Hiroshima." This was the movement's most characteristic picture, the world as Hiroshima. Many speakers and writers went on to offer science-fiction images as sober truth. For example, a physicist talking over loudspeakers to a hundred thousand people at a German rally predicted that only a few youths would struggle through to rebuild society, "perhaps beginning in the Stone Age." In short, like earlier groups the movement spread powerful images with the aim of promoting nuclear fear. Few citizens escaped hearing the message—as a British leaflet put it, "Act Now or Perish!"[2]

The author of this leaflet was the movement's patron saint, a very old man with a shock of white hair and the grim look of a schoolmaster whose charges have been up to dangerous mischief: Bertrand Russell. When the philosopher preached the imminent end of all life on earth, he was drawing on an unusual personal concern with death. All through his youth he had considered suicide, frantic to escape an unbearable loneliness. In his most suicidal period Russell had written a story of a scientist who invented an atomic gadget that could destroy the universe, and who found politicians so appalling that he decided to press the fatal button. The despair this story reflected may have begun when Russell was two years old, for his mother had died then, and he would have been unlike other humans if that loss had not left him desperately lonely. When the news from Hiroshima arrived, Russell predicted world doom and fell into what he later called "a very much exaggerated nervous fear." He found that millions shared his feelings.[3]

Russell and his fellow campaigners focused not only on abstract future danger but on present-day bomb tests, until the antibomb movement became inseparable from the fallout controversy. A typical SANE advertisement hitched the two issues in tandem: "NUCLEAR BOMBS CAN DESTROY ALL LIFE IN WAR . . . NUCLEAR TESTS ARE ENDANGERING OUR HEALTH RIGHT NOW." This was the mingling of fears that left many of the public worrying chiefly about radioactive contamination and even displacing their fear of war onto strontium-90 in milk.

The movement drew its strength from a new style of culture and politics. The CND, SANE, and their fellow organizations attracted not so much mainstream politicians and business leaders as ordinary people who were not entirely comfortable with the established order. Adherents came mainly from the middle classes; among the marchers schoolteachers and civil servants were far more numerous, if less visible, than radical Marxists. But a sociologist who studied the CND found that these apparently solid bourgeois suffered vague social discontents. They questioned not only nuclear weapons but everything from the religious establishment to unfettered capitalism.[4]

The backbone of the movement was a very large social category with an increasing distrust of authority: women. Hardworking housewives had been invaluable organizers ever since the early 1950s, when the only protests had come from small church groups and the like (especially among the *hibakusha*). Perhaps one reason these tens of thousands of women took the lead in opposing weapons was that females in many societies tended to suffer more anxiety, and to avoid risks more openly, than males did. One explanation offered was that they held less power to control their fate. "Men have always played irresponsibly with human life," some of them declared, "while women have always protected it."[5]

Most visible of all were the young. For the first time in modern history, young people made up a large part of a mass campaign. To many students the marches were only an open-air diversion with guitars, part of the new youth culture. But this was an increasingly alienated and rebellious culture. Young people began to call nuclear bombs a prime example of everything they distrusted in adult society. Indeed, most of the movement's followers believed they were fighting against coldhearted authorities and the entire aggressive culture that produced bombs. As the sociologist who studied CND supporters explained, they "tended to translate the general anxiety arising from the dangers of a nuclear war into a full scale critique of contemporary society."[6]

Foremost among these social critics was J. B. Priestley, whose article had helped start the CND. Reflecting in his old age, he came up with a clue to his way of thinking: his mother had died soon after he was born. "There were areas of dark bewilderment," he wrote. "Something was missing that should have been there." This is not to say that his mature character reflected only infantile thinking. When Frederic Soddy and Philip Wylie were left motherless they grew up lonely and bitter, talking as if the world almost deserved to be blown up. From the same loss Priestley, and Bertrand Russell too, fought their way through pessimism to a sharpened perception of human failings and a courageous refusal to despair. They threw themselves into battle on behalf of life, taking their personal longing for a trustworthy world and transforming it into a call for a better world for everyone.[7]

Priestley decried not only nuclear armament but every modern, mean-spirited form of conflict. He called on his countrymen to rise above partisan selfishness and start a moral "chain reaction" that could sweep the world. Any noble cause would do, but renouncing the bombs would be best of all. Many other campaigners agreed that in fighting nuclear weapons they were fighting a deeper moral corruption. It was not so much the thought of being killed themselves that motivated them, many declared, as the fact that their government was poised to slaughter innocent people in Moscow by the million.

Pamphlets, articles, and books debating the morality of possessing nuclear weapons arrived in a flood, more in a few years around 1960 than in all the years before or since. Some preachers and authors argued against nuclear arms on traditional pacifist grounds, adding only that war was now more hateful than ever. Others argued that the new weapons were particularly evil because it was impossible to use them in a just and measured way. Many simply took it as self-evident that nuclear bombs were uniquely blasphemous. With vigils, fasts, pilgrimages, and sermons, the movement took on the air of a religious revival. After all, nuclear weapons posed questions such as only religion had once addressed. A British protestor, arrested at an illegal demonstration, wrote from prison: "Our position is not dissimilar to that of the early Christians. With us, as with them, the last things have arrived."[8]

The campaign was starting to resemble a millenarian movement—the kind that has turned up in many human societies stressed by change.

They range from messianic "cargo cults" in tribes confronted with Western civilization to modern urban revolutionary sects. According to historians and social scientists who studied such movements, the followers would typically be convinced that their society was in the last throes of decay, and that they could be midwives of a world rebirth to something far better. Often a millenarian group simply projected onto society each member's hopes for personal transmutation. But some problems of an individual are in truth social in origin, and cannot be solved unless the distressed individual joins with like-minded people to reform society. Nuclear war was such a problem.[9]

The question posed by nuclear fear raised a creative answer: world community. This was no longer a platitude, but the realization that all humans truly depend upon one another. Fallout had begun the process, thanks to the sensitive instruments that could detect a handful of dust carried thousands of miles on the wind. Magazine drawings of ballistic missiles arcing around the globe drove home the point. The idea of world community became the movement's greatest strength not only morally but practically, as groups of various nations joined in mutual support. Compassion for all life, from the victims of Hiroshima to the citizens of Moscow to generations unborn, was the most profound answer that could ever be given to the nuclear question.[10]

But what policy was a government supposed to actually adopt? The movement's rallying cry was "unilateral disarmament": a refusal to own nuclear devices, no matter what other nations did. Some admitted that this policy might involve yielding to Soviet conquest. The idea, simplified into "Better Red than Dead," was accepted by a substantial minority in some nations. Few foresaw that nations led by Communists would have a penchant for attacking one another, so you could be red *and* dead. Nobody noticed a tremendous fact: well-established democratic governments, alone of all types of government, do not make war upon one another.[11]

Only a few supporters of the antibomb movement analyzed the problem of state aggression. Some who did attacked the very existence of nation-states. They pointed out that states had originally been found necessary mainly to fight off invaders—but there was no defense against nuclear missiles. Although few went on to demand the abolition of governments, many translated their abhorrence of the bombs into abhorrence of all governmental authority. After all, among everything else nuclear weapons stood for, they were a supreme expression of the state's possession of supreme power.

Rational argument became less and less prominent in the controversy. One reason was that the means of persuasion available to political outsiders—demonstrations and hand-painted banners, appeals to passersby to sign a petition, occasional newspaper advertisements or thirty-second radio spots—did not allow much space for sophisticated exposition. Another reason was that many of the movement's supporters tended from the outset to reject abstract reasoning.

In talking about nuclear weapons, the people most likely to use detailed logic and numbers were the men professionally involved with them, such as generals who had to decide how many bombers to assign to each target. From the mid-1950s on, the best-known center for such calculations was a flat, nondescript building fenced off among the palm trees of Santa Monica, California. This was the RAND Corporation, devoted to rational analysis. Sometimes the analysts felt that their worst enemy was not so much the Soviets as their own sponsors, veteran Air Force generals who sneered that logic could never analyze the chaos of war. Such disagreements were only to be expected in an organization trying to weld together two very different groups, scholars and military officers—a new linkage that was altering both sides.

The most famous of RAND's scholars was a young physicist, Herman Kahn. He seemed the opposite of a military man with his exceedingly round stomach and thick glasses, but he could pour out neatly packaged historical anecdotes and daring ideas (the Doomsday Machine was a favorite), keeping senior officers fascinated through hours of lectures. When Kahn published a book in 1960, controversy erupted over his unfamiliar logic.[12] Did it make sense to base national strategy, as Kahn would have it, on the distinction between a war that killed 80 million citizens and one that killed only 10 million?

Poles away from that mode of thinking stood people like Pat O'Connell, a housewife who was arrested again and again in the early 1960s for leading civil disobedience demonstrations. According to a writer who interviewed her, she saw things "simply in terms of human beings." People like O'Connell felt as a personal blow each individual story, each maimed survivor of Hiroshima, each child who might die of leukemia from fallout. Much later, in 1982, a survey confirmed that the people most active in opposing nuclear war were motivated not by technical statistics but by concrete personal images, charred bodies of people and the like.[13]

Nuclear strategists like Kahn complained that protestors, carried away by emotions, were hiding from the hard facts. Kahn insisted that nuclear weapons must be analyzed only with the coolest detachment. It began to sound as if the choice between building bombs and banning them was equivalent to the choice between rational logic and human feelings. Some protestors reversed the equation, declaring that they were the reasonable ones. Like some military officers, they found RAND's clever logic and calculations at heart irrational, meaningless in face of the turmoil of real war. Some protestors went on to insist that any talk of using nuclear bombs, for war or even for deterrence, was plain madness. The CND titled its newsletter *Sanity*, while SANE's name spoke for itself.

Who specifically was not sane? The RAND theorists in general and Kahn in particular made the most obvious target. The portly analyst was first bewildered, then outraged, when critics called his analyses not rational but lunatic. A stereotype was at work, one that Priestley had elaborated back in 1955 in a notorious essay. He had sketched the personality of "Sir Nuclear Fission," supposed to represent a breed of modern man. As a youth, said Priestley, Mr. Fission had isolated himself in abstract scientific studies, disregarding the claims of his senses and emotions. Warped by scarcely recognized frustrations, "brilliant but unbalanced," even as he made scientific discoveries Dr. Fission became divorced from the real world. Because he lacked friends and pleasures, the eminent government adviser Sir Nuclear undervalued human life and was all too ready for war. Priestley was not, of course, describing real scientist advisers, lively and often sentimental people whose careers were rich in personal interactions. He was describing the traditional mad scientist ready to blow up the world—a stereotype he had portrayed already in a 1938 thriller.[14]

There was at least one real nuclear authority who seemed much like the stereotype. Since his student days Hyman Rickover had been unsocial and obsessed with knowledge, turning himself, according to *Life*, into "a tough intellectual." The magazine added that Admiral Rickover even looked like an incarnation of relentless thought, his body no more than a slender appendage "utterly controlled by the head, not permitted to engage in frivolities." The missile submarine officers, selected and trained by Rickover, had to be equally impervious to sentiment, if only because they literally held in their hands the keys to catastrophe. Magazine articles explained that the submariners endured months locked in steel corridors, forbidden any contact with their wives; "after a while," a reporter wrote, "even their talk of sex stops."[15]

The same stereotype covered the creator and chief public symbol of the Strategic Air Command (SAC), General Curtis LeMay. He was the picture of toughness, with his bulldog face and scowling eyes, a cigar clamped between compressed lips; and anybody who doubted that the general was ready to destroy cities had only to recall that he was the man who had designed the Tokyo firebomb raids. Magazines described LeMay, and by implication SAC, as "more machine than man," a creature of "irrefutable logic."[16] It was the logic of deterrence, after all, that made SAC's mighty armament desirable: the airmen's paradoxical motto was "Peace Is Our Profession."

Popular movies told tales of marital conflict when a SAC pilot showed more dedication to his deadly airplane than to his wife (of course in the end the wife patriotically acquiesced). Blind emotion was unacceptable in men who could drop fearsome bombs at will. The public was not told that LeMay's airmen and Rickover's submariners suffered from widespread psychological problems and a high divorce rate. The acceptable image was of officers who had made themselves into logical machines, the better to protect the nation from war itself. As a historian later explained, many citizens clung to a culture of stability, relying on the nuclear predominance of the United States to keep the world safe. This culture was the enemy of the new, chaotic, dissenting culture that likewise centered its concerns upon nuclear weapons.[17]

The stereotype of nuclear officers took its purest form at the missile bases the United States built in the early 1960s. *Life* showed men in underground capsules, wearing white coveralls like laboratory technicians. One told a reporter, "We're like robots in a way." The only thing these men seemed to have in common with military heroes of the past was their suppression of sentiment in the name of duty. The image of the soldier as steely defender of his homeland was starting to look . . . creepy. A physicist remarked that tales of soldiers heroically interposing themselves between their loved ones and the desolation of war had become "fairy tales, and not nice ones at that."[18]

Links between officer and scientist were being forged in real places like the briefing rooms of RAND and the burgeoning weapons laboratories.

This linking would have happened to some extent even if there had been no nuclear weapons, but the bombs did more than anything else to label military officers as masters of awesome technology. In wide regions of popular culture it was getting hard to distinguish the military officer from the mad scientist.

14

SEEKING SHELTER

While the movement for nuclear disarmament was getting under way, people who believed in tough-minded logic and military strength laid plans of their own. In 1958 Eisenhower was taken aback by intelligence reports claiming that the Soviet Union was building missiles at a furious pace: it could soon have ten times as many as the United States! (In fact the early 1960s would see just such an imbalance . . . in the Americans' favor.) The worst-case estimates quickly found their way into the press, and Democrats began to attack Eisenhower for allowing a "missile gap." If the Soviets got far enough ahead, wouldn't they believe they could attack with impunity?

This sounded like an echo of a debate of the mid-1950s, when American generals and senators had cried "bomber gap" while demanding funds for the Air Force. In both cases people deliberately fostered a feeling of insecurity so the public would approve buying more weapons. Of course nobody could expect conscientious military officers and their supporters to say that the nation had enough arms to guarantee perfect security. Like many other actors in this story, they felt duty bound to remind people of mortal dangers.

What if deterrence did fail? Most people had not even tried to think beyond that possibility. But now the prospect of a real missile war began to preoccupy governments. One essential element would be an effective civil defense program. That opened the way for a new assault of nuclear imagery on the public mind, striking closer to home than anything so far.

The advent of hydrogen bombs had thrown Eisenhower's civil defense program into disarray. As a historian later remarked, after 1955 "discussion of civil defense in the cabinet . . . assumed a disconcertingly surreal character. Policy makers seem to have lost their footing." When Kennedy became president he promised a revived program. Civil defense officials, starved for public respect and funds, hopefully bombarded the new president with memoranda. Imitating his campaign slogans, they said that a shelter program would "awaken the country" to defense needs and "show the world that the U.S. means business." Still more persuasive to Kennedy was an argument for shelters as "insurance" in case of war.[1] A family that stayed beneath a few feet of dirt for a few weeks after bombs fell would greatly improve its likelihood of surviving fallout. How could anyone refuse a chance to save millions of lives? Kennedy was nevertheless unwilling to ask for the enormous multibillion-dollar program that civil defense experts recommended. The one thing he could do easily was warn people to take care of themselves, and he began advising Americans to build shelters.

The matter became urgent when Khrushchev threatened war over West Berlin and Kennedy made an equally belligerent reply. A nuclear scare built up, worse than any before, frightening the public around the world. It came to a climax in a tense speech the president gave over national radio and television in July 1961, implying that the world was on the brink of war. Kennedy warned citizens they must be ready to protect their families, and said he would ask Congress for funds to stock shelters with food, water, and first-aid kits (an indelicate mention of "sanitation facilities" was cut from the text at the last minute).[2]

Kennedy's speech shocked the nation. The federal civil defense agency got more than 6,000 letters a day asking for information, more than it used to receive in an entire month. In December a Kennedy aide canvassed the nation and reported that shelters had become the chief domestic concern, a fad verging on hysteria. Kennedy had announced that he would send a civil defense booklet to every household in the nation, but his aides, taken aback by the uproar, kept delaying the publication for fear of setting off complete panic. Other people hurried to fill the vacuum. Swimming pool contractors became experts on shelter construction, and manufacturers of everything from biscuits to portable radios discovered that they were already making survival equipment. Fly-by-night shysters and respected merchants advertised on television, encouraging the war anxiety. (One bank, apparently not excessively worried that war was imminent, offered loans to build shelters with "up to 5 years to repay.")[3]

The antibomb movement, seizing the issue as another opportunity to spread frightening images, vehemently attacked fallout shelters. Those would save only a few people, they said, who would anyway emerge into a world of rubble and would envy the dead. Shelter advocates argued that shelters would help dissuade the enemy from launching the dreaded First Strike.

Opponents replied that civil defense would give Americans false confidence, make the Soviets more nervous, and so render a First Strike more likely.[4]

The debate turned bitter when critics noted that there would never be enough completely safe shelter space for everyone. Novels and films featured desperate fights at shelter doors—the old theme of people reverting to savagery, like "naked wild animals," as one television drama suggested. In fact most shelter owners were willing to let in neighbors and even strangers, but about a quarter of the owners did say they would fight off intruders, and a few displayed rifles. Reporters exaggerated the issue, drawing peculiar statements from clergymen as to whether or not Christians had a duty to shoot someone who was trying to break into the family shelter. An aide warned Kennedy that the spirit of "do it yourself" that had inspired the home shelter program was slipping into a mood of "every man for himself."[5]

After all, what could be more American than the image of a father defending his homestead with a shotgun in a lawless world? Home shelters fitted nicely with the image of a new frontiersman who "could venture forth" after the bombs, as *Time* said, "to start ensuring his today and building for his tomorrow."[6] Teenage boys fantasized about using a nuclear attack as an opportunity to take a girl into a shelter and persuade her to give up her virginity (at one school assembly a teacher warned the girls not to give way to such end-of-the-world arguments).[7] A few critics noted sardonically that a shelter somewhat resembled a womb, a safe place to await rebirth. Indeed, the long wait in uncertainty, the cramped conditions, and the darkness all sounded suspiciously like the traditional rite of passage into a new life, resembling the initiatory trials that young people in many primitive tribes were made to undergo in order to enter the adult world.

More commonly the shelter furor conveyed grimmer images, bringing back air-raid fears with redoubled force. Most frightened of all were children, like a little girl who, upset when her parents decided not to build a shelter, picked out a closet to hide in when the bombs dropped. In many such cases the dark shelter beneath the ground represented victimization, separation, and pure death.

To the majority, shelters did not clearly signify panic and savagery, or heroic survival and rebirth, or separation and the grave, or anything they could picture at all. Most of the photographs in magazines and the sketches in American or Soviet civil defense booklets showed shelters as architecture, either empty or housing deliberately bland families. The shelter debate drove home the idea of nuclear war as an indescribable catastrophe, while reinforcing murky associations with fantasies of victimization and survival, but it did little to bring the vague imagery into focus.

One definite result of the shelter debate was to center attention more than ever on radioactivity. By 1962 most Americans believed that if a bomb dropped on their city, fallout would kill or injure them; barely half feared the bomb's blast; fewer still feared death by fire. The logic of this response was exactly backward, for in fact a hydrogen bomb would do most of its damage with heat and next with blast. Fallout would be the greatest killer only in a limited attack that fell on remote military targets but generously avoided cities. Yet there was nothing you could do to defend yourself against a fireball. So here, as in the controversy over testing, talk was dominated by anxiety about radioactive dust.

In a few months the Berlin crisis cooled off, and talk of civil defense began to settle back toward a moderate level. The whole squalid controversy had been confined almost entirely to the United States. In other nations (except Sweden and Switzerland, busy digging massive shelters), people felt that attempts to defend against hydrogen bombs were futile. Everywhere, mention of shelters did less to reassure people than to remind them of a rising danger.

"The very existence of mankind is in the balance," exclaimed the secretary general of the United Nations. It was October 1962. Kennedy had appeared on television to announce that Soviet missiles were being placed in Cuba, and hinted that if the missiles were not promptly removed he would launch an attack. In a secret message to Khrushchev, the president warned that the Soviets risked "catastrophic consequences to the whole world." He told his advisers that he was not worried about the first step but about climbing the staircase of escalating threats to the fourth or fifth step— "and we don't go to the sixth because there is no one around to do so." The Soviet premier wrote back that the risk was "reciprocal extermination." He had not meant to provoke war, Khrushchev exclaimed: "Only lunatics or suicides, who . . . want to destroy the whole world before they die, could do this." In the crisis the world's leaders abandoned the logic of strategic missile exchanges. Rather, they called up images of uncontrolled men blowing up everything.[8]

Among leaders and the public together, nuclear fear reached a higher peak during the Cuban missile crisis than at any other time before or since. As Soviet ships approached the American blockade fleet, a considerable number of people from London to Tokyo thought they might not live to see another dawn. In some cities food-hoarding panics stripped supermarkets bare. Those who had thought hardest about deterrence were especially worried. Undersecretary of State George Ball told his wife to turn their basement into a fallout shelter, and Herman Kahn carried a transistor radio everywhere he went. Leo Szilard, most sophisticated of all, flew to Switzerland. Years later Secretary of Defense Robert McNamara, holding thumb and forefinger an inch apart, said, "Rational individuals came *that* close to total destruction of their societies."[9]

In the end the Soviets packed their missiles back home in exchange for American promises to leave Cuba alone and to withdraw missiles placed in Turkey. A great nuclear war had been avoided only by difficult self-restraint—imposed by nuclear fear, the sheer terror of hydrogen bombs. Otherwise the bombs had been worse than useless. The solution had come through old-fashioned diplomacy. Nuclear warheads had only created the unnecessary crisis in the first place, and then carried it from a bothersome local confrontation to the edge of catastrophe.

Something would have to be done to reduce the deadly tensions. Since 1945 world leaders had sworn over and over that they were determined to bring about nuclear disarmament. Newspapers had given acres of front-page space to ingenious plans offered by one authority or another. These diplomatic ploys had inevitably interacted with nuclear imagery. In particular, an obsession with "control," especially of secrets, dominated all nuclear negotiations. In reality the possibility that some fraction of weapons or bomb tests might go undetected was only one of many complex issues in arms control. But missiles hidden in mine shafts, bombs exploded undetectably within underground caverns, endless talk of "inspection"—these were the questions that continually blocked progress. The unending flood of news about negotiations reminded the public again and again that nuclear energy was connected with dangerous hidden things.

The problem was circumvented when, in August 1963, the United States, the Soviet Union, and Britain signed a treaty promising not to test nuclear weapons in the atmosphere. It would be easy to tell if someone violated the promise. And there would no longer be fallout from tests drifting on the winds. The treaty did nothing in practice to slow the nuclear arms race, for tests continued in underground shafts at a faster pace than ever. But out of sight was out of mind. The space the *New York Times* gave to arms control abruptly dropped to less than a third of the peak level. There followed years of détente, with neither the crises nor the vehement rhetoric of the 1950s.

When bomb tests went underground, the nuclear disarmament movement's tactic of concentrating on fallout lost its point. And the growth of détente robbed the outcries about war of their urgency. The CND, SANE, and similar organizations from Germany to Japan had all failed to go beyond slogans to a political program that a majority of citizens would vote for. Increasingly dominated by radical young people, after the Limited Test Ban Treaty was signed the movement abandoned the nuclear issue in favor of agitation for black civil rights in the United States and South Africa, for peace in Vietnam, and for other causes.

The worldwide collapse of interest in nuclear war showed up in every indicator I have studied, including bibliographies of magazines, indexes of newspaper articles, catalogs of nonfiction books and novels, and lists of films. From their peak around the time of the Cuban missile crisis, all these measures plunged by the late 1960s to a quarter or less of their former numbers. Even comic books with "Atom" in the title faded from the newsstands. In short, after the Cuban crisis and the ban on open-air testing, people turned their attention away from nuclear weapons as swiftly as a child who lifts up a rock, sees something slimy underneath, and drops the rock back.

What caused this astonishing event, the only well-documented case in history when most of the world's citizens suddenly stopped paying attention to facts that continued to threaten their very survival?

The coming of détente and the apparent success of arms control and deterrence satisfied many people on a rational level. Yet reason could also point out that the bombs not only remained, but grew more numerous year after year. Close observers believed the precipitous drop of interest reflected less rational factors. Some said the Cuban crisis had served as a catharsis, but what did that mean?

The public had been bombarded again and again by war scares and hydrogen-bomb stories, by warnings against fallout and fantasies about monsters, by antibomb outcries and shelter debates, with the Cuban shock only the most terrifying in a long series. Yet awareness of peril did not spur action. Scarcely one in eight Americans actually took precautions during the Berlin and Cuban crises, and only about one in fifty built any sort of fallout shelter. Even RAND analysts and civil defense officials rarely built shelters for their own families.

A panel of social scientists that Kennedy quietly assembled in 1962 inspected the psychological consequences of such decisions, and concluded that citizens faced "cognitive dissonance." This referred to a theory that was popular among academic psychologists, and that moreover had common sense in it. The theory began with the observation that most people held a confusion of half-formed beliefs that might not logically agree with the way they acted. If they were forced to pay attention to the contradiction, to the dissonance, people would strive to adjust either how they acted or what they believed.

In nuclear affairs, ever since 1945 many people had admitted to deep confusion. The results were compatible with cognitive dissonance theory. A small minority of Americans made a deliberate decision not to protect themselves, and these people held the corresponding conviction that nuclear war would be so ghastly that shelter was pointless: they had brought action and belief into accord. Others thought survival was possible, and built shelters. Most Americans, however, said they thought that civil defense would be of some use, yet did nothing about it. The panel of social scientists noted that the worst dilemma was faced by people in areas liable to be wiped out by bombs—and polls found a majority of the population believed that they lived in target areas. The only way people in such areas could survive a future war would be to uproot themselves and go live elsewhere, at once. Very few families had done that (although millions of city-dwellers had taken a "vacation" during the Cuban crisis). Thus the majority could keep their beliefs compatible with their actions only by deciding there was little chance that a nuclear war could actually happen.

Ever since the early 1950s, roughly a third of the world public had admitted strong fear of nuclear war, with most of the rest at least somewhat worried, and the numbers did not change markedly once the Cuban crisis was over. In fact, into the 1980s responses in polls about nuclear war showed little change, shifting by scarcely 10 percent one way or the other. In the few years after 1962, when published attention to nuclear war dropped to a quarter or even a tenth of its previous level, this drop was not because of any great change in the public's beliefs and concerns. People still admitted their nuclear fear if asked about it, but they no longer brought it up spontaneously. Brushing the whole issue aside was the easiest way of all to ward off cognitive dissonance. Would the real tomorrow, the one you actually prepared for, reveal a limitless Hiroshima, or would it hold familiar scenes of everyday life? The question had been posed ever more insistently, beginning in 1945 as a problem for the misty future and becoming by October 1962 a demand about what the next morning might reveal. Most people chose to live with the image of peace.

The polls showed some other curious features. From 1945 on, sociologists could not find their usual correlations. Some people seemed desperately afraid of war and others completely uninterested, but the difference had little to do with social class. A study that addressed the matter directly found that the amount of knowledge a person had about fallout had no relation at all to the amount of anxiety the person felt; anxiety did not come from a special knowledge or from a special ignorance of danger, but from somewhere else. On the average, women were somewhat more worried about the bombs than men were, and poor or uneducated people somewhat more worried than wealthy or educated ones; but studies have found that those groups tend to express more insecurity on almost any issue. The differences were greatest not between groups but between individuals within a given group. Attitudes and images had taken root far below normal social categories, in levels of the mind that simple questions were ill suited to probe.

What difference did it make to have this appalling danger as a normal part of daily life? Although it is difficult to measure, the confrontation with the real possibility of nuclear war probably led many Americans to realize that their government could not defend them in the manner remembered from the Second World War. During the Cuban crisis, supporters of Kennedy's presumed tough stance had clashed on college campuses with proponents of disarmament; at peace demonstrations there were catcalls, eggs thrown, fistfights. It was an early sign of what would become a great divide between those who trusted the government and those who would never trust it again.[10]

Sensitive observers believed that young people were especially disturbed. The first postwar generation had tended to be conformist "organization men," accepting the existing stereotypical gender roles and other social relationships. But already in the 1950s a distinct youth culture was forming, characterized by a sense of isolation and withdrawal. Outright distrust of established ways blossomed during the 1960s. A study of incoming college freshman between 1948 and 1968 found a steady decline in attitudes of cooperation. There were many reasons for these changes, but everyone agreed that exposure to nuclear fear—the bomb drills, the civil defense instruction, the war scares, the movies, and all the rest—was playing an important role.[11]

Some time after the Cuban crisis a poll found 40 percent of adolescents admitting a "great deal" of anxiety about war, more than twice the rate found in older groups. A survey that said nothing about bombs, but only asked schoolchildren to talk about the world ten years ahead, found over two-thirds of the children mentioning war, often in terms of somber helplessness. In the 1960s observers reported talks with young people who said it was pointless to study or save up money when the world might end tomorrow (I heard some say this myself). In 1982 a psychiatrist, summarizing decades of studies, said that the nuclear problem had left many young people with "a sense of powerlessness and cynical resignation."[12]

But among the young, as among their elders, nuclear fear came to be expressed less and less openly. A college teacher in the 1960s asked his students every semester about such feelings, and noticed the shift. Those who had entered adolescence before the bombing of Hiroshima frankly admitted anxiety, but the next generation did not. The teacher felt that all the students were nevertheless acutely aware of danger, and he concluded that the danger had become so frightening that they denied it.[13]

Denial: one of the few mental defense mechanisms whose existence is accepted even by severe critics of psychological theory. Doctors found that some patients, when told their disease was incurable, simply forgot the intolerable fact; psychiatrists knew patients who were so disturbed by certain things that they literally became blind rather than face them. Many children in Second World War air raids rigidly insisted that their homes were perfectly safe. As a young adult said in 1965, "If we lived in fear of the bomb we couldn't function."[14] And images of nuclear war were still more liable to be repressed because the imagery was mixed up with fantasies involving primitive fears and desires that many people hid even from themselves.

To test denial directly, in the mid-1950s a psychologist showed people a set of drawings ambiguously related to nuclear war and asked them to make up stories around the drawings. The more explicit and potentially frightening the pictures were, the *less* some people would allow the idea of war into the stories they told. Those who failed to see war in the pictures made up stories that were nevertheless gloomy and vague. This was the phenomenon that psychologist Robert Lifton and others later called "numbing." Somebody who denied an idea would also thrust away everything reminiscent of the idea; refusal to feel, to think, or to contemplate action could spread outward indefinitely. One of the great unanswered questions of our age is how far denial of nuclear dangers has promoted such numbing.[15]

Denial was supported by more elementary psychological processes, particularly "habituation." Animals exposed to a loud noise that is never followed by physical harm will eventually behave as if they do not hear the noise. The principle was applied with notable success to cure people of phobias, and one of the classic papers explaining this therapy, published in 1963, is of special interest.

A salesman was afflicted with morbid fear of nuclear war. His terror had become so acute that he avoided the radio or anything else that could remind him of the international news, until he lost his job and spent most of the day hiding with covers over his head. He believed that his real fear was of losing his wife, on whom he was childishly dependent—a state in accord with suggestions I offered earlier about thoughts of world doom—but his psychologist chose to treat the phobia directly. The patient was told to put himself in a relaxed and pleasant state of mind and then imagine glancing briefly at a newspaper. After many repetitions he could do this without anxiety. He then progressed to the radio and so forth, until he was cured.[16]

How many people in the early 1960s were similarly "cured" of nuclear fear? There can be no sure answer to such a question, but I believe that what happened in an extreme case to one patient could also happen in a milder way to the entire populace, all exposed to war scares and other stimuli that were never followed by actual harm. Unfortunately, as in the story of the boy who cried wolf, growing accustomed to outcries was not the same as getting rid of the wolf.

Reinforcing the move to apathy was another elementary process, later named "learned helplessness." Psychologists found that a caged dog given electric shocks at random, with no way to avoid them, would eventually cease trying to save itself. Later, put in a different cage in which a normal dog could learn to avoid shocks, the creature that was taught to feel help-less would only lie down and whimper. The corresponding human state was fatalistic apathy, which could be experimentally induced simply by presenting a person with a series of insoluble problems. Bertrand Russell complained of this phenomenon on a larger scale in 1964, noting the waning of the CND and the growing silence about bombs. The trend did not mean people were truly indifferent to the prospect of catastrophe, he said. Rather, they were overtaken by a "sense of helplessness . . . most people feel utterly paralyzed by the vast impersonal machinery of war and state power."[17]

These were not just matters of personal psychology but also of social forces, of propaganda in the most general sense of the term. Propaganda does not always aim to excite particular opinions and actions; most societ-ies have a ceaseless background propaganda pushing toward an *absence* of opinion, a silent acquiescence. For example, the perennial publicity about arms control talks and about new types of weapons tended to persuade people that national authorities were working hard to solve the nuclear problem. It seemed that only the authorities could understand the intri-cate diplomatic and technological details. This was not learned helpless-ness, but taught helplessness.

Whether people rationally decided that nuclear war was not an urgent matter because deterrence and détente appeared to be successful; or avoided cognitive dissonance by choosing to believe that it was unnecessary to flee their home; or found the images so terrifying and disturbing that they consciously or unconsciously denied them; or became accustomed to warn-ings and no longer felt fear at all; or felt so helpless that they turned to other matters; or accepted the authorities' claims that they could be trusted to handle the problem, the result was the same. The images of nuclear war remained a mixture of the fantastic and the abstract, never developed into conceivable practical actions. Like the denizens of an enchanted Sleep-ing Beauty's castle, each image and each concept kept its silent place while nearly two decades passed. It was a stagnation of military, political, and moral thought without modern precedent.

Yet subtle changes in thinking were under way. People had become sus-picious of the control that authorities held over nuclear energy, and some began to wonder about controls within technological society in general. That questioning did not subside into silence after the missiles left Cuba.

15

FAIL/SAFE

W hat exactly had doomed the human race to slow death? According to Fred Astaire, playing a drunken physicist in *On the Beach*, it was "a handful of vacuum tubes and transistors—probably faulty." In the book that inspired the film, however, the end of the world had begun with nations deliberately attacking each other. Between 1957, when the book appeared, and the 1959 film there began a shift of emphasis, not only in this story but throughout public thinking. The threat of catastrophe was beginning to seem less a question of international politics than one of technology. As Astaire's character put it, systems were proliferating until "we couldn't control them."

Before 1945 scarcely anyone had imagined a war set off by a technical accident. If writers spoke of inadvertent war they meant an error not of electronics but of national leadership. Beginning in 1946 atomic scientists warned that automated war systems could increase this danger: What if a general should panic and push the buttons too hastily? It sounded like the prewar fantasies about an experimenter who inadvertently blows up the planet, and at first few people took time to worry about it. But by the mid-1950s there was good reason to worry.

In deep secrecy, American warplanes were regularly flying across Soviet territory to take photographs and test defenses. The flights annoyed Khrushchev, and in 1957 he hinted discreetly about practices that might set off an accidental war. The old image of a scientific accident destroying the world began to creep into serious military discussions.

Real failures took a hand, for during 1958 the Strategic Air Command's new B-47 bombers suffered seven crashes and other accidents. The most spectacular came in March, when a nuclear bomb fell from a B-47 over South Carolina. The chemical explosives that were part of the warhead went off, injuring five people and scattering about radioactive material.

Radio Moscow complained that a similar mishap could panic SAC into attacking Russia.[1]

At the end of the 1950s SAC was in fact on a hair trigger. Theorists calculated that in a Soviet First Strike, America's bombers could all be blasted on the ground like sitting ducks. So SAC made it a rule to put a small fraction of its fleet into the air at the first suspicious signal. In practice such signals included an unidentified group of airplanes, a radar error, even a meteor shower. The issue came into the open in a 1958 news story. Introducing the public to some engineers' jargon, it explained that "Fail Safe" rules would always turn the bombers home if anything went wrong.

The reality was that American nuclear weapons, at first guarded jealously under the AEC's civilian control, had gradually come into the hands of military field commanders. By the end of the 1950s some individual pilots were authorized to drop bombs on their own. Those facts were secret, but, as often happened with crucial matters in the United States, plain hints reached the public. After 1961 the Kennedy administration installed safeguard devices and also halted the flights over the Soviet Union, but it was too late to avoid the darkest fears. Further bomb accidents, such as one that scattered radioactive material around Palomares, Spain, in 1966, continued to darken the public image of "the atomic enterprise generally," as an Atomic Industrial Forum spokesman lamented.[2]

If nuclear war ever came, it would most likely come, like every other war, in a crisis in which the leaders of both sides believed they had less to lose by fighting than by yielding. Yet few could forget the flawed transistor or—what seemed much the same—the rigidly dedicated bomber pilot. However implausible, these images of fatal innocence had a peculiar fascination. For they masked more intimate thoughts that were far from innocent.

The mask was partly lifted by a thriller titled *Fail-Safe.*[3] Serialized in the *Saturday Evening Post* in October 1962, it chilled millions of readers (with some help from the Cuban missile crisis that same month). The book was the only nuclear novel ever to make the list of the top ten best-sellers for a given year. The story revolved around the failure of a minor electronic device that caused a coded signal to send an American bomber group into Russia. Air Force spokesmen complained that only a voice command could launch a bomber. But obviously the chance of *some* kind of accident, however unlikely, was never zero. Some argued that the consequences of failure would be so dreadful that even the most implausible chance must be treated as if it were bound to happen eventually. If you accepted that logic, the only solution was to dismantle the entire apparatus for nuclear war.

The movie producer Max Youngstein, a member of SANE, filmed *Fail-Safe* as a contribution to the disarmament cause. In 1964 his movie put millions into a cold sweat—a visceral reaction of horror to the thought of losing control over the bombs. If the story had no robots running amok, it did have humans who would have been at home in a robot tale. For example, the movie showed a strategist who icily advised a mass attack when he calculated that war was inescapable. Most of all it was military officers who were robotic. The pilot of the bomber heading for Moscow refused to heed radio commands to return, since his orders told him that voices could be enemy ruses. The audience saw the pilot's wife at the microphone, tossing her head in a frenzy as she begged her husband to turn back; he stolidly switched her off. The real danger came not from automated machines but from automated men.

Even more penetrating and influential imagery came out of a 1958 British novel that few people read, Peter George's *Two Hours to Doom*. This author, too, wanted to take a stand against hydrogen bombs. Perhaps he also drew on more personal emotions: a few years later, shortly after finishing a novel in which bombs reduced humanity to warring tribes, he killed himself. The plot in *Two Hours to Doom* openly addressed something akin to suicide, for here it was not electronic breakdown that launched bombers into the Soviet Union but human mental breakdown, a psychopathic general.[4]

Stanley Kubrick found the idea plausible. The veteran film director had thrown himself headlong into studies of nuclear war, pondering how to send the world a message. He happened upon Peter George's little-known thriller and found the plot he had been seeking. The outcome in 1964 was the most important of all nuclear films, *Dr. Strangelove*.

At last the Man Who Started the War showed his face, and it turned out to belong to a SAC officer. But the film's message was only beginning when General Jack D. Ripper, convinced the Communists were poisoning his "precious bodily fluids," privately arranged to attack Russia. Kubrick turned the whole grand apparatus of authority and technology into a haywire clown's contraption. Dr. Strangelove himself was a scientist-strategist in a motorized wheelchair, scornful of human feelings. As a critic noted, he "comes to us by courtesy of a Universal horror movie. In his person, the Mad Doctor and the State Scientist merge."[5] Even Strangelove's crippled, black-gloved right hand was a sly reference to the similar feature of a robot-making scientist in the classic 1926 film *Metropolis*. That film in turn had lifted its imagery from the Frankenstein story and the paraphernalia of medieval sorcerers. At last the secret that had been hinted at for decades was proclaimed aloud. The psychological problem of nuclear war was precisely what the traditional mad scientist had always stood for: the problem of human authority losing its self-control.

The stories of accidental war resulted, finally, only in further catharsis. Like the Cuban crisis but on another level, the novels and movies could help people to bring up their secret anxieties, and to partly resolve the dissonance of incompatible beliefs and actions—if only by deciding to shrug the whole business aside. That particular resolution was easier to reach because these novels and films drew attention away from real politics, focusing instead on cracked transistors and cracked minds. What was the point in fretting over remote possibilities lying beyond anything the citizen could affect or even understand? There was an edge of truth in *Dr. Strangelove's* arch subtitle: *How I Learned to Stop Worrying and Love the Bomb*. The 1958–1965 spate of films, novels, and magazine articles about accidental war brought down the curtain on the long series of important nuclear war fiction and nonfiction published since 1945. Debate over accidents was the final burst of serious argument about nuclear war before attention turned elsewhere.

A faint echo lingered, for there was one way nuclear accidents could plausibly connect with news of international events. In 1960 France had exploded its first atomic device, and China followed suit in 1964. Everyone understood that more nations with bombs meant more chances for inadvertent catastrophe from a faulty transistor, a psychopathic officer, or (more realistically) the escalation of some local conflict. Policymakers called the problem "proliferation"—a biological term, as if bombs were prone to reproduce all by themselves. Once nation A got bombs, surely its enemy B would decide to get its own, followed by its neighbor C, and so forth, as uncontrollable as the spread of a poisonous weed.

Responding to these fears, in 1968 many nations signed a Non-Proliferation Treaty that established the long-sought corps of international inspectors to control the production of plutonium. But the International Atomic Energy Agency inspected only reactors; a determined nation might find another path. Much expert attention went into hypothetical connections between bombs and the rising nuclear power industry, but the real question of proliferation lay elsewhere. Would more nations decide to build bombs?

To everyone's surprise, over the decades scarcely any nations felt that such a choice would be to their advantage. Public debate over proliferation therefore grew desultory, and nobody wrote an extended fictional treatment of the subject. Public imagery wandered off in a direction of its own, back to tales of fanatics building or stealing bombs. Genuine national leaders were forgotten while stereotypical scientist-criminals brandished infernal devices. That tradition was carried on, for example, in 1962 by the first immensely popular James Bond movie, *Dr. No*, in which the agent defeated an evil genius (complete with black gloves) on his nuclear-powered island stronghold. Unlike other nuclear fiction, such stories did not disappear from the paperback book racks and movie theaters after the mid-1960s, but remained roughly constant in numbers to the end of the century. Time and again James Bond or his imitators saved the world from secret criminals who wielded hydrogen bombs. These movies won the widest audience of any tales relating to nuclear energy.

As archetype, Bond was of course the hero who confronts the monster. Typically the villain aimed to wield nuclear weapons for personal gain, perhaps even intending to set off a world conflagration. And typically the secret agent thwarted him—like the hero-survivors who faced nuclear war in the dreams recorded by Michael Ortiz Hill—not with countervailing force, but through a combination of ingenuity and simply surviving attacks from the villain's minions.[6]

If there was any social message it was that ordinary citizens had no role but to watch, trusting in the wily hero-survivor with his special abilities and technological gadgets. As one film critic remarked, scriptwriters had domesticated the hydrogen bomb, transforming it into "just another technological toy" in the perennial spy fantasy.[7] But as a theatrical prop, nuclear energy was peculiarly fit to spice up the fantasy. For nuclear energy brought with it associations with key components of spy tales: aggressive violence, secrecy, sexual adventure. The novels and movies reinforced these linkages in return. Thinking about nuclear war had gone as far as it could go in a direct line; it was reverting to more primitive and hidden levels.

As people spoke less and less about controlling nuclear weapons, some began to turn to the problem of controlling nuclear reactors. In practical terms, an accidental launching of bombs could scarcely have been more different from an accidental explosion of a reactor. In terms of imagery, they could scarcely have been closer. Talk of reactor accidents had been muted during the early barrage of Atoms for Peace publicity, but an image of dreadful risk nevertheless gradually arose. It was developed out of public view by the very men who were responsible for building reactors.

Most nuclear authorities had always felt confident that sound engineering could prevent disasters. Nevertheless, since the Manhattan Project days they had insisted that reactors should be handled with a care beyond anything known in other industries. Overlooking how many thousands of lives were routinely sacrificed in ordinary coal mines and chemical factories, scientists and other nuclear industry leaders insisted, with a perverse pride, that reactors were "by long odds the most dangerous manufacturing process in which men have ever engaged." That idea dominated the thinking of reactor builders, and eventually of everyone else.[8]

The first widespread publicity came as usual from American scientists, at the 1955 Geneva Atoms for Peace congress. The most impressive paper, by Teller and two colleagues, discussed the radioactive fission fragments that would build up inevitably in any reactor. They believed—almost boasted—that these isotopes were a million times more toxic, gram for gram, than ordinary industrial products. (It was only later that the comparable hazards of certain chemicals were recognized.) The authors warned that running a reactor was like "conducting both explosive and virulent poison production under the same roof."[9] Most reporters ignored these and other hints from the experts, but a few journalists mentioned the matter, not to deny that Atoms for Peace was a wonderful idea but to call for caution during its development.

Some of the nuclear industry's public relations experts sensed that they were on thin ice. After all, people were familiar with science-fiction stories about fission plants causing spectacular accidents. Teller warned that a single serious accident, the kind that happened now and then in almost any industry, could be a "psychological disaster" that would impede the advance of the peaceful atom. Engineers must proceed with great caution to protect not only the public, but the public image of nuclear energy.[10]

Nuclear leaders created formal institutions to perpetuate their concern about reactor dangers. The first and most important had been set up by the U.S. Atomic Energy Commission in 1947. Its new Reactor Safeguard

Committee was only an ad-hoc group of a half-dozen experts serving as part-time advisers, but within a few years they would force the entire nuclear industry to think as they did, and eventually their way of thinking would impress itself upon the public at large. Teller was the committee's chair for its first six years, the formative years. Dedicated to a particular view of the safety problem, he would dominate all decisions.

On the spectrum of opinion about safety, ranging from nonchalant over-confidence to apprehensions of disaster, the Reactor Safeguard Committee inclined toward disaster. That inclination accorded with the chairman's personal preoccupation with every possible sort of nuclear doom. More im-portant, the committee members worried that the careless attitude of cer-tain reactor builders might cause accidents that would spur the public to retard the industry's growth. The Safeguard Committee took its job so seri-ously that before its first year of work was done it was at odds with many of the AEC's other experts, forbidding them to build reactors anywhere near a populous area. The Reactor Safeguard Committee was sometimes called, even to Teller's face, the Reactor Prevention Committee.

Their solution was to locate pilot reactors in a desolate part of Idaho. The reactor design group at General Electric in upstate New York ob-jected. They were not eager to commute to a desert thousands of miles distant, and besides, commercial nuclear power would be costly if reactors had to be located many hundreds of miles from the cities they served. The GE staff therefore proposed to enclose their reactor within a tremendous steel shell, so that if an accident did happen the radioactive substances would be safely contained. This idea the Safeguard Committee accepted.

The GE architects now raised a fateful question. Exactly how thick should they make the containment shell? Up till then everyone's attention had gone into preventing accidents, and nobody had tried to calculate the details of what an accident might look like. But the architects needed precise numbers. The reactor designers therefore had to describe a catastrophe—in fact the very worst that could plausibly be imagined. The container would then be built to withstand even that. It was a new idea to plan such mas-sive precautions for such a remote possibility. "We basically went through the whole nuclear controversy," one of the GE staff recalled. "We ourselves played the devil's advocate."[11]

Soon the Reactor Safeguard Committee itself began to insist that any group that wanted to build a reactor must describe the worst failure—what came to be called the Maximum Credible Accident. The committee members would review that scenario and try to conceive an even more dreadful failure. Engineers would calculate in detail what the worst case would look like and then design something to contain it. Teller's committee made all thinking about reactor safety center upon the Maximum Credible Accident. In short, reactor safety meant dealing with unspeakable calamity.

That was not how engineers in other industries addressed safety. They gave primary attention to the modest accidents that from time to time actually killed people, groping toward reliable practices through experience with hundreds of small-scale tragedies. Experience in many industries showed that big accidents stemmed from the same causes as little ones, so if you understood and prevented every ordinary failure, there was little more to worry about. The nuclear industry did not adopt this approach, and at first it could not. In the 1950s engineers were developing reactors so swiftly that before experience could reveal the drawbacks of one design they were building the next, with a different design and several times larger. Furthermore, the 1950s reactors suffered few accidents serious enough to serve as lessons.

The Maximum Credible Accident was like the First Strike of nuclear strategists: something that had never happened, spun out of pure thought, yet a thought so horrible it was studied more painstakingly than any actual event. This was the logic of the U.S. Air Force's "worst case" guesses about Soviet capabilities, guesses that led Teller and others during these same years to campaign for ever more weapons to remedy supposed bomber or missile gaps. All such worst cases were based less on technical facts than on fear of the unknown, of somehow losing control. Indeed, the Reactor Safeguard Committee was particularly worried that a vicious or unstable person might set off a reactor accident on purpose, much as SANE worried that such a person might start a war.

Nuclear energy not only overturned conventional warfare, but likewise overturned conventional industrial safety engineering. For the contents of a reactor, if scattered across a countryside, could render thousands of square miles uninhabitable. You could afford to have a coal-fired power plant fail in the worst possible manner, but not a reactor. And if you planned to build thousands of reactors, each operating for tens of years, then something that had only one chance in 10,000 of happening—an unprecedented earthquake, a fiendishly clever saboteur—was almost certain to strike eventually. Your reactors must be prepared to survive it.

Concern about maximum accidents became institutionally embedded in reactor laboratories. The first formal organization outside of Teller's group came in 1952, a Reactor Safety Review Committee set up to oversee work on the American pilot reactors in Illinois and Idaho, that is, to oversee roughly half of all reactor research in the world. The committee spent long days and weekends interviewing engineers and reactor crews and going over every piece of every reactor with a fine-tooth comb. The committee could and did get careless people fired. The other AEC laboratories working on reactors, as well as General Electric, Westinghouse, and other private firms, likewise set up their own in-house safety committees. These, too, reviewed reactor designs and reported their findings straight to top management.

Most other nations followed the American pattern, with rigorous safety committees and massive containment shells around reactors. The exception was the Soviet Union and its satellites. Authorities there publicly insisted that there would never be a reactor calamity, and they probably believed their own publicity. Besides, to criticize a technology that was controlled by the state was tantamount to criticizing the state itself.

In the type of reactor the Soviets built most abundantly, with uranium rods embedded in graphite, the chain reaction could run away if the temperature rose. This reactor design was not an arbitrary choice, for the laws of physics have a role in history. One result of these laws is that a graphite reactor is particularly suitable for producing plutonium, that is, bomb material. Moreover, for that purpose the reactor should be refueled frequently. Soviet civilian reactors were directly descended from ones built to produce military plutonium, and they used many of the same design shortcuts. Moreover, Soviet reactors built before the 1980s—such as the one that later became famous at Chernobyl—had on top nothing more than a screen through which refueling took place, a screen literally full of holes. Doing without a full containment shell saved money and made frequent refueling easy. In short, for Soviet reactor designers, safety was less important than building civilian reactors that could produce military plutonium if desired, and building them cheaply.

Not even the most dedicated effort could make a flawless reactor. The American program had a number of minor mishaps during the 1950s. In 1952 and 1958 Canadian reactors caught fire, and in 1959 and 1961 came the world's first fatal accidents as reactor operators died in Yugoslavia and in Idaho. Each accident caused a stir, drawing newspaper comments on the problem of reactor safety. But these were all experimental reactors, where sensible people expected problems. The public showed little concern.

The response was different when a reactor released a significant amount of radioactive material into the open. To get plutonium for bombs the British had constructed large reactors at Windscale in northern England, looming above verdant meadows where sheep and cows grazed. In October 1957 some of the uranium fuel rods caught fire. Black headlines exploded: traces of radioactive iodine had been discovered on the pastures for dozens of miles around Windscale, and officials were confiscating milk from the area. The press furiously criticized nuclear officials while frightful rumors spread. The accident may have slightly increased the risk of cancer in the region and thus resulted in perhaps a dozen deaths over the next several decades, but at the time nobody was visibly injured. Most people believed that the affair ended without harm, and the press turned to other matters. Yet if the outcry was brief, it had been strong enough to show that the public remained extremely wary of any escape of radioactive substances.

The first case of sustained public opposition to a reactor came in the United States during 1956, and it revealed much about the roots of opposition. The central issue was political—a feature that would later be half hidden under layers of seemingly nonpartisan arguments. Another fact that was later obscured was that the negative imagery and evidence that supported the opposition originated within the nuclear community itself, where they served quite different purposes. More revealing still, these negative factors first emerged into public view just where positive imagery was greatest. The fears were the dark twin of an enthusiasm evoked for the most utopian of all reactor types.

Around 1945 Manhattan Project scientists had become excited by the fact that the plutonium produced in a reactor, transmuted from the uranium used as fuel, could itself be used as reactor fuel. Enthusiasts soon began to speak as if the process was a philosophers' stone producing endless new gold. A Soviet scientist said that it sounded like burning coal in a stove and getting back more coal. Leo Szilard, an ardent promoter of the process, dubbed it "breeding," as if it captured the limitless reproductive powers of life. The Russians frankly termed the process "reproduction," and Americans said the uranium-238 was a "fertile" material. As a Westinghouse pamphlet later put it, "breeder" reactors promised "what might be called perpetual youth."[12]

The breeder attracted some of the most audacious minds in industry, and one of these was Walker Cisler. The sober-faced executive with a social crusader's soul became interested in the device first as an engineer, second as chief of the Detroit Edison Company and founder of the Atomic Industrial Forum, and last but not least as the creator of a new private consortium, whose goal was to break the government monopoly on nuclear power. Cisler wanted to prevent the government from strangling free enterprise and freedom itself; at the same time he expected that the breeder reactor would add wealth to his company, his nation, and eventually the whole world. Since Lewis Strauss and others on the AEC shared this vision, Cisler struck an agreement with the agency for help in building a prototype breeder. It would go up at Lagoona Beach, Michigan, twenty miles outside Detroit.

The "fast" breeder Cisler planned would be more sensitive to mistakes than almost any other type of reactor. There was a chance of a runaway chain reaction that would heat the device so rapidly that it would explode roughly like an ordinary TNT bomb. That possibility worried the Advisory Committee on Reactor Safeguards (as the Reactor Safeguard Committee was now named) when they studied the project. What if the fuel somehow melted, and what if that somehow caused an explosion, and what if that somehow broke the containment shell, and what if the wind were just then blowing toward Detroit? After days of debate the committee advised that such a breeder should not be permitted until more research was done.

The Atomic Energy Commission rejected the Safeguards Committee's advice. Strauss felt sure that by the time the reactor was completed, the research needed to keep it safe would be in hand. That decision appalled Clinton Anderson, the chairman of the Joint Committee on Atomic Energy. The Democratic senator saw Lagoona Beach as the worst move yet in Republican plans to give nuclear power away to private business. He stirred up opposition in Michigan, starting with the United Auto Workers (UAW), who were happy to take up cudgels against the Eisenhower administration in an election year. Joining with other unions in the Detroit area, the UAW took the breeder to court. The unions drew on the AEC's own ideas about Maximum Credible Accidents, tracking down reports prepared as part of the safety review process, publicizing calculations that the worst breeder failure might kill thousands. To many in the public, it was as if an atomic bomb were to be placed near Detroit.

The unions and other breeder opponents were by no means opposed to nuclear power in general; they would be happy with reactors in government hands, an atomic TVA. What each side chiefly feared was that the other would seize a monopoly over what promised to be a centrally important new industry. At Lagoona Beach, however, this question of control within society became associated with the technical safety issue. The Maximum Credible Accident broke out of the obscurity of engineering reports to become an actor on the political stage.

The move from concern over social control to concern over dreadful technical failures was reinforced by Congress in its most ambitious attempt to ease public worries about reactors: the Price-Anderson Act of 1957. Some key provisions of the act came straight out of the Lagoona Beach fight. Senator Anderson insisted the AEC must hold public hearings before giving any license to build a reactor, for he expected that the process would ultimately strengthen public confidence in nuclear power. The results were precisely the opposite. Over the next decade many hearings were held, each one conducted by the AEC only after its staff had fought out a reactor's safety questions with the designers and decided to give approval. With the engineers' doubts resolved out of sight, most hearings were manifestly nothing but a legal charade. The hearings therefore reinforced feelings that the AEC was arrogant and secretive. Moreover, the AEC used every available legal and bureaucratic maneuver to brush off anyone who questioned its decisions, converting modest skeptics into embittered enemies.[13]

Another provision of the Price-Anderson Act did still more to distress rather than soothe the public. Cisler and other nuclear spokesmen complained to Congress that reactors did not have enough underwriting. Insurance experts and even the Atomic Industrial Forum told Congress that the "catastrophe potential" of reactors was "more serious than anything that is now known in any other industry," and some companies said they might flatly refuse to develop nuclear power for fear an accident would bankrupt them.[14] Only the government itself could underwrite reactors. The Joint Committee, determined to push Atoms for Peace with all speed as part of national grand strategy, requested half a billion dollars of protection, and as usual Congress passed with little debate what the committee requested.

The press scarcely noticed, for reactor catastrophes seemed a remote problem. But Congress had put an official stamp of plausibility on the Maximum Credible Accident.

Meanwhile the AEC inadvertently made the idea more concrete. At first the Joint Committee had not been convinced that the nuclear industry needed a special indemnity, and if it did, they wanted dollar numbers. The AEC therefore asked its laboratory in Brookhaven, Long Island, to prepare a report. The Brookhaven study group took a worst case: what if half the contents of a reactor were turned to powder and hurled into the sky? Nobody could say how that could possibly happen with a typical American reactor, but then, nobody could prove it was impossible. The Brookhaven group proceeded to assume that if it did happen, the radioactive contents (not some of it, but all of it) would somehow escape the containment as a powder, and at a time when the wind was blowing straight toward a nearby city, but not so fast as to disperse the cloud before it got there.

The odds against all this ever happening were astronomical, but never mind. The scientists had provided a cautionary tale. They were willing to encourage Congress to set a high level for insurance on reactors, and meanwhile prod engineers to be scrupulously careful. The scientists felt sure that if everyone took risks seriously, all the safety problems would be worked out.

The most important critique of the draft report came from Teller, who warned that it "understates the actual dangers." He could imagine still worse circumstances, for example, that just at the moment when a radioactive cloud from an accident reached a city, rain would fall and sweep the powder to the ground. As a result of such criticism the final Brookhaven report, designated WASH-740, estimated that damage from a reactor accident could amount to as much as $7 billion, while immediate deaths—not even counting later cancers—could exceed 3,000. Some experts felt that WASH-740 had gone far over the border into the incredible, but most people accepted this new set of authoritative "facts" about reactors. The idea of a Maximum Credible Accident, from its beginnings as the science-fiction tale of an experiment gone awry, had been condensed into numbers.[15]

WASH-740 was supposed to be a dry technical report for the eyes of engineers and policymakers. But the public took up the part of the scientists' thoughts that matched familiar images of doom. Opponents of the Lagoona Beach and subsequent reactors made much of the story of deaths by the thousand. Other AEC studies in later years had similar effects. For example, a 1965 calculation found that in a worst case, temporary agricultural restrictions such as the confiscation of milk near Windscale might affect an area the size of the state of Pennsylvania. Opponents used those calculations to imply the kind of picture that *Astounding Science-Fiction* had offered in 1940: a vast territory left as barren as the moon. The old undercurrent of anxiety was working itself free from the countervailing confidence that experts would keep everything safe. A few citizens began to think of organizing protests against any reactor proposed in their locality.

Reinforcement arrived from an unexpected quarter. After David Lilienthal had retired as chairman of the AEC, he had grown increasingly dismayed by the actions of his Republican successors. In the privacy of his diary he scoffed at the "zealots" of Atoms for Peace with their "orgasms of promises." In early 1963 he went public with his worries about reactor accidents. The fundamental problem, he said, was that engineers "are human beings who are fallible."[16]

In November 1963 Lilienthal met his former colleagues on their home territory, a meeting of the American Nuclear Society, where industry people gathered to trade notes and make speeches praising the atom's future. He warned that unless the nuclear community became more open to criticism they might let loose a "wave of disillusion" against all of science and technology. The audience booed him. But as Lilienthal remarked, the louder the industry tried to shout down its critics, the more the public would wonder. The political maneuvering over nuclear reactors was being submerged in a larger question, the *Fail-Safe* question: Could anyone be trusted to control nuclear energy?[17]

16

REACTOR PROMISES AND POISONS

In 1958 the Pacific Gas & Electric Company (PG&E) decided to build a power station north of San Francisco at Bodega Bay. This was pristine coastline with immense vistas, ocean waves surging against a headland, fields of grass bending in waves of their own under the wind. In 1961 PG&E announced that the proposed plant would use a nuclear reactor—a bold step toward the clean and well-ordered atomic White City of the future. But already local residents were organizing opposition. They warned that any kind of power plant would "ruin the scenic value of Bodega" and "deflate real estate values."[1] Centuries-old worries about the ravages of industry were beginning to zero in on reactors.

Aesthetic scruples had never done much to stop industry, and they were not going to stop PG&E. Nuclear anxiety was more effective. Deliberately invoking the anxieties of the bomb test debates then under way, critics warned that an earthquake in the nearby San Andreas Fault might split the plant open and bring "fallout." Opponents went to Bodega and released batches of balloons marked "Strontium-90."[2] Meanwhile geologists discovered that a small fault ran right through the site. The AEC staff, never convinced that the reactor would be safe, now vetoed it, and in 1964 PG&E gave up. The reactor was defeated not so much by local opposition as by general misgivings among experts. But the details of this battle involved a new image: with reactors as with bombs, people were beginning to fear not just that the things would explode but that they would vomit forth dreadful contamination.

Concern about pollution from a civilian nuclear industry, like most nuclear ideas, started among a minority of experts at an early date and only gradually reached the public. Since 1945 a few scientists had warned against careless leakage of radioactive industrial by-products. The critics first attracted public attention in 1955 when they brought up the subject at the Atoms for Peace conference in Geneva. In the next few years they got into the newspapers now and then, as when a National Academy of Sciences panel insisted that radioactive wastes were more hazardous than any other industrial material and must be handled with extreme care. The press and the public gave the matter only passing attention, preferring to leave nuclear wastes to officials. Officials left it to nuclear experts, and most nuclear experts left it alone. Wastes were messy, unglamorous stuff, and even the expert critics agreed that there were many feasible means of disposal. The AEC and its counterparts in other nations studied the matter at leisure, meanwhile finding temporary expedients for storing unwanted radioactive materials.[3]

But as the controversy over bomb tests intensified, more and more people noticed a similarity between reactor by-products and fallout. For example, in January 1957 *McCall's* magazine ran a banner on its cover, slashed across a picture of a baby, declaring that "RADIOACTIVITY Is Poisoning Your Children." The featured article began by exclaiming that fallout from bombs could kill most of the human race, but gave about the same amount of attention to the way the nuclear industry's wastes "polluted the earth." The first book in English critical of the industry, *Our New Life with the Atom*, published in 1959 by a pair of veteran conservationists, devoted about half its length to bombs and mixed talk of fallout with talk of reactor by-products. A British movie about a monster lizard engendered by radioactivity put newsreel clips of bomb blasts alongside talk of nuclear industry wastes. The poisonous associations of pollution were leaking from military to civilian activities. "After all, 'radioactivity' is a frightening word," remarked a newspaper editor in 1959—showing how much things had changed since the start of the century.[4]

As I noted in Chapter 11, radioactive fallout carried all the elements that studies of risk perception identified as important. Industrial radioactive effluents likewise were new and mysterious; they carried a stigma of horrible imagery, which movies and media controversy made readily available in the back of the brain; they seemed liable to poison the whole world and future generations; they were managed, or perhaps not managed, by institutions that were coming to be widely distrusted. It was small wonder that the studies found nuclear risks evoked greater aversion than any others—more even than house fires or riding in a car, whose grave rates of mortality were blunted by familiarity.

Thoughts about waste had unmentioned resonances. In reality the world's most dangerous pollutant by far was not nuclear waste but human excrement, a carrier of disease that killed tens of millions of people every year; during the 1960s the greatest waste problem even in the United States was uncontrolled sewage. When some critics called radioactive wastes "sewage" they were suggesting thoughts of excrement. The nuclear industry itself described its wastes as products of the "back end" of nuclear fuel processing, a transparent metaphor, and some industry workers described radioactively contaminated objects as "crapped up." A French psychologist who ran association tests on the word "wastes" found that it evoked disgust and sometimes a direct mention of feces. This association was hard to discuss in a culture that suppresses such topics so strongly that the common words for human wastes are used mainly in anger.[5]

Excrement was often associated with hostility, and therefore, curiously enough, with bombs. Clinical psychologists noticed that small children thought of their feces not only as poisonous but sometimes as explosive, and fantasized about using the stuff to blow things up. Second World War bomber crews occasionally remarked that they were symbolically defecating on the enemy, and some flew warplanes like the "Privy Donna," decorated with a painting of a flying outhouse depositing bombs. I would not go so far as Robert Oppenheimer, who one day despondently told Szilard that their weapons were of no use for any good purpose: "The atomic bomb is shit." I do suggest that on every level of human thought, radioactive wastes—in association with weapons—were seen as filthy insults against the proper order of things.[6] This undercurrent of physical disgust at the thought of nuclear wastes could easily transfer into moral disgust, a belief that the nuclear officials responsible for wastes were unreliable or dishonest.

It was almost impossible to see the genuine, straightforward technical problems of the nuclear industry amid the images on the one hand of worldwide pollution, and on the other hand of an endless supply of cheap power and medical marvels. By the mid-1960s a few people held the horror in the foreground of their thinking. The great majority of citizens, however, kept waste problems in the background alongside science-fiction exaggerations, all obscured behind the dazzling visions of Atoms for Peace.

After feeling a flick of the whip of opposition, the nuclear community redoubled its efforts to soothe the public. During the five-year period

1963–1967, American agencies and corporations made available over twice as many films about reactors and three times as many about nuclear safety and the environment as they had offered during the five years preceding. But now the intention was less to excite the public about nuclear power than to calm them. The new productions often diverged from the 1950s publicity by toning down utopian promises, focusing on ordinary electricity production rather than medical and agricultural fantasies, and giving more space than ever to prosaic reassurances.

The occasional criticism of reactors grew fainter, perhaps less because of public relations work than because the forces behind the opposition slackened. After the ban on atmospheric bomb tests, with scientists and street demonstrators no longer lashing out against radioactive dust from weapons, people were less often reminded of radioactivity from reactors. Moreover, in the United States the proponents of private nuclear power decisively won their battle against government ownership, so the political dispute was put in cold storage, and both sides got on with promoting reactors.

Public interest in the nuclear industry waned still more because it became clear that the promised atomic industrial revolution was not coming soon. The economics of electricity from reactors looked increasingly shaky. New technologies such as offshore drilling and strip mining were holding down the costs of oil and coal, while nuclear power plants ran afoul of costly technical surprises. Around 1960 almost every nation cut back its goals for the amount of electricity to be produced from reactors in the next decade, settling for a half or a quarter of the levels in the first ambitious plans.

Around the world, the total space magazines gave to military and civilian nuclear energy reached a peak around 1960 and then steadily dropped, falling to little more than the fraction occupied by radium in the early decades of the century. Articles on peaceful atoms peaked in the second half of the 1950s and then died away. As I noted earlier, interest in military uses was sustained through 1964 by bomb tests, war scares, and the proliferation and fail-safe debates, but after the mid-1960s discussion of weaponry fell back to about the same modest fraction of magazine space as civilian uses.

At a few localities in the United States small bands of citizens kept on opposing proposed reactors, supported by a handful of hardy critics who kept the fight alive on the national level. But the nuclear industry, having learned to be wary, tried to choose reactor sites where the neighbors would feel that tax advantages and new jobs outweighed any risks. Even accidents aroused little interest. In 1966, while the "Fermi-I" breeder reactor that Cisler had built at Lagoona Beach was going through its shakedown tests, a piece of metal came loose and partly blocked the flow of coolant. The result was a melting of fuel roughly comparable to the Maximum Credible Accident, that is, the sort of thing the engineers had designed containment to withstand. No radioactive material escaped. The press scarcely noticed. As far as polls can show, a majority of people around the world were as willing to accept nuclear power, and a minority as suspicious of it, in 1966 as they had been in 1956 when the Lagoona Beach fight began, or for that matter in 1946.

Underneath the expressed opinions, however, something had changed: the nuclear industry had become plausibly associated with drastic error, mass death, and dreadful pollution. As the visions of wondrous progress receded into the haze of a distant future, these anxious images remained in place. At the peak of the Atoms for Peace campaign, around 1956, fully half of the articles on civilian uses listed in the *Readers' Guide* had optimistic titles, with most of the rest displaying no particular tendency. Only four years later, less than a quarter of the titles sounded positive. Articles with negative or fearful titles, and about as many more with titles that noted questions about nuclear risks but did not take sides on the issue, were climbing in proportion to the rest as gradually and regularly as the advance of a tide. Back in 1957 a World Health Organization study group had remarked that educated people, faced with official confidence about reactor safety, hesitated to voice their inner worries lest they appear ridiculous.[7] Five years later it had become respectable, if not yet popular, to voice such worries.

As long as most of the public showed little interest in the industry, nuclear officials could remain confident. In February 1969 a leader of Congress's Joint Committee on Atomic Energy gave a pep talk to nuclear engineers at Oak Ridge. He admitted that there were "still some critics who like to get their names in the paper by making muckraking charges or fabricating half-truths." But he said that because the nuclear power industry had taken enough care to forestall any serious accident, it had "largely solved the public acceptance dilemma it faced just a few years ago."[8]

It was time to move ahead boldly, and the Oak Ridge staff meant to be in the lead. In their "City of the Atom" they already lived contentedly alongside reactors and other nuclear installations, scattered among the rolling hills of Tennessee. This was TVA country, with a local tradition of visions of high technology blended with rural self-reliance. The Oak Ridge scientists felt challenged to work still harder toward a utopian community, and they had a technical word for it: the "nuplex."

A nuplex, a nuclear complex, would be a town centered on reactors. Confident about safety, engineers figured the reactors ought to be right in the middle so that their output of heat could serve factories. Virtually independent of fuel supplies, nuplexes could be dropped into jungle or tundra. For some people this was not just a technical idea but a move toward social rebirth. Oak Ridge leader Alvin Weinberg proudly explained that this was precisely the dream of a world set free that he had learned from reading H. G. Wells. The nuplex, a monumental cluster of white hemispheres amid houses and lawns, was the most specific plan ever devised for a futuristic town. But the image had at least one drawback. As Lilienthal later noted, the social corollary of the nuplex was "an invisible cadre of experts" doing as they thought best in isolation from the larger community. Such a nuclear "priesthood" (as Weinberg approvingly called it), safeguarding their perilous and wondrous powers, would appeal to experts like Weinberg more than to ordinary citizens.[9]

The main promoter of nuplex imagery during the 1960s was another atomic scientist, Glenn Seaborg, who served as chairman of the AEC for a decade. Tall and rangy, austere in his personal life, cautious and orderly in his thinking but with a romantic streak when his mind turned to the future, dedicated to improving the world through science, Seaborg seemed to embody the values of H. G. Wells's virtuous technocrats. Since his career was based on his Manhattan Project work with plutonium, it was only natural for him to believe the metal would prove supremely valuable. Seaborg told the American public that they would soon have a "plutonium economy." The element would become the basis of world prosperity, perhaps even replacing gold as the monetary standard; alchemical gold was nothing next to the "large-scale alchemy" of reactors. Moreover, Seaborg and his colleagues expected economic difficulties unless they had breeders in hand by the 1980s to stretch out the supply of uranium fuel. For the number of uranium reactors had begun to climb swiftly.[10]

The breakthrough had come in 1964, when General Electric announced that it would build a reactor in New Jersey under a contract that guaranteed to deliver electricity more cheaply than a coal-burner. Soon GE and

Westinghouse were furiously bidding against each other for orders. The new reactors ran into every obstacle that could be imagined for an untried technology, and when the dust cleared it appeared that between them GE and Westinghouse had lost a billion dollars. But the cost seemed worthwhile to make nuclear power a permanent fixture of the American economy. Other nations followed suit, returning to the ambitious plans they had laid aside a few years earlier.

The new plants would each generate close to a thousand megawatts of electrical power, but so far nobody had any experience in operating a reactor larger than a few hundred megawatts. An observer later remarked that it was "as if the airline industry had gone from piper cubs to jumbo jets in about fifteen years."[11] A few safety experts grew so concerned that they left the mainstream of optimism.

After 1965 the AEC's Safeguards Committee began to write confidential warnings about something worse than their former Maximum Credible Accident. If most of the fuel in one of the gigantic new reactors ever melted, there was a chance that it could burn its way down through the floor of the containment building. If the molten mass hit groundwater and set off a steam explosion, the containment shell might be bypassed. An engineer dubbed the problem "the China Syndrome" for the direction the reactor core was heading. Like the deep shafts in mad-scientist movies, the phrase embodied feelings of uncanny dread.

Safety researchers began to focus on ways to prevent such a meltdown, as usual giving the most attention to the worst conceivable case—a main pipe carrying coolant somehow wrenched apart as if by a giant hand. Engineers were challenged to calculate what would happen as steam rushed away from a white-hot set of fuel rods while cold water flooded in. The work offered researchers fascinating insights into the nature of violent fluid behavior, and also offered them (as reactor manufacturers wryly noted) years of funding. The research was closer to fantasy than to anything seen in real life.

As a few experienced engineers pointed out, and as events would later show, what caused actual accidents in any technological system was more modest. Nearly always there was some combination of three factors: a hidden flaw in the design of one of the thousands of components, along with inadequate instrumentation that gave a poor picture of what was going on, along with mistakes by inadequately trained workers. But that sort of combination was less straightforward, less liable to bring total catastrophe, and, in short, less interesting. Although the AEC did require reports on every little failure, the reports piled up unnoticed while studies of spectacular imaginary disasters forged ahead.

There was one official at AEC headquarters who fought against the preoccupation with calamity. Milton Shaw, head of the division that developed new reactors, said that studying emergency cooling was like studying how an aircraft disintegrated during an accident; he thought the money would be better spent making sure that accidents never happened in the first place. In 1965 he began to reshape safety work with an iron hand.

Shaw was a Rickover man who had joined the AEC after working on naval reactors, and like the admiral he was a tough, tireless administrator. His favorite word was "disciplined": more disciplined organization and more disciplined control over manufacturers, just as in the Navy—that was the route to safety. The AEC's laboratories were strangers to this approach. They had been purposely set up with a loose structure to attract top scientists, the sort of people who chose their own tasks. Shaw slapped down such attitudes. Soon research assignments were dictated from Washington, and people who did not toe the line might find their jobs yanked away. Shaw became not only the hardest working but also the most despised administrator in the history of the AEC.

In one way Shaw's strategy succeeded. From the early 1970s on, every valve and switch on every American reactor was forced to meet rigorous tests. If the least pipe failed there would be records to trace back who had inspected it and even where the steel had come from. People in the industry hated the truckloads of paperwork and the skyrocketing costs that Shaw's methods imposed, but in the end they had to obey.

Such controls had kept submarines reasonably safe. But submarines had a uniform design and rigorously disciplined crews. American civilian industry had a bewildering variety of reactors each unlike the next, designed by different manufacturers, built by a variety of contractors, and run by all sorts of utilities. That manner of proceeding did not fit so easily into the harness of a centralized bureaucracy.

A strict yet flexible and realistic control over American reactor safety had been achieved in the 1950s by the independent safety review committees that had appeared spontaneously in each laboratory. But now engineers who had to meet Shaw's demands begrudged taking further months to work through the problems again with the in-house committee. And if the local committee had its own ideas about safety methods, they were swept aside by the AEC's rigid specifications. One by one the laboratory safety committees either withered away altogether, or became mainly concerned with previewing designs so the AEC staff would find no embarrassing flaws. The primary concern of engineers was no longer how they would look in front of a jury of their colleagues who reported to the organization's bosses; now they had to measure their work against shelves of written standards issued from AEC headquarters.

Most fateful of all, nuclear experts had become divided. A few researchers at Oak Ridge and other laboratories, backed by the Safeguards Committee, kept discovering further possibilities for bizarre accidents. They begged for more funds so they could study whether the scenarios were really possible, but Shaw allocated only a little. Some experts began to feel that the AEC leadership had made an alliance with industrialists to deliberately suppress every awkward safety question. These frustrated feelings, if they ever came into the open and mixed with the public's latent anxieties about all things nuclear, would make an explosive combination. The spark that set off the explosion would arrive from an unexpected quarter: some startling new ideas about nuclear war.

17

THE DEBATE EXPLODES

Thousands of warheads would slant down from the direction of the North Star. When they hit the upper atmosphere they would burn like meteors, drawing lines of fire pointing to their targets. Deep in caverns at the end of those lines the defenders would activate their computers and await the outcome; the next acts would be too swift for human reflexes. Electronically guided rockets would dart upward like meteors in reverse to thwart the attack, nuclear fireballs by the thousand would illuminate the air, and then . . . and then what? Nobody knew how it might end, this warfare of the 1970s.

The potential threat brought an actual attack of imagery, a renewed eruption of hallucinatory visions across the landscapes of the mind. Missiles were not the only disruption. By the 1970s many people were longing for a more human scale in every technology, for a world that was nearer to the cozy villages of children's books and farther from the incandescent puzzles of computer screens.

When technologies came into question, nuclear energy would stand at the head of most critics' lists—and by nuclear energy they would mean reactors. Yet public criticism of the civilian nuclear industry became important only after other problems had stirred the public. The first radically new technology to run into widespread and effective opposition was not the reactor, but the antiballistic missile.

In the late 1960s a new game began in the rooms where military strategists assembled. Between 1960 and 1966 the United States had quadrupled the number of vehicles that could reliably drop a bomb on Soviet territory, leveling off at more than 2,000 B-52 bombers, land-based and submarine missiles, and the like. Half a dozen years behind came the Soviet reply, giant rockets whose numbers doubled every two years. Meanwhile engineers on both sides were designing Multiple Independently-targetable Reentry Vehicles (MIRVs), each of which could divide up to destroy several enemy bases. No doubt a few missile silos would survive and launch their multiple warheads in retaliation. But now American and Soviet engineers claimed they could build a rocket to shoot these down—an antiballistic missile, or ABM. Would that make it possible to launch a First Strike with impunity?[1]

Poring over the photographs sent down by spy satellites, in 1966 American analysts discovered primitive ABM bases under construction around Moscow. This time it was the Republicans who were out of power, and they cried that an "ABM gap" threatened the nation. President Lyndon Johnson soon announced that the United States, too, would defend itself with amazing inventions.

Some knowledgeable scientists were dismayed. They suspected that starting to build such defenses might frighten the Russians into attacking while they still had a chance, or might even delude Americans into thinking they could safely launch a war themselves. Their technical analysis showed that no defense system, faced with a barrage of many warheads mixed together with decoys, could offer salvation: missiles would always get through. Physicists who had done little since the atomic scientists' movement of 1946 began to organize a new campaign. They were middle-aged now, university professors with high prestige and inside knowledge of weapons. Yet their lobbying within the government had failed against other experts who thought that ABMs were worth a try. The opponents began to publish magazine articles, but at first these did little to arouse the public. It took the U.S. Army to do that.

In November 1967 the Army announced it was choosing sites for ABM bases near Boston, Chicago, Seattle, and seven other major cities. The news hit a nerve. Suddenly in the outskirts of each city there was a particular location singled out as a bull's-eye for Soviet attack. That neighborhood would also be a home for hydrogen bombs on the tip of each ABM rocket—and might one explode by accident? By early 1969 local groups ranging from Audubon Society chapters to town boards were straining to shove the fearful things away. People's feelings were caught by a newspaper headline: "Anywhere Except Near Us."[2]

The scientists raised these feelings from a local to a national issue. A few hundred dedicated physicists crisscrossing the country gave speeches, debated generals on television, distributed bumper stickers, and marched on the White House, raising the first strong outcry of nuclear anxiety that the world had seen since bomb tests went underground. A college student remarked that he had tucked the threat of nuclear war out of consciousness for half a decade, but the ABM debate made it all "real to me once again." A SANE newspaper advertisement, reproduced as a popular poster, featured a caricature of generals slavering with goggle-eyed glee over a model ABM rocket. "They're mad," said the text. "They're absolutely mad . . ."[3]

SANE labeled these generals "The People Who Brought You Vietnam"— the stereotype of untrustworthy nuclear warriors was linking up with larger issues. Many people around the world had begun to turn not only against fighting in former colonial regions, but against the economic damage done by the arms race and against "militarism" in general. Earlier debates had concentrated on nuclear bombs while scarcely ever addressing the social factors that underlay decisions about armaments. But when critics in the late 1960s looked anew at nuclear war, they examined the bombs within their larger context.

The first precise questions had been posed in 1967 by no less a personage than the American secretary of defense, Robert McNamara. Disgusted by the pressures that were forcing him to build a useless ABM system, he declared in a speech to news publishers that a "mad momentum" was pushing the nation to arm itself above any sensible level. Soon many were explaining that the purchase of ever more weaponry was driven by simplistic ideologies, political ambitions, and plain greed for contracts. A few authors even considered how these forces combined with the imagery of fear. They described the extensive public relations efforts that encouraged military spending by playing up foreign threats. And they studied how national leadership groups harbored poisonous suspicions that were only loosely connected to the actual behavior of other nations. For a few years after 1968 a crowd of books and articles about the "military-industrial complex" rushed into print, urging citizens to strike down the ABM and other weapons at their economic, social, and ideological roots.[4]

In March 1969 the new Nixon administration announced that it would not put H-bombs in the backyards of cities but would only defend remote missile bases. In 1972 the ABM was abandoned by mutual agreement with the Soviet Union—the first true victory for arms control. As nuclear weapons once again receded from view, and as the Americans retreated from Vietnam, study of the military-industrial complex faded away as rapidly as it had arisen.

The brief ABM debate had agitated only a small fraction of the public, but it showed that nuclear fear had not disappeared. All that was needed to bring opposition roaring forth was a scenario in which some neighbor-hoods were singled out to face a threat and felt they had a chance to push it away. Meanwhile the debate reminded citizens that government decisions about nuclear technology could be scientifically suspect, and politically vulnerable too. The Atomic Energy Commission and the Joint Commit-tee on Atomic Energy were only peripherally involved with the ABM. But as that battle died away it gave birth to another, which would destroy both institutions.

The opening gun was an article in the September 1969 issue of *Esquire* magazine by a physicist, Ernest Sternglass. Sternglass had known about the problems of radiation since childhood, for his father, a physician, had often used X-rays. In 1947 the physicist's infant son became desperately ill, and he worried that he had passed on a defective gene originating in his father's exposure to X-rays. Sternglass began to worry about radiation from bombs too. He became famous with his *Esquire* article, titled "The Death of All Children."

Sternglass's argument, which he repeated in other magazines, news-paper interviews, and appearances on major radio and television shows, began with the fact that infant mortality had not declined recently in the American South as it had done in other regions. He said this had hap-pened because the South lay in the path of fallout from the Nevada bomb tests. Extrapolating his results to the whole world, he figured that one out of every three infants who had died in the 1960s had been killed by bomb test fallout. He concluded that in a war, even if ABMs destroyed every missile with nuclear explosions high in the atmosphere, the ABMs' own fallout would doom every baby born for decades in both America and Russia. In short, the ABM was a "doomsday machine."[5]

If using weapons would mean the end of humanity, then no govern-ment would dare to deploy them. For people anxious to abolish bombs, it was almost a disappointment when Sternglass's calculations turned out to be worthless. It was not even true that southern states were downwind from bomb tests. The reason infants in the South died disproportionately was the region's poverty.

While the experts were dismissing Sternglass, ordinary citizens were less sure. He was not only a physics professor but a good performer, making his points clearly and persuasively. When he and an AEC expert each had a few minutes to present opposing views on a television show, it was impossible to tell who was right. Most effective of all was the way Sternglass shifted ground. When his bomb test study was thoroughly discredited, he came up with statistics about infant mortality in the vicinity of nuclear power plants.

Others had been worrying inconclusively about the fact that all nuclear reactors release slight traces of radioactive materials into nearby air and water. Into these debates strode Sternglass, insisting that radiation from existing reactors had already killed babies by the thousand. Worried public health experts scrutinized his figures and discovered that they were full of holes. The regions where Sternglass pointed to high death rates had typically received a far more serious dose of poverty than of radiation. However, by the time one set of statistics was refuted, the professor had a new set that he found even more persuasive. Not one other scientist was ever convinced by his figures, but there was always some newspaper reporter who would publish Sternglass's latest horrific calculations.[6]

Although Sternglass was distressed that reactors, as he believed, slaughtered babies, his main motive was his original one of discrediting bombs. "The military is behind the entire nuclear reactor program," he declared. He believed that evil officials would do anything to make people think that radioactivity from reactors was safe, because otherwise nobody would tolerate nuclear weapons either.[7]

Sternglass's attacks dismayed nuclear experts. They believed that reactors were going to be much cheaper than other sources of energy and therefore would reduce poverty, the real slayer of babies. The AEC encouraged its own scientists to refute him, and that assignment turned out to be a mistake.

Nowhere in the AEC's domain was enthusiasm for nuclear energy higher than at the Livermore Laboratory. Fenced in among the hills of central California, the laboratory was Teller's creation, built to rival Los Alamos in designing weapons. During the fallout controversy Livermore had set up a health research program under John Gofman, a respected radiation medicine expert with a fringe of graying beard and a friendly smile. Despising government authority in general, and Teller's enthusiasm for explosions in particular, he remained apart from the Livermore spirit. Gofman later explained that he found all officials nauseating: "They can only think how to obliterate, control, and use each other."[8]

When the AEC sent Sternglass's *Esquire* article around its laboratories for comment, Gofman asked another Livermore scientist, Arthur Tamplin, to look into the matter. Tamplin, another opponent of the ABM, found himself in a dilemma once he understood that Sternglass's argument against the weapon was nonsense. He solved his problem by recalling Pauling's argument about fallout: stray radiation might well have killed some thousands of babies even if that could not be proved. When the AEC tried to prevent Tamplin from publishing this speculation, he and Gofman turned their attention to reactors as well. However slight the amount of radioactive material a reactor released, they pointed out, there was at least a *possibility* that it would cause somebody harm. The pair toured the nation, giving talks and interviews to warn against materials routinely emitted by reactors. Meanwhile they wrote a book titled *"Population Control" through Nuclear Pollution.*[9]

Gofman and Tamplin built upon the worries that Sternglass had originally provoked. For example, all three were given generous time in a 1971 television documentary broadcast in Los Angeles. Viewers were left with the thought that, as narrator Jack Lemmon put it, "nuclear power is not only dirty and undependable . . . it's about as safe as a closetful of cobras."[10]

While the AEC was losing the battle of images, it also fared poorly in the scientific dispute. Gofman and Tamplin refuted the claim, upheld by some of the nuclear industry's friends, that there was solid proof of a threshold below which radiation caused no damage. There was no proof of the no-threshold hypothesis either, but gradually the majority of radiation experts came to accept it, if only because it was more conservative than the alternative. Gofman and Tamplin's most effective argument took as a model Pauling's large numbers for fallout, which assumed no threshold. They said repeatedly that if everyone in the United States got a dose at the maximum level the AEC permitted reactors to emit, there would be 32,000 excess cases of cancer every year.

Nuclear industry spokesmen replied that the calculation was grossly unfair, since most members of the public got only a minute fraction of the maximum permitted dose. Very well, said Gofman and Tamplin, then why not set the rules so that public exposure would remain low as reactors multiplied across the land? In 1971 the AEC reluctantly bowed to this logic, imposing far stricter regulations on the amount of radioactive material that a reactor could legally release. Now even Gofman and Tamplin's calculations showed less damage to the public from reactors than from various everyday hazards. (In fact by far the worst exposure the public has always had to dangerous radiation is from the traces in tobacco, inhaled with cigarette smoke.) The radiation controversy faded away. But it left behind a new image: reactors insidiously slaying people not just nearby but anywhere.

Real events reinforced the idea. The first visible problem came at uranium mines in the American West, poorly regulated by the responsible state agencies and thick with radioactive dust. In 1967 the secretary of labor announced that about a hundred miners had perished needlessly of lung cancer, with hundreds more probably doomed. Over the next few years the federal government imposed tight standards on uranium mines, much as it began clamping down during the same years on deadly practices in coal mines. Meanwhile the uranium miners' tragedy had revealed that nuclear work was not somehow set above other industries: it really could bring a health disaster.

A less visibly deadly problem disturbed people even more: What was to be done with the radioactive by-products of the nuclear industry? Worrisome news came in the early 1970s from Hanford, Washington, where intensely radioactive liquids from AEC reactors were temporarily stored in underground tanks among the barren hills. Since the late 1950s some tanks had leaked, releasing hundreds of thousands of gallons of wastes into the sandy soil. The AEC had kept silent about the leaks for over a decade, so when the news came out, few trusted the official reassurances. Was any place safe?

Researchers who had been studying reactor by-products in a leisurely way had concluded that the problem could be solved. It was only necessary to mix wastes with cement or glass or the like to make a rock-hard substance, seal that in thick canisters for extra protection from groundwater, and bury the canisters. After all, in some parts of the world natural radioactivity in rocks, buried near the surface with no precautions, had lain for billions of years without straying out of the rock bed. Under pressure from criticism building up over the Hanford leaks and several similar cases, the AEC hastily announced in 1971 that it would solve the problem for good. It would bury wastes in an underground salt dome in Kansas. Such rock formations, after all, were perfectly dry and self-sealing.

Kansas officials soon found that the region was riddled with boreholes into which salt miners had injected water. The AEC had hit upon a remarkable rarity: a wet and leaky salt dome. Faced with an uproar among geologists and the Kansas public, the AEC abandoned its plans. Now there was no place to put wastes permanently, and little confidence that the government could be trusted to find a place. Further government efforts over the next decade led only to paralysis, as the experts' reassurances of perfect safety met a determined cry: "Not in my back yard!"[11]

According to polls in the United States and Europe, at the start of the 1970s few people would bring up the issue of nuclear wastes. But by the mid-1970s a majority saw the wastes as a major problem, and many worried about it more than about any other nuclear hazard. Many felt that reactor wastes were somehow unique, more dreadful than any other industrial danger.

The most persistent nuclear disputes of the 1970s would center on such by-products, particularly in France and Germany, where bitter opposition arose against installations for reprocessing used nuclear fuel to extract the plutonium. No other nuclear issue brought out such a wide spectrum of passions. At the highest level were moral objections. What gave anyone the right to contaminate the earth, to endanger innocent animals, to make money at the expense of future generations? At the other extreme the imagery of wastes continued to remind some people of infantile anal concerns. American and Japanese critics likened building reactors without a permanent waste repository to building houses without toilets. When public relations experts in Germany sent an "information bus" into villages near a proposed nuclear reprocessing plant, protestors trimmed the bus with toilet paper and smeared it with dung. A French tract called nuclear wastes "shit-heaps."[12]

In fact reactor wastes are the only type of industrial waste with a volume small enough to bury all in one locality with scant risk to people in other places. Even experts who from the start had criticized the nuclear industry for carelessness mostly agreed that risks from the waste were relatively minor and local, compared with many industrial pollutants that were only loosely controlled. But their calculations could not outmatch the image of nuclear objects buried in the earth—what people had associated for decades with wicked violations and dreadful dooms. Some citizens believed, as one of them told polltakers, that reactors threatened a "nuclear infection eventually destroying life on this fair planet." The "silent bomb" (as one writer called it) offered the same threat of contamination as nuclear weapons.[13]

Characteristic was a 1972 British television movie about radioactive waste dumped in the ocean near a coastal village. Fish grew to a great size while the fisherfolk became shambling monsters with a penchant for insane violence. As if to demonstrate the links with tradition, a cheap horror film about oversized ants engendered by nuclear wastes was produced by the same man who had made one of the 1950s films about monstrous insects engendered by bombs.[14]

From the early 1970s on, whenever a nuclear reactor was proposed in the United States a local group of housewives and students, reinforced by a few professional people, would spring up to oppose it. Most nuclear experts felt the opponents didn't have a leg to stand on, in terms of hard science and engineering. But in 1971 the opponents found support from deep within the AEC itself.

The Union of Concerned Scientists (UCS) was a group of students, scientists, and others, centered on the universities around Boston, who had come together in 1969 during the ABM battle. They had broad ideological aims, reflecting fears that science and technology would only degrade and endanger people, unless they were controlled in a more democratic way. As the ABM battle died away the UCS moved on to other issues; its membership declined. Drawn into a local controversy over a nuclear reactor site not far from Boston, they cast about for information and discovered . . . the China Syndrome.

The problem was no secret, but neither had it been publicized. In July 1971 a UCS report thrust the question into newspapers and television news broadcasts. UCS members had been talking with scientists from the reactor laboratories and were surprised to find that some AEC experts were unsure that the system for cooling the reactor in an emergency would work. For the next half-decade the UCS would focus its efforts on emergency cooling and other aspects of reactor safety, along the way gaining a sharp increase in membership and funding.

The UCS and its allies showed that it was not yet proven that a reactor could be prevented from ejecting radioactive materials if it happened to lose all its cooling water at once. Whether that particular, spectacular accident would ever happen in a typical American reactor, and whether research should concentrate on that rather than on more mundane routes to failure, was less discussed. Everyone from student opponents to nuclear engineers, from television commentators to AEC officials, became more preoccupied than ever with the most ghastly conceivable possibilities.

———

A 1977 television movie, *Red Alert*, had the same title as the American edition of the novel that foreshadowed *Dr. Strangelove* and *Fail-Safe*. Like those movies, the television production was built around forbidding technical systems, for example, a control center where anxious experts studied a computerized wall map. The story was again one of nuclear doom threatened by electronics gone out of control and by a rigid official who himself acted like "a robot." However, the threat in the 1977 production was not nuclear war but the simultaneous explosion of every reactor in the United States. And the official was no general but a reactor expert. The film showed how the image of reactors could borrow wholesale from the image of bombs—and even displace the weapons as a focus of public anxieties.

Not only Sternglass, Gofman, Tamplin, and the UCS, but many others who started out disgusted with military authorities began in the 1970s to turn the same thoughts against reactor builders. For example, one of the most effective activists was Helen Caldicott, a lively and tireless pediatrician. As a youth in Australia she had been deeply impressed by reading *On the Beach,* and as an adult she helped lead a campaign against French bomb tests in the Pacific. When she moved to the United States and found nobody there interested in bombs any longer, she began to fight reactors. Entire organizations took the same path. For example, one of the most effective groups was in St. Louis, inspired by the biologist Barry Commoner. They had begun work in 1957 with scientific studies of fallout, but in 1971 (as the Scientists' Institute for Public Information) they brought a lawsuit against the AEC's reactor program. In the coffeehouses where students gathered in the early 1970s, bulletin board posters opposing the Vietnam War and the ABM were covered over with notices of meetings to protest reactors.

Polls found a third of the world public held the false belief that a reactor could blow up exactly like an atomic bomb. Even among better-informed people, the image of a poisoned territory following a China Syndrome accident was not very different from the bleak science-fiction landscape of an after-the-bombs Forbidden Zone. And for almost everyone, radioactive reactor products seemed as alive with unspoken horror as if they were pieces of a bomb.

There was a core of truth in this last thought, for reactors do make plutonium. In the mid-1970s the press reported a new idea: somebody, anybody, might steal a few kilograms of the metal from a civilian plant and craft his or her very own atomic bomb. For example, a 1975 American television documentary showed that an undergraduate student could design some sort of fission explosive.[15] Experts tried to explain that the requirements were so stringent that any plutonium device that could be constructed without a cadre of top experts would yield a blast more like a large ordinary bomb than a Hiroshima cataclysm. But the news reports sounded as if the decades of fiction about a city blown up by a terrorist could come true at any moment.

Far more realistic was the risk that a nation might steal from itself, secretly diverting plutonium from a reactor to make bombs. Beginning around 1970 there were authoritative hints that India and Israel were doing exactly that, and in 1974 India dismayed the world with an atomic bomb test. Although both nations were drawing plutonium not from civilian industry but from "research" reactors they had built with explosives distinctly in mind, critics warned that the nuclear power industry could tempt a nation into making plutonium bombs.

Most experts on both sides of the nuclear debate had agreed ever since 1945 that this risk of weapon proliferation was the most worrisome of all liabilities of civilian nuclear industry. In the mid-1970s high government circles gave it more attention than any other nuclear hazard, getting into arguments over the sale of reactor materials to Brazil and so forth. The press took only passing notice of these arguments. A survey of American magazine and newspaper articles of the mid-1970s found that only 8 percent of all items on nuclear power focused mainly on weapon proliferation, and a minuscule 3 percent on theft or sabotage. Similarly, when poll-takers asked the public in America and Europe to explain their worries about nuclear power, the replies centered on health, safety, and wastes. Unless asked about it specifically, scarcely anyone brought up the fact that reactors could be used to make the stuff of bombs.

It was remarkable how the world public continued to avoid thinking about nuclear weapons. Nothing could have been more spectacular than images of missile warfare, the rockets blasting up from their hiding places, the warheads falling like shooting stars, yet the imagery remained un-developed. Aside from one or two almost unknown exceptions, movies and television and popular fiction of the time simply did not show what missile warfare would look like. As the 1970s wore on there erupted passionate protests about the chance that some reactor might explode despite every effort of its designers, but scarcely any protests against the chance that thousands of devices built to explode might fulfill their purpose. In various nations emotions flared over shipments of civilian reactor fuel along highways, while few mentioned the far more extensive travels of warheads. No agency dared to bury industrial fission products permanently, but masses of the same isotopes were casually implanted underground in every bomb test. Anyone reading the newspapers might have supposed that as soon as governments crafted plutonium into a bomb it ceased to be of interest.

The concentration of nuclear anxieties on reactors disturbed the few people who continued to struggle against weapons. Lilienthal, for one, though critical of reactors was far more concerned that citizens had apparently given up hope of halting the arms race. Since nuclear reactors did seem politically vulnerable, he said, people attacked them as "a surrogate for bombs."[16]

Lilienthal was saying that fear and hostility had been displaced from weapons onto civilian nuclear power. All the elements of classic displacement were indeed present. There was a persistent anxiety about nuclear war. There was an inability to dispel the anxiety in the only genuine way, by getting rid of bombs. Finally, there was a target toward which the frustrated feelings could redirect themselves, and all the more easily because of the many associations between bombs and reactors. If you had spent your life in a room with a grimacing Russian who kept a flamethrower pointed at your head, you might well feel upset when somebody struck a match.

Probing for social mechanisms, surveys around the world found that with reactor fears, as with war fears, the most remarkable result was no result: only minor differences were found between the standard social groups, a sign that all varieties of nuclear fear had deep roots in individual personalities. A closer look showed that the same groups that worried a bit more on average about nuclear war—the young, the lower social and economic classes, and especially women, or, in short, the relatively powerless—also worried more about reactors. Perhaps both varieties of anxiety reflected a more comprehensive concern that extended beyond reactors—and beyond bombs too.

———

Anxiety about science and technology was cropping up all over, from scholarly discussions to student demonstrations. Polls showed that the number of Americans who felt "a great deal" of confidence in science declined from more than half in 1966 to about a third in 1973. A main reason for misgivings about science, according to a poll that had studied the matter in detail, was "unspoken fear of atomic war." In the 1970s atomic bombs remained the one technology that large majorities frankly distrusted. But public confidence in every form of authority was waning, with the military, business, and government all increasingly under suspicion. Nuclear affairs were caught up in a larger social tide—which nuclear fear had played a significant role in raising.[17]

People were noticing that new technologies such as offshore oil drilling and strip-mining of coal could harm broad regions, and they began to protest. Groups like the Sierra Club, which traditionally had devoted themselves to local affairs such as preserving stands of redwood, began to question industry on a national or even global scale. They were joined by new and more militant groups such as Friends of the Earth. They won impressive press attention that reflected and encouraged a strong rise in public concern. Governments responded with strict laws and new bureaucracies to enforce them. In 1970–1971 alone, major environmental agencies or programs were founded in the United States, Britain, France, Japan, and even the Soviet Union.

A historian of the movement, Walter Rosenbaum, found the new mood had three main components—ones with a familiar ring. First was a sense of dire crisis, a fear that humanity was headed for self-imposed catastrophe. Nonfiction works along with novels and movies warned the public against new ice ages, stupendous earthquakes provoked by injection of liquid wastes, and dozens of other agents ready to poison, choke, or otherwise destroy civilization. Predictably, many of the fictional apocalypses led through social disintegration to a new and better world.[18]

A second theme of the environmental movement that Rosenbaum noted had less to do with traditional fantasies. It was a concern with whole

systems, with the ways in which pollution could affect ecological relationships around the globe, and the ways in which our entire modern culture
and economic system encouraged pollution. A third theme was disillusion
with these modern values and authorities.

These themes—dismay with technological authorities and with systems
that seemed about to doom the entire world—had first been thrust upon
the public by hydrogen bombs. Had nuclear fallout taught anxiety about
all technology? Barry Commoner, most prominent of the 1960s ecologist
critics, said so: "I learned about the environment from the United States
Atomic Energy Commission in 1953." The publication that his group
began during the fallout debate, *Nuclear Information*, turned into *Environment* magazine, one of the new movement's main vehicles. But popular environmentalism had really started with a book attacking chemical
pesticides, *Silent Spring*, begun by Rachel Carson in 1958. She wrote privately that in earlier years, despite scientific evidence about harmful
chemicals, she had clung to the faith that "much of Nature was forever
beyond the tampering reach of man . . . the clouds and the rain and the
winds were God's." It was radioactive fallout, she said, that had killed this
faith. Her book, published in 1962, opened with a fable of a town dying
from a white chemical powder that snowed down from the sky, a powder
that Carson likened to fallout. Many other environmentalist writings likewise used images taken over directly from nuclear protests.[19]

Nuclear fear was only one of many forces behind the environmental
movement, but it held a special place. It raised emotions earlier and on a
more visceral level than any other issue. And it served as a banner that everyone could rally around. In return, environmentalism would provide a
solid base for the opposition to reactors, lending organization, seasoned
leadership, masses of followers, and a set of ideals.

As early as 1965 a poll showed that the great majority of environmentalists, unlike most other Americans, opposed nuclear power plants. A study
of a coalition that battled the AEC over emergency cooling systems and
the China Syndrome found that two-thirds of the leaders had been in
environmental groups. These leaders were usually mature, middle-class
liberals, the sort of people who had distrusted the AEC ever since the
campaign against fallout. Many of them had begun with no specific dislike of civilian nuclear power, and during the 1960s some environmentalists had accepted reactors as less polluting than coal. At first, when such
people opposed a reactor at a particular site such as Bodega Bay, they had

done so for reasons that would have led them to oppose any industrial plant there. But raw nuclear fear added a further driving force.

A process of conversion spread through the environmental movement. Typical was the evolution of Ralph Nader. As a crusader against the hidden evils of industry and government, in 1970 he had been less concerned about nuclear power than about air pollution from oil and coal. But Gofman and Tamplin's warnings caught his attention, and so did the China Syndrome controversy. Convinced that in nuclear power the public faced another indecent official "cover-up" of hazards, in 1974 Nader convened a national conference that brought together more than a thousand reactor opponents. A reporter said the assembled militants, wearing everything from backpacks to business suits, "had the grim flavor and messianic fervor of the movement to end the war in Vietnam." Nader institutionalized this spirit the following year in Critical Mass, a coalition of hundreds of local and national groups.[20]

The most powerful organization to join the fight was the Sierra Club. The club had traditionally been wary of all power plants, including hydroelectric dams. In 1974 a spontaneous wave of feeling at the grass roots prompted a heated debate over the environmental pros and cons of reactors (weren't they preferable to hydroelectric dams and coal-fired plants?). In the end the club's directors took an official stand against the technology.

Other environmental causes were faltering. Polls in the mid-1970s found that the public had become less worried about ordinary industrial pollution, if only because the new laws really were doing some good. The press turned to newer topics. Yet articles critical of nuclear power kept trickling out as one problem or another—wastes, reactor accidents, proliferation, homemade bombs—had its day in the news, looming ever larger in proportion to other technological concerns.

Nuclear opposition absorbed back into itself much of the driving force that nuclear fear had lent to environmentalism; along the way the force was concentrated upon reactors. The public might be growing used to alarms about oil spills and pesticides, as it had grown used to bomb threats. However, nuclear reactors represented not just a new anxiety but a new *type* of anxiety, one that addressed the goals of all industrial society.

18

ENERGY CHOICES

October 1973: the first oil crisis. Shortages and a huge leap in fuel prices stunned the world. Economies staggered as irate motorists idled their cars in long lines at gas stations. A few decades earlier, most advanced nations had been nearly self-sufficient in energy; now nearly all of them relied heavily on imported oil. As the flow faltered, governments laid frantic plans to safeguard national prosperity and security. These plans included ambitious long-range programs to build hundreds of reactors and make them a linchpin of the economy.

The programs greatly stimulated controversy. Up to then the reactor debates had mainly involved on one side nuclear officials and on the other side small local groups backed by a few full-time opponents, arguing technical questions at government safety hearings and the like. After 1973 the opposition became a mass movement. Would the world's nations be allowed to build reactors as planned? The outcome would depend on some combination of public imagery and technical facts; neither by itself was sufficient to prevail.

If we could not power civilization over the long term with oil, was there any alternative better than reactors? Most reactor opponents pointed to sunlight. The nineteenth-century dreams of a utopian solar-powered civilization persisted, as seen for example in a 1940 *Astounding Science-Fiction* tale by Robert Heinlein: a scientist fought industrialists who wanted to suppress his discovery of solar electric-power panels. Heinlein took it for granted that his readers would understand that energy from the sun would mean a decentralized civilization with riches and freedom for all.[1]

Large majorities of people, particularly the younger and more highly educated ones, whether or not they favored reactors, had far more enthusiasm for the sun as an alternative than for any other energy source. Solar technologies, according to a prominent reactor critic, were "ideally suited for rural villagers and urban poor alike," and so forth.[2] Only a few historians noticed that the promises of prosperity and freedom made for the smiling sun had been made two decades earlier on behalf of nuclear reactors.

Scientists and engineers had been working on solar power for a century, and in the 1970s many more set to work, backed by abundant government research funds. Every year some experts spoke of imminent breakthroughs. The informed consensus was less sanguine: solar energy would not begin providing truly important quantities of electricity until the early decades of the twenty-first century—and perhaps never. As the history of reactors showed, no matter how good an industry appeared on paper, it could not be built up to full scale in less than a generation, and then it might come up with unexpected problems.

Environmentalists came up with another proposal: conservation. For example, a major "energy resource" turned out to lie in giving houses better insulation, and another in giving automobiles better fuel economy. Yet conservation had limits. After 1973 the advanced nations took strong measures that did reduce their total energy use, yet their use of electricity slowly rose; electricity seemed to be more indispensable than other forms of power. Worse, it turned out that when people saved money by using less power, they might just spend the surplus on another power-eating gadget. In any case the population in many regions was rapidly climbing. When people in a city came home at night and turned on their lights, something would have to furnish the electricity.

For most of the "atomic age" that something had been oil, but the oil feast was coming to an end. By the mid-1970s informed people realized that half of all the oil that could easily be taken from the ground in the United States was already gone, and the world's halfway point would be reached within their children's lifetimes: the world was beginning to glimpse the bottom of the oil barrel. Natural gas, tar sands, or oil shale might stretch out the supply for another generation, but that prospect was entirely uncertain.

For most utilities responsible for lighting a city, the choice came down to a nuclear reactor or coal. The use of coal by American utilities increased 50 percent between 1975 and 1982 as oil burners were converted and new coal-powered plants were built, while similar sharp rises took place in other nations. Reality was posing a blunt question: Which would the public rather draw its electricity from, a coal plant or a nuclear reactor?

Just how good or bad was the nuclear industry—and compared with what? While experts fretted over proliferation of weapons, the public concentrated on matters closer to home. Was your life, and your children's, in danger? Some pointed to the hundreds of deaths among uranium miners as an indicator of peril. But neither reactor critics nor proponents remarked that other technologies regularly provoked tragedies on that scale—failures at hydroelectric dams had killed hundreds and even thousands at one time, and so forth. Few thought that nuclear problems could be comparable with the problems of less magical industries.

There were exceptions among engineers, members of a profession dedicated to the best use of limited resources. They pointed out, for example, that hundreds of millions of dollars were being spent on improvements in reactor pollution controls that would save a few lives at most. Yet only a few million extra dollars given to a city's health department could save hundreds of lives, by attacking infant mortality among the poor. Nothing could make an engineer's blood boil like resources wasted on an inefficient way to achieve a given result. No doubt everyone ought to look in this holistic way at all the problems and benefits of every technology together. Yet a comparison between things as unlike as reactor construction costs and a city budget made little impression.

There was one area in which nuclear power could be and indeed had to be compared face-to-face with a familiar technology. Coal, too, had its perils. Well known in mining and industrial regions, the hazards became obvious to everyone after a week in 1952 when a "killer smog" of coal smoke settled over London, destroying 4,000 lives. Might a switch from coal to uranium save more lives than it endangered?

The rise of the antireactor movement made the comparison with coal more urgent, and beginning in the early 1970s a few people worked out the relative risks and benefits with increasing precision. It was the birth of a new type of analysis, soon applied to many other industries; nuclear issues became the paradigm for detailed "risk-benefit" analysis of technology. By the end of the 1970s the key facts regarding coal and uranium were clear. Some of the conclusions can help us see where the fear of nuclear reactors relied upon facts and where it did not.

Analysts agreed that the nuclear industry's chief health risk, not only in popular belief but in fact, was radioactive wastes. The worst problem came from the mining and milling of many thousands of tons of uranium ore. What was the comparison with coal? That would be clear to anyone who visited areas where uranium was mined and then visited coal country. Thousands of square miles of barren, misshapen land, thousands of miles of creeks where the water ran yellow and stank—those were the marks of coal mining. In contrast, uranium mining areas were scarcely altered from their original state. The same disparity could be found at the far end of the process. The materials removed from reactors were compact and tightly monitored, but there was no federal regulation over the many tens of millions of tons of coal ash dumped each year into landfill and waste ponds.

Specifically, the nuclear industry's most severe problem turned out to be radon gas that seeped from excavated uranium ore, perhaps to cause lung cancers. Radon-producing uranium was not found only in concentrated ore, however, for minute traces are everywhere in rocks and soil. American coal mining brought about 20,000 tons of uranium to the surface each year, mixed with dirt and a billion tons of coal. Taking into account the regulations that required uranium mine tailings to be covered over, but not coal residues, the harm to health through radon from the two sources of energy turned out to be on more or less the same level.[3]

With coal, unlike uranium, the worst problem was not radon. One material found in coal was arsenic, and there was good evidence that traces of arsenic could bring cancer and genetic damage. As with plutonium, experts debated whether there was a threshold below which no damage would be done. But for both plutonium and arsenic the conservative assumption held that even a single atom might cause a cancer or mutation. Unlike plutonium, however, arsenic did not undergo radioactive decay to another element, with a half-life of some 20,000 years: it was a stable element and would persist in the environment forever. Similar problems held for mercury and lead, also found in coal and not easily removed from a power plant's smoke by pollution control devices. And whatever the devices did remove would be dumped in the open anyway, as part of the billions of tons of sludge and slag that piled up each year. The arsenic, lead, and mercury in many coal-ash dumps leaked into water supplies at levels deemed unsafe, yet nothing was done.

These metals were not the worst problem of coal. Scientists had never studied most chemical compounds with anything remotely like the care that they applied to radioactive isotopes, but in the 1970s some began to study the fly ash that wound up either in the air or in sludge piles. It turned out that these complex chemicals, too, might damage genes or cause cancer.

Radiation does not bring harm through some uncanny magic, but by breaking apart molecules within the body, that is, by causing chemical changes. That is why cancers and mutations caused by radiation are essentially indistinguishable from the same ills caused by chemicals.

Complex chemicals were still not the worst problem of coal. The limited knowledge available by the 1980s suggested that the gravest health hazard came from simple chemicals and microscopic particles in smog. Their effects, like those of radioactive isotopes, were too diffuse for scientific proof, and some estimates pointed to only hundreds of illnesses. But equally plausible estimates suggested that where coal was heavily used, as in the central and northeastern United States, much of Europe, and in China and India as their economies developed, it could be causing tens of thousands of premature deaths from cancer and lung disease. (It was only much later, in the early twenty-first century, that studies showed conclusively that the microscopic particles in coal smoke did cause roughly 10,000 premature deaths a year in the United States, and far more worldwide.)[4]

None of this meant that burning coal was unconscionable. The advanced nations reduced the total of their air pollution in the 1970s, and if the political will had remained they could have continued not only to increase their use of coal but to decrease its pollution for several decades more. Besides, even the worst industrial contamination added only a small fraction to the total of human sickness and death, while poverty was killing wholesale; the most industrialized countries had by far the healthiest populations. It is clear that, by every sensible measure, as a health and environmental issue coal deserved many times more attention and care than nuclear power.

There was one category of environmental danger, however, where the comparison was harder to make, and it bothered some people most of all: the maximum accident. Nobody could estimate the chances of a catastrophe for which there had never been a single example. But the AEC tried, organizing a lengthy study under the engineering professor Norman Rasmussen. The Rasmussen Report, completed in 1974, was designed to reassure the public. Its heavily publicized conclusions said that a citizen was more likely to be killed by a meteor than by a reactor accident. The report immediately came under fire not only from antinuclear groups but also from independent scientists who showed that the conclusions were riddled with methodological problems and in some ways far too optimistic. Evidently the whole subject was a swamp of uncertainty.[5] Nuclear engineers and their critics continued to focus single-mindedly through the 1970s on the spectacular events of the China Syndrome. Meanwhile there were as usual hundreds of little mishaps each year. If several of those ordinary events, recorded but scarcely attended to, should happen one on top of another . . .

In 1979 an elaborate chain of failures, any one of which would have been harmless by itself, came together to cause a medium-sized accident: partial melting of the fuel in the Three Mile Island reactor near Middletown, Pennsylvania. Thanks to the containment vessel, scarcely any radioactive material escaped; nobody was injured, although the reactor was ruined. But it was not clear for days whether the containment would hold. The press screamed, schools closed, authorities warned pregnant women to stay indoors, citizens for hundreds of miles around were alarmed almost to the point of panic. I will say more about these public reactions in the next chapter. But what did the accident look like to engineers?

Later, calculating from the detailed analyses buried in the Rasmussen Report, engineers found that some such medium-sized accident might have been expected somewhere in the world, with fair probability, before the end of the 1970s. The exact types of failures that came together to wreck the Three Mile Island reactor had been occurring at one or another reactor for years without happening all at the same time, but reports on those failures had been overlooked amid the storms of debate over maximum disasters and the blizzard of regulatory paperwork. Taken aback by the Three Mile Island accident, American officials and engineers at last began to pay closer attention to the real problems of reactor safety. Just as hundreds of minor individual failures had been observed for each troublesome incident, and hundreds of those for each serious accident like Windscale or Three Mile Island, so among reactors with containment shells there should be many such cases of partly melted fuel for each great disaster in which a still longer chain of failures let a radioactive cloud break into the open. For typical American and Western European reactors, many accidents like Three Mile Island could be expected before the first dreadful catastrophe. Long before then, of course, the owners would either make reactors more reliable or abandon them, if only because each ruined reactor cost them more than a billion dollars.

The problems would become plainer after a far worse accident: in 1986 a reactor at Chernobyl in the Soviet Union burst into flames. Once again the immediate fault lay in a series of small errors, which were routine in the slapdash Soviet industry. The mistakes were magnified out of hand by the Soviet design that made it easy for the reactor to experience runaway heating. It was a Maximum Credible Accident—without a containment shell. As I will describe later, the results would come close to what experts had warned such a result might bring: immediate deaths by the dozen, health problems affecting millions, contamination extending hundreds of miles. However, prior to 1986 the public did not notice that Russian reactors lacked containment. Most of the world's reactors were far safer.

The Chernobyl catastrophe was about as close as one could get to the worst conceivable result of a nuclear industry accident, and anything similar was very unlikely for any reactor with a containment shell. For coal, the possibilities for disaster were unavoidable and on an immeasurably greater scale. Burning fossil fuel increases the amount of carbon dioxide gas in the Earth's atmosphere. Since the late 1950s a few scientists had been predicting that the resulting "greenhouse effect" would heat up the planet. If the emissions continued decade after decade until a good fraction of the available coal was burned, the global warming would bring problems that would certainly be very serious, and might be catastrophic for civilization. The scientists predicted, however, that the warming would not become clearly detectible until around the start of the twenty-first century (they turned out to be correct); in the 1970s that seemed very far away. Besides, many scientists doubted that global warming would happen at all. Most people shrugged off the matter—if they were even aware of it. The risk would not be widely publicized until the late 1980s, and would not be taken seriously by the world's governments until the first decade of the twenty-first century.[6]

In fact the public heard little about the comparative risks of coal and nuclear power plants. When Glenn Seaborg and his colleagues remarked that a typical coal-fired plant routinely emitted more radioactive material into the air than a nuclear reactor, and when other experts estimated that using coal instead of uranium might cost hundreds of lives each year for each power plant, they spoke with restraint. Most scientists and engineers, along with a majority of the public, did not despise any form of civilian technology. After all, over the long run technology saved many times more lives than it cost; whoever favored reactors usually favored coal as well. Government officials and utility executives felt no urge to insist that their nations' use of coal was poisoning people for thousands of miles downwind.

Environmentalists were even less inclined to publicize comparative risks. Many who fought against reactors fought with equal conviction (if less success) against coal plants, finding no need for comparisons when they condemned both alike. Others claimed that coal was "far cleaner" than nuclear power. Widely read authors ignorantly assumed that radioactive isotopes were the only chemicals that had long lifetimes or could cause genetic damage. Others simply took it for granted that nuclear things were more horrible than any alternative. Besides, committed opponents of nuclear power had little incentive to study, still less to advertise, comparisons that favored reactors.[7]

In 1978 a group of antinuclear militants had a private meeting with André Gauvenet, an elderly gentleman, trained in engineering, who had worked his way up within the French Atomic Energy Commission to become its head of nuclear safety. Calm and refined, intensely civilized, Gauvenet had at his fingertips a rebuttal to every technical criticism his opponents raised. The militants quickly tired of the argument. Particularly irritated was a leader of the local Friends of the Earth, a young man who had left his family and studies to commit himself to life in a commune and antinuclear campaigning. He raised his voice to insist that the real problem was not in technical safety games: "It's political!"[8]

The confrontation between the silver-haired authority and the frustrated militant took place on many levels. The two were poles apart not only in their views on reactors but in their deepest social and political attitudes. The nuclear debate had exposed a rift that ran straight through the core of twentieth-century society.

The young militant, like most other nuclear protest leaders in the 1970s, owed many of his views to the worldwide countercultural movement that had burst forth a decade earlier. The movement's many causes ranged from the advent of television to the huge postwar baby-boom cohort of educated youth. But of special importance was something closely associated with reactors: the shadow of nuclear war.

As one protestor put it, recalling civil defense drills, "In many ways, the styles and explosions of the 1960s were born in those dank, subterranean high-school corridors near the boiler room where we decided that our elders were indeed unreliable." A psychological survey of young people in the mid-1960s confirmed that in their thoughts of imminent nuclear bombing, reality was reinforcing adolescent fantasies about inadequate and destructive adults. Most people who observed the young radicals agreed with the famous 1962 manifesto of the Students for a Democratic Society, which said they were "guided by the sense that we may be the last generation." Some eminent social commentators said that fear of nuclear war was not just one force, but the primary force, behind the rebellion.[9]

In the political sphere the first signs of a youth rebellion had come with the movement against fallout. The most revealing case was in Germany, where the struggle had given birth to a Campaign for Disarmament (not just, as in Britain, a Campaign for *Nuclear* Disarmament). The increasingly young, increasingly radical protestors demonstrated by tens of thousands against militarism in general, government police powers, the American presence in Vietnam, and the "fascist" aspects of modern capitalism. In other countries, too, the antibomb movement of the early 1960s was a main source of both practical lessons and experienced leaders for the later, more general protests.

During the mid-1970s the countercultural movement concentrated a large part of its attention on nuclear energy. It was not being diverted but was returning to basic concerns, for even the early bomb protesters had addressed issues beyond war; now the general disaffection with modern technology and the officials who controlled it was becoming plain. Nuclear officials were reluctant to recognize that they faced something new. At first they believed, as Seaborg put it, that reactor critics "have just not considered the facts." Surely they could be won over by sound information! The opponents of the 1970s, however, were different from the people who had worried over nuclear energy back in the 1950s, on average younger and better educated. The nuclear industry began to call them spoiled children and Marxist radicals. There seemed to be an entire type of person opposing reactors, a type with strong values of its own.[10]

The antinuclear people agreed with that. They called themselves democratic, warm, spontaneous, frank, and free; they called their enemies a hierarchical, cold, and secretive establishment of cramped bureaucrats who cared little for humans. Many went beyond environmentalism to a fundamental critique of modern society. The critique was pioneered by intellectuals such as the economist E. F. Schumacher, whose 1973 book *Small*

Is Beautiful argued for restricting industries to a scale that could be controlled by an individual or a small organization. Schumacher envisioned a network of diverse, modestly sized, quasi-autonomous groups, each using only technologies that were readily understood, each giving close attention to human and spiritual values. People like Schumacher scoffed at the TVA dream of using central government organization to promote a modern Arcadia. They found hope for economic prosperity and justice, and even world peace, only through complete decentralization.[11]

Especially persuasive was a 1976 article by Amory Lovins, the very picture of an intellectual, with a broad forehead, small chin, and mild eyes behind dark-rimmed glasses. He pointed out that reactors necessarily required highly centralized power systems, which by their very nature were inflexible, hard to understand, unresponsive to ordinary people, inequitable, and vulnerable to disruption. A German journalist, Robert Jungk, went still further. In *The New Tyranny*, internationally the most widely read antinuclear book of the late 1970s, Jungk described the fight against reactors as an apocalyptic battle: the nuclear power industry was driving us into a robotic slave society, an empire of death more evil even than Hitler's.[12]

Jungk and many others saw the nuclear industry as the supreme paradigm of decision making by remote and heedless authorities. Indeed officials had often patronized or ignored any nonexpert who worried about reactor safety. Jungk and others told how nuclear authorities—stirred by public threats of sabotage along with a few actual bombings of their offices and installations— ordered surveillance of opponents. Were they not preparing the way for a garrison state? Once again we were at a crossroads. The choice between nuclear and solar power, one writer declared, was "a choice between corporate rule and social democracy, between suicide and survival."[13]

Opposition to reactors was developed most fully in Western Europe. The breakthrough came in 1974, when governments announced that they would meet the oil crisis with hundreds of reactors. Like their American counterparts, most of the European groups that pioneered the campaign against nuclear power could trace their origins to indigenous struggles against weapons or environmental pollution, but often the immediate spark that ignited opposition came from the United States. Europeans translated and repeated much of the American polemics, beginning with the Sternglass, Gofman, and Tamplin warnings. If Europeans borrowed ideas and tactics from the United States, they would soon return the favor, bringing a new level of political sophistication.

Their greatest early success was in the French and German Rhineland, where they won over citizens from housewives to town officials, all marching behind a new slogan: "Better Active Today than Radioactive Tomorrow!" A climax came in 1975 at Wyhl on the German side of the Rhine when 20,000 people advanced on the site of a proposed reactor and tore up the fence surrounding it. Despite police resistance people swarmed onto the site, setting up barricades and putting up tents in the forest. Over the next months students and other activists from all over Western Europe came to the encampment at Wyhl, mingling with local citizens and organizing study groups at a "people's school." The discussions covered not only reactor technology and ecology but also the police state and similar political topics. Here and at demonstrations elsewhere, the legacy of the student countercultural movement of the 1960s was in evidence—the flowers and songs, the rejection of bourgeois society, the spontaneous informal organization almost like a rural commune. But communard attitudes mingled with the solid conservatism of German farmers who only feared that a nearby installation might spoil their vineyards. The battle moved into law courts, and the Wyhl reactor was never built.[14]

Subsequent years saw further attempts to occupy nuclear industry sites in Europe and, by imitation, in the United States. In particular, a reactor at Seabrook in New Hampshire was opposed by a "Clamshell Alliance" that borrowed the Wyhl tactics of civil disobedience, and passed them along to other groups in turn. The climactic battle came in France, where the government had put its greatest hopes in nuclear industry.

If the Wyhl campsite was a living model of anarchist-Arcadian hopes, its counterpart was visible in the offices of Electricité de France and the French Atomic Energy Commission. These were located in prosperous sections of Paris, along quiet boulevards where strollers could linger at an occasional café or fountain—boulevards deliberately modeled on the nineteenth-century vision of the White City. The offices themselves were in perfect order, with polished glass, a guard in uniform at the door, efficient secretaries, silent elevators. Of a piece with this social order was the ingenious device going up at Malville in the Rhone Valley. Because France had only a modest amount of uranium as of all other fuels, the government had determined to build a complete breeder reactor economy, staking the nation's long-term future on a prototype at Malville. The device was named "Super-Phoenix" after the mythical bird reborn from its ashes, another of the countless symbols of transmutation.

July 31, 1977, was a rainy day at Malville. Some 50,000 protestors advanced on the fortified reactor site, slogging ahead in columns, silent in the dismal weather. The antinuclear movement with its environmentalist and pacifist background was by nature nonviolent, but also by nature it lacked strict organization. Small bands came with iron clubs, helmets, red or black flags, and Molotov cocktails. The police met the marchers with truncheons, high-pressure water hoses, and tear gas grenades. When the cries and confusion ended there were over 100 injured, and one demonstrator was dead. Construction of the Super-Phoenix and dozens of other French reactors went ahead on schedule.

The opponents could not win in the field, and they could not win in the French courts either. Unlike the American and German systems for licensing reactors—mazes of hearings in which a determined group of opponents could raise endless obstructions against a builder—French licensing was a straightforward matter of working things out among government experts. The French governing system did not leave as many openings for opponents as did the American and German systems, with their numerous rival local and national authorities, any of which might be turned into a roadblock. French reactors would go ahead so long as a bare majority of the nation's citizens continued to trust their leaders.

In many nations the opponents' demands were more effective. The American movement was the first to achieve a major political goal: the destruction of the Atomic Energy Commission. Promoting and regulating a technology at the same time, as the AEC did, was common among government agencies, but it seemed less and less acceptable for the special case of nuclear energy. In 1974 Congress tried to satisfy critics by splitting up the AEC. One part became the hapless Nuclear Regulatory Commission, while the rest, demoralized, eventually joined the new Department of Energy. Meanwhile Congress stripped the Joint Committee on Atomic Energy of its unique powers. Nuclear energy had lost its strongest institutional supports.

More direct political gains followed. A 1980 referendum left Sweden with a nuclear program scheduled to end; a 1978 referendum in Austria stopped reactors there forthwith. Similar results were achieved in other democracies such as Italy, the Netherlands, several individual states in the United States, and wherever else political and judicial systems gave a determined minority a veto. The mechanics of the process were most visible in Germany, where a whole new political party, the Greens, rose up with opposition to nuclear power the keystone of their program. The major German political parties, scrambling to regain votes, grew hesitant about reactors. Around the world, by the early 1980s far fewer reactors were on order than had been planned a decade earlier.

It was not only politics that slowed the deployment of reactors. Largely because of the rise in oil prices, the world economy stagnated and interest rates soared, undercutting any industrial project that had to borrow large sums of money. Reactor construction was further wounded by errors and delays. These came partly because the technology of giant reactors was new and untried, and partly because a stream of government safety regulations required seemingly endless design changes. Costs climbed far above the cost of reactors finished in the 1960s. Some utilities that had staked their future on reactors, for example the TVA, wound up in deep financial trouble. On the other hand, there were some American utilities, and foreign ones such as Electricité de France, that built reactors of proven design and were highly pleased with the results. In economic terms nuclear power could be either a bust or a boon.

When utility executives tried to choose among coal, uranium, and other alternatives strictly in terms of money, they remained subject to pressures from imagery. Planning an electrical system meant peering thirty or forty years into the future. Yet nobody could predict even a few years ahead what would happen to fuel costs, interest rates, or demand for electricity. Estimates about reactor economics varied wildly from one expert to another. By the mid-1980s American utilities were shying away from nuclear power, only completing reactors already begun or even abandoning them halfway. The French were meanwhile building an electrical system that by the end of the century would draw nearly four-fifths of its power from reactors—and export some of it to neighboring countries at a tidy profit. Most nations hesitated in between, planning a gradual and wary increase in their use of nuclear energy.

One thing was certain: every decision on whether to build a reactor would depend critically on guessing what requirements the national government and the local public might someday impose. Therefore, even economic projections depended on the attitudes of everyone who held a piece of power, from leading politicians to patchworks of neighborhood groups. The great energy debate turned largely on public images—and on the little-known forces that promoted images.

19

CIVILIZATION OR LIBERATION?

When proponents and opponents of reactors stood up to debate face-to-face, forces long obscure came into the open, startling onlookers with their vehemence. Many nuclear opponents frankly displayed their anger and anxiety, and if advocates of reactors made a show of calm rationality, on that side, too, anyone listening closely could detect anxiety and anger. With growing exasperation each side accused the other of peddling disgusting nonsense—and sometimes of being disgusting people. In formal debates nobody seemed able to refute anyone else's statements; arguments flew past each other almost without touching.

The longer the controversy went on, the more each side was subject to what sociologists call "social amplification"—the tendency of members of a group to coalesce around a set of shared opinions. There is a tendency for this position to move toward an extreme, to distinguish itself from whatever the group's opponents maintain. The tendency is solidified by "confirmation bias" (something easy to see in others, harder to see in ourselves): give opponents the same set of facts, and each side would pay respect to the facts that supported their position, while ignoring or disbelieving the facts that tended to contradict them.[1]

Observers began to notice that the two sides held widely divergent assumptions about not only reactors but many other things. Sometimes the debaters spoke what amounted to different languages, using different words and expressing themselves in different modes. The fight pointed to a major division that reached into the very core of modern society.

By the mid-1970s many nuclear opponents were saying that their battle was not just against the reactor industry but against all modern hierarchies and their technologies. The lines of battle were mapped by polls that found a seeming paradox. People who identified themselves as "conservatives," supporters of the established order, were more likely to accept new technology than were "liberals," who said they favored alterations in society. This was not really a paradox. Technological change had become so embedded in modern civilization that to uphold such change was to uphold the existing social authorities and ways of thinking.

More specific surveys found, among many factors, only one that clearly predicted whether a person would be for or against reactors: it depended on whether or not the person had confidence in basic social institutions. This would reflect the individual's political ideology, no doubt, but it would also be shaped by how the person felt about the risks and benefits of various technologies. Confirmation bias could make this a circular, self-reinforcing process. In the end, people would form a deep emotional response to things like reactors as fundamentally "good" or "bad," attractive or horrible. Another study found that this "affective response," more even than an explicit worldview, determined whether people thought reactors were safe or risky, and even whether they were economically sound. It was with good reason that people on both sides of the controversy worked to affect the public's affective response.[2]

A few activists hoped the movement against reactors would serve as "a mechanism for raising the political consciousness" of people who had originally feared only for the environment. Reactor dangers would teach them the necessity of a "revolutionary movement" against capitalism. Ordinary opponents of reactors did not share such a radical political critique, yet the critique revealed much about the thoughts that helped sustain the movement.[3]

Perhaps the sharpest analysis was made by feminists who saw nuclear energy, both military and civilian, as a form of male domination. A few women dragged into the open the sly messages long hidden in journalistic talk about atoms, accepting the phrases as disgracefully true. Mastery of nature, they said, was only a variant of male suppression of women. Some were offended to see "the earth raped by the insertion of missiles." Others claimed that a reactor spilling radiation was equivalent to incestuous assault, pollution inflicted by father figures: "The patriarchy . . . has created the rapist-energy of radiation-ejaculating nuclear power." Most of the antinuclear movement would not follow the analysis that far, but they all responded to images of victimization at the hands of authorities.[4]

For antinuclear people, the ambiguous structure that I have described earlier in this book—scientist and victim, monster and hero, promiscuously interchanging places—resolved into a simple pattern. They saw all atomic affairs through the lens of a simple bipolar structure that might be written in shorthand as *authority/subject*. On one side stands the scientist or technologist with his dangerous devices, the domineering male (especially a government, industry, or military official), the authoritarian father, the entire generalized threat from an overregulated technological society. On the other side stands the guinea pig, the enslaved worker, the dominated woman, the rejected child, the individual crushed by modern society.

The evil atomic scientist stereotype, which had already partly infiltrated the public's images of strategy theorists, Air Force officers, and other government authorities, now infiltrated the images of private industry too. The heartless and calculating businessman, another centuries-old stereotype, had originally had little to do with science, but he fitted the same psychological pattern as the cold-blooded scientist. As industry increasingly connected to science, stories began to show technologists as masters over industrial serfs and even as owners of a runaway atomic power plant. That was the role played, for example, by Ming the Merciless in the 1936 Flash Gordon serial and by Dr. No in the 1962 James Bond movie. The stereotype took a more overt meaning in the 1970s as opponents accused industrialists of using perverse technology to pollute both nature and individuals.

The emotional force became clear in 1974, when the antinuclear movement announced its first martyr, Karen Silkwood, a nuclear industry worker. She was exposed to radioactive contamination from the processing plant where she worked and then died in a peculiar automobile accident. Because she had been a union activist digging for secrets about mishandled plutonium, many believed that Silkwood had been murdered by order of some anonymous official. Much that was written about her portrayed her as the complete victim: a woman, a suppressed proletarian, and even a childlike innocent, who learned the evil secrets of her masters, was polluted by unwholesome technology, and was then struck down by malign authorities.

The most striking images came in a television movie modeled on the Silkwood legend, *The Plutonium Incident*, widely seen in 1980 and later. When the naive young heroine stumbled across wrongdoing in a fuel processing plant, officials deliberately contaminated her and finally killed her with a burst of radiation. Along the way the movie showed the heroine forced down struggling and crying on a bed while technicians inserted a "womb monitor."[5] Outright symbolic rape at the hands of the nuclear industry . . . facts were of no importance where imagery took charge. In a popular 1979 movie, *The China Syndrome*, the heartless male authority, possessor of perilous nuclear forces, was shown explicitly as the chairman of the board of an electric utility.

These movies and many other antireactor productions included something more subtle. Much of the tension came from a conflict between the smooth and reasonable-sounding assurances of officials and the protagonist's gut feeling that something was dreadfully wrong. Here the division between authority and its victims was placed in parallel with another division, which I would summarize as *logic/feelings*.

As noted in earlier chapters, human thinking is in fact shared among these two modes, the rational and emotional; neither is "better" than the other, and both are important for making good decisions.[6] Social commentators had long recognized the distinction. The success of engineers relied upon rational systems from which everything personal and emotional was excluded; at an opposite extreme worked artists and poets, preoccupied with individual human emotions. The difference between personal approaches had social meaning. There is evidence that the rational mode prevails especially in large hierarchical organizations (corporations, for example), whose members must think in terms of categories and logical relationships. Thus if nuclear engineers stressed pure cognition, their preference reflected not only their training and, perhaps, their inborn inclination, but also their social position.[7] Such a mindset looked like a defect to people who trusted their feelings. Not only scientists and engineers but industrialists, military officers, and government bureaucrats were increasingly called mechanical and unfeeling. Contrasted with them were groups traditionally described as closer to their emotions and instincts, more attuned to spontaneity and intuition: women, children, primitive peoples, peasants, manual laborers. These were precisely the groups excluded from power, and especially from power over technology.[8]

In fact peasant women could be as rational as anyone, while the people who in fact held power often cared little for scientific rationality. Nevertheless many took it for granted that authority and rationality were inseparable, and together opposed to human feelings. The countercultural rebellion and most people in the antinuclear movement took the side of intuition. Starting with attacks on established institutions, they went on to criticize the laboriously established structure of science and verified knowledge itself.

The lines of division were not always clear. Many who wanted a less hierarchical society were friends of particular technologies, devoting effort to promoting wind turbines or solar heating. Opponents of nuclear reactors often deployed reasoned arguments backed up by an impressive array of facts, and their ranks included a number of experts. A 1975 French petition questioning nuclear power was signed by 4,000 scientists and engineers. Was this really, as its enemies charged, an antirational movement?

A French sociologist sympathetic to the reactor opposition studied some of the activists, including a few of the petition-signing scientists, and uneasily concluded that they never saw themselves as champions of rationality. Although they marshaled technical arguments as a debating tactic, at heart they were profoundly skeptical of expertise. The antinuclear scientists seemed especially concerned about the bureaucratic organization of their laboratories. Other committed opponents disavowed modern science altogether in favor of intuition or a frank belief in magic.[9]

Still, many reactor opponents had mastered a variety of technical arguments and presented them objectively. Indeed they were sick of the scornful way in which nuclear authorities called them "irrational" and "emotional." The problem was that, after decades of secrecy, evasions, purportedly "scientific" statements that turned out to be dubious, and the occasional outright lie, they no longer trusted the facts and reasons put forth by the authorities.[10] Trust is fragile: once lost, it is not easily regained. Most people on both sides of the controversy quite properly believed they were rational, in the sense that their views were commonsense deductions from what they believed to be true.

However, people might believe false things. In particular, studies found that the average citizen was severely mistaken about the frequency of various ways in which people might die. Compared with actual statistics, most people tended to greatly overestimate the likelihood of spectacular disasters (whether nuclear or other), while underrating the risk from commonplace causes of harm such as heart attacks.[11] This tendency fitted with the normal urge to insist that things one does every day, such as overeating, are not really too dangerous. Moreover, the availability of images of catastrophe in the press exerted a pressure that separated "common sense" from the conclusions reached by experts.

Common sense is another way of describing beliefs that are formed quickly and intuitively, based on associations with the experiences that have left residues in our memories. Besides this fundamental and primitive process, which I have been discussing throughout this book, humans have a second process for reaching decisions (this was not clarified by experiments until the 1990s). Whereas animals must base everything on as-sociations among their direct experiences, humans also have language: our decisions can be influenced by things that we are told. This is a pro-cess of linear, rational deduction that uses different regions and pathways in the brain from the process based on emotions. It is not immediate and hardwired; it must be learned. The primitive process is designed to make a snap decision—valuable indeed for our animal ancestors. Then reason-ing comes along at leisure to evaluate and perhaps correct that decision.

This second path does not draw on images and emotions, but reaches its culmination in logical rules, charts, statistics, and even equations. It comes naturally only to scientists, engineers, and the like who have spent arduous years training themselves in its methods. Small wonder, then, if they evaluated many risks differently from people who built their beliefs from television news items, movies, and editorial cartoons.[12]

Cognitive processes aside, many people were simply not interested in mortality rates and all the rest of the apparatus for calculating risks. To people who frankly preferred intuition and human feelings above logic, statistics of accidents were just not the point. What really mattered was the issue of values.

Reactor builders tended to keep their feelings to themselves, but in pri-vate conversation they revealed a growing bitterness and sense of isola-tion. Indeed, both sides sometimes displayed the paranoid style common in millenarian groups: each felt besieged by evil forces that conspired to impede their program to bring peace and plenty. The most ardent debat-ers on each side tended to see themselves as members of a community engaged not just in a political dispute but in a mission to drive back death and transform the world. The difference between the two extremes came down to just what sort of world transformation they desired: toward Arca-dia or a White City?

Of all the dichotomies people have used to organize their thinking in debates over the future of society, one of the most powerful has been the bipolar structure *culture/nature*. In Western thought this grew into polarities such as city versus countryside or civilization versus wilderness, polarities embedded everywhere from folktales to social philosophy. In personal terms "nature" was often associated with intimacy rather than formality, with instinctive impulses rather than self-control and planning, and, in brief, with feelings as opposed to logic, the individual as opposed to hier-archical authority. None of these associations was necessarily valid, but in

Western tradition the pattern of association was so pervasive that most people took it for granted. Nature was to culture as feelings were to logic, female to male, liberty to order, freedom to security, organic to mechanical, and so forth indefinitely.

Sometimes culture and nature were seen as equally capable of good or ill. However, another way of thinking was widespread: distrust of the wild as a haunt of ravening beasts and outlaws. Man's duty was to tame and civilize, to transform the useless disorder of desert and forest (and "primitive" colonial natives) into a structured landscape. By the nineteenth century this idea of "progress" was identified with the growth of large, rationally structured organizations in government, industry, and elsewhere.

Values could also be assigned in the opposite direction. Already in the eighteenth century some intellectuals began to praise untouched wilderness along with spontaneity and freedom. By the late twentieth century such critics of "progress" had become numerous, for the balance between nature and culture had reversed. Whereas wilderness had once surrounded and menaced our frail villages, now the world-city surrounded and menaced whatever unspoiled nature remained.

The antinuclear movement was thus taking its stand on ideas that were becoming widespread. An American survey found that antinuclear people, as compared with pronuclear ones, gave more emphasis to such nature-related values as "a world of beauty," "inner harmony," and "equality," while giving less emphasis to values of civilization such as "a comfortable life," "national security," and "family security."[13] But the central question was more explicitly related to power structures. A 1996 study found that people with an egalitarian ideology, who thought wealth and power should be widely distributed, were anxious about environmental risks in general and nuclear power above all. Less concern was found among people who believed in a hierarchical social order. This was the deep division, only approximately "left vs. right," that we saw earlier in the battle over fallout (Chapter 11).

Many on the antinuclear side did argue, in the leftist tradition, for a greater government role—but only to regulate and restrict industry in order to advance social justice within a legal framework. And their opponents preferred a minimum of government regulation, however much they might admire hierarchical leaders like a father, a captain of industry, or even a specific political leader. They felt that unfettered individual enterprise was essential so that everyone could win his or her appropriate place in the social order.[14]

Age, gender, education, income, and even explicit political orientation were not the keys to differences in attitudes toward the risks of nuclear power measured in a detailed 2008 study. What did matter was values (this study called the distinction one between "traditional" values and "altruistic," i.e., egalitarian, ones). The only other factor that really mattered in this study was trust in government nuclear institutions, a trust that correlated with acceptance of reactors. This is not to say that Republicans and white males did not especially favor nuclear power—they did tend to favor it— but the bias reflected these groups' more traditional values and greater trust in hierarchical authorities.[15]

A study of attitudes toward a nuclear waste facility again found only weak correlations with the usual social-demographic variables, and even correlations with ideology and values had only a modest effect compared with the effect of one variable. This crucial factor was, again, the degree of trust in the nuclear experts and organizations to do their job competently and honestly.[16] Of course that was connected with values, insofar as they affected attitudes toward authorities. To egalitarians, a loathsome hierarchical authority meant loathsome contamination—and more. From the mid-1970s forward the controversy became an explicit battle for legitimacy.

Each side, whether they knew it or not, supported the beliefs and values that would add legitimacy to their side. After all, hierarchical officials and experts would be expected to support the values of rationality and the corresponding ideology that decisions should be made by experts and officials—that is, by themselves. The fact that nuclear reactors would always require a large, well-paid corps of engineers and administrators was not a drawback in their eyes.

What about the opponents of reactors and of large-scale technology in general? If power was taken away from the traditional hierarchies, then decisions would still have to be made—but by whom? With remarkable unanimity, leaders of the antinuclear movement spoke up for local initiative, small "affinity groups," interdependent networks, the populace at large. In such a situation the power to make choices for society, including technological choices, would rest with those skilled in the communication of ideas and feelings. Journalists and novelists; television producers and moviemakers; artists and actors; schoolteachers and professors; lawyers;

ministers; community activists and organizers: in a society without centralized authority, these were the people who would ultimately determine what was acceptable.

These groups made up a "New Class," according to some of their critics, although it was neither exactly new nor exactly a class. The fastest-growing professions were remote from the struggle of industrial production, resolutely independent professions that manipulated ideas and symbols. This was far from the traditional sort of class distinction; a television film director and the vice president of a manufacturing corporation might live side by side, having the same income and sending their children to the same schools, yet stand poles apart in their values.

Going straight to the point, the sociologist Stephen Cotgrove separated out two viewpoints in British polls. On one side he set people who distrusted science and technology, saw environmental problems as extremely serious, and were against industrial growth; on the other side he set people who felt the opposite way. Those who favored economic growth and so forth typically turned out to hold jobs close to industrial production: engineers, industrial scientists, and managers. Their environment-minded opponents were usually remote from production and even from the market economy: research scientists and other academics, teachers, clergy, social workers, and artists. Similarly, a poll among influential Americans found none so pronuclear as professional engineers and none so antinuclear as university social scientists.[17]

New Class professions seemed to be unusually numerous in the rank and file of the antinuclear movement. Such people were certainly the most visible, here as in many other movements that opposed authorities. And by the very nature of their skills they were the most influential people in such movements.

Industrial officials remained firmly in power overall, yet the "feeling" and communications professions were gaining leverage. Their ideas appealed also to farmers, independent tradesmen, students, and housewives, all of whom were numerous among signers of antinuclear petitions. Such people felt that once the technocrats were discredited, once decisions were made in community networks, their ideals would have a better chance to guide the world. Since of all authorities the most distrusted were nuclear officials, for all the reasons I have laid out in earlier chapters, it was natural to make them a primary target.

Given that the antinuclear movement was commonly identified with the political left, some observers thought it strange that institutions such as labor unions and the European Communist parties remained largely pronuclear (or at least their leaderships did). The explanation was that union and Communist leaders were firmly on the side of both rationalized organization and the established pattern of economic growth. Meanwhile Republican housewives could march alongside radical poets, students bent on transforming society could make common cause with conservative farmers who sought only to safeguard their crops, united under the banners of victimization, feelings, and nature.

The history of attitudes toward nuclear energy is a sensitive indicator of something larger: attitudes toward all rational knowledge, technological progress, and organized decision making. The nuclear debates make particularly clear just how the trend toward confidence in these things, a confidence that had been growing sporadically for a millennium, faltered in the last third of the twentieth century.

Nothing in the controversy was so obvious or so important as the differences in rhetoric. A political scientist who studied 1973 congressional hearings on reactor safety found that reactor advocates spent four-fifths of their time talking about technical facts or administrative expertise, while the opponents spoke of such things in less than one-fifth of their testimony, devoting the rest to warnings against careless officials. Outside such formal settings, the nuclear industry filled volumes with numbers and charts, marshaling painstaking evidence, while their opponents communicated with stirring phrases and slogans, cartoons, and tuneful songs.[18] Of course there existed antinuclear calculations and even a few pronuclear poems, but few people spent much time reading such things. A reactor opponent like Helen Caldicott would tell of watching one particular child die painfully of cancer; a reactor advocate would deploy statistics of illness, waving away emotional talk about any single case.

The opponents considered it a virtue to feel emotions and used emotive imagery deliberately. The most widely read polemic of the movement's early years, the 1969 book *Perils of the Peaceful Atom*, had a foreword titled

"In Defense of Fear," in which the authors said they hoped that "some of our distress is communicated to the reader."[19] Cartoons about nuclear wastes showed two-headed babies or giant rats. Antinuclear writers applied the word "nuke," which originally meant nuclear bomb, to reactors as well, making the devices seem interchangeable. Particularly popular was the greatest of all antecedents of the evil scientist: opponents called nuclear power a "Faustian bargain" with the devil. In novels and cartoons the archfiend himself appeared, embracing reactors. Some opponents exclaimed that they were engaged in a battle against Evil and Death itself; in street demonstrations they wore skull masks, carried coffins, and set up mock graves.

The antinuclear movement's mode of expression clearly reflected a reliance on the associative, emotional cognitive system, but it also reflected the movement's social character. Outsiders trying to change the course of policy had more reason to use emotion-laden statements than did officials who could silently take for granted the established values of industrial progress. Within the antinuclear groups themselves, relying as they did on volunteers and direct-mail fundraising, passionate cries of horror would do more than statistics to keep up spirits and to attract new members and money.

Advocates of reactors also attempted to mobilize symbols. In the United States a Committee on Energy Awareness mounted a large industry-funded advertising effort in the late 1970s, while some individual companies undertook their own campaigns. These typically relied on the logical cognitive system. A steady stream of news releases couched in technical jargon that only an engineer could understand, combined with flat statements that all was well, might reassure citizens, but it did little to establish a trust based on real understanding. Trust had been weak for decades, undermined by events from the Oppenheimer affair to the fallout battles. To more and more people, a show of information by nuclear officials no longer conveyed any confidence at all.

Pronuclear advertisements often aimed more directly at the public's attitudes, appealing to patriotism and hopes for progress. But such campaigns were mostly couched in abstractions, as when an American advertisement stated that nuclear power, by forestalling war over foreign oil, would bestow "Energy Independence and World Peace."[20] Some publicists used more concrete images, tending to follow the established modes of industrial propaganda. Films, advertisements, and exhibits extolled progress by showing impeccable mechanisms such as gleaming uranium fuel rods or a white power plant set in a meadow, or fully equipped kitchens enjoying the benefits of nuclear electricity. Colorful invective did not come easily to senior officials in government agencies and industry, and still less to nuclear scientists and engineers. After all, the whole aim of the engineer and scientist was to replace the shifting sands of personal emotion with the solid ground of rules and logic.

The two different modes of expression each appealed to only one of the two basic human modes of evaluating risks and reaching decisions, and they had very different impacts on the debate. Most of the articles explaining that nuclear industry was safe and beneficial were printed in the recondite technical journals of physics, health physics, and engineering. On the paperback racks in bookstores, most of the books about nuclear power were lurid and openly hostile. And that was nonfiction; the contest was still more unequal in fictional works. Roughly a dozen ephemeral thrillers about nuclear reactors were published in the 1970s, and their titles reveal their viewpoint: *The Accident, Meltdown, The Nuclear Catastrophe,* and so forth.[21] Typical plots used the idea that Robert Heinlein had invented for "Blowups Happen" in 1940: an advanced reactor, officials stiffly determined to keep it going, an individual striving to avert doom. There were—there could not possibly be—equivalent thrillers in support of the everyday operation of nuclear power. It was a simple matter, this bias toward colorful talk of danger, but it gravely influenced the image of all technology.

One of the most respected scientists in the field of medical radiation, Lauriston Taylor, reported that since the 1950s editors had asked him pointblank to give alarming statements, telling him there would be no audience for calming explanations. He indignantly complained that whenever a radioactivity story came along, the media would eagerly take up the "willfully deceptive" statements of a half-dozen or so antinuclear scientists, always the same ones, people whose reputations with their colleagues had long since vanished; "collectively they account for more news lines than the hundreds of reliable professionals accepted by their peers."[22]

Of course it was normal for the press to play up the unusual—there was nothing exciting in an expert supporting nuclear power. Journalists could also be expected to play up attacks and human emotions that would sell publications. But beyond the sensationalism that inevitably attends a free press, nuclear advocates began to accuse journalists of frank antinuclear bias.

Detailed studies of American newspapers, magazines, and television, and briefer surveys of the European press show that in the 1970s antinuclear presentations came to outweigh pronuclear ones by two to one or more. No other technology was so despised in the press. Yet polls in the same period found the public at large slightly favoring reactors. The press, then, was not simply mimicking the public's views. Neither was it reflecting the views of government officials and business leaders, who were largely pronuclear. Still less was the press reflecting the views of scientists and engineers—for such professionals, even those who had no connection with the industry, supported nuclear power by the strongest majority of all. When the press gave the few antinuclear experts substantially more space than the many pronuclear ones, the implied picture of the scientific community was highly inaccurate.

Direct surveys in the United States found that not only the articles, but the journalists who wrote them, were far more antinuclear on average than the groups they supposedly drew upon for their information. At the extreme were television journalists, openly antinuclear. The slanted press coverage had consequences. For example, after a heavily biased 1985 ABC television special that suggested that the nuclear power industry could end life on earth, a survey of the viewers found a significant and lasting shift against nuclear power. In sum, there was good evidence that the reactor wars reflected a profound division between key groups in modern society, and that this division influenced the public debate.[23]

In the last several chapters I have focused on articulate and active minorities. What did everyone else think? In most nations, and at most times since polls on nuclear power began in 1945, it would have made sense to divide the public into five groups of roughly equal size. The first group were strong advocates, confident in the ability of experts to keep reactors safe and impressed by the benefits of economic growth. Directly denying all that and about equally numerous were the convinced opponents, at first silent but increasingly vocal. In between were those who leaned one way or the other: a group that favored reactors even while harboring some misgivings about their safety, and a group that frankly feared reactors but were not convinced they should be banned. Both these middle groups would usually go along with whatever the authorities decided. In polls and referenda from the 1940s through the 1970s, as governments pushed their reactor programs, these groups usually joined with convinced advocates to give a solid majority in favor of nuclear power. But whenever authorities themselves were divided the middle groups hesitated, giving a majority attitude of wait and see. The fifth group consisted of people with no position, and often no interest, in the whole question.

The antinuclear campaigns had some effect. Polls in Germany and France showed that around 1975, when the issue of nuclear safety began coming into the headlines, support for reactors fell off from a dominant majority to a bare majority. Support among Americans similarly declined. It dropped below a majority after the Three Mile Island accident, which brought a significant loss of what trust in nuclear authorities remained. These were modest shifts rather than major swings; negative attitudes, like positive ones, had been present everywhere over many decades.

Among the ordinary people who made their voices heard only in opinion polls or at the voting booth, advocates and opponents of reactors (as of nuclear weapons) were spread quite evenly through the traditional social groups. Even education made little difference. Nuclear experts often suggested that the antinuclear movement was based on ignorance of facts, while the opponents, too, felt they would win support if only they could teach people the truth. Neither side was correct, for studies showed that the way people felt about nuclear power was mostly independent of how much they knew about it: this was confirmation bias at work. (In fact most people had only rudimentary knowledge, mixed with various bits of misinformation.)

Gender was the one area in which a strong divergence turned up in every poll, in every country, at every time. For example, a survey conducted near several American reactor sites asked people to complete the sentence: "When I think of the nuclear power plant, I feel ——." Twice as many men as women came up with talk of progress and economic benefits; twice as many women as men spoke anxiously about dangers. Careful analysis of polls suggested that the phenomenon was connected with a tendency, when any technology was mentioned, for women to think in terms of safety and their children. But why did women diverge from men more on nuclear energy than on any other technological risk? The only explanation I can find is that to many women nuclear energy stood for the worst in technology, because it had become most specifically associated with aggressive masculine imagery: weapons, mysteriously powerful machines, domination of nature, contamination verging on rape. On no other technological issue was the sexual imagery so thoroughly developed and, from a woman's standpoint, so viscerally disturbing.[24]

Images remained at the center of all public thought, for the old myths kept their vigor into the 1980s. Many people in Eastern Europe, in South America, and even in the United States continued to visit spas that discreetly advertised the healing powers of their radioactive waters. Meanwhile antinuclear writers told of weird sea creatures found near nuclear facilities (unfortunately no specimens of these marvels were preserved). Most commonly the creatures were said to be abnormally enlarged, if not quite so much as in the 1950s monster films. Novels and popular movies continued to display every sort of transmutation imagery, and anyone curious to see fantasies of rays with uncanny powers needed only to watch a few hours of American children's television on a Saturday morning.

As ever, the strongest fantasies dealt with human transformations. Nobody was so successful here as Stan Lee, creator of the characters that brought Marvel Comics to sales of over 100 million copies a year, and not only among children. Lee's first popular creations were the Fantastic Four, whose powers were induced by cosmic radiation in 1961; the next year came Spider-Man, bitten by a radioactive spider; meanwhile, with Dr. Frankenstein and Dr. Jekyll explicitly in mind, Lee created the mighty Hulk from a physicist struck by nuclear rays. Nuclear energy remained the leading manufacturer not only of superheroes but of monsters. For example, in a 1981 comic book Superman and Spider-Man teamed up to quash a monster—a nuclear industry worker transformed by exposure to an experimental isotope. With scarcely a pause the heroic pair then quashed a super nuclear reactor, built by a villain to control the world, which threatened to reduce the planet to ashes.[25]

The outcome of the whole history of images was reflected in a word-association test conducted in the mid-1970s by a French psychologist. Asked to respond to the word "atom," many people gave only banal descriptions, as if reluctant to bring up emotions. Of those who did give emotion-laden responses, only a small minority included such hopeful words as "progress" or "future." The rest were fearful: "Hiroshima," "disaster," "death." Some people came up with what the psychologist called "personal preoccupations centered around the problem of origins," for example, the word "father."[26]

Other interviewers found that the pronuclear and antinuclear public together shared a distinct image of reactors. The white buildings behind a fence stood for industry and science, an immense and mysterious force that was barely contained. Neighbors of one reactor believed that it had vast subterranean installations, like the laboratory of the comic-book genius. In short, whether one saw a reactor as good or evil, or more likely as a mixture of both, it stood as a full symbolic representation of attempts by the forces of order to master nature. In terms of prospective hazards to our daily lives, reactors were not obviously worse than, for example, the chemicals industry. What reactors did offer that nothing else could match was a unitary image tying together everything involved in the confrontation between "nature" and "culture."

The Three Mile Island reactor accident provided a litmus test of how the public in general, and the press in particular, saw reactors. Hundreds of reporters converged on the accident, which dominated the headlines and television news reports for a solid week. The reports were so frightening that roughly half the residents of the region, some 200,000 people, moved away temporarily. Scholars later studied these reactions in detail. They concluded that the press could not be blamed for raising fears, for these had already been alive in the people the reporters interviewed— including nervous staff members of the Nuclear Regulatory Commission itself. At each stage as information passed from reactor operators to officials, from officials to reporters, and from reporters to the public, attention had concentrated on the most frightening possibilities.

But there was also, of course, frank sensationalism. Journalists sought out the most worried people for interviews, while on national television Walter Cronkite philosophized about Frankenstein and man's "tampering with natural forces." The world press made numerous references to monsters, robots, and the devil, as well as to Hiroshima, nuclear war, and the end of humanity. On top of all that, the *China Syndrome* movie was just then playing in movie theaters.[27] The press, adopting a narrative prepared by the antinuclear movement, covered Three Mile Island with an intensity far beyond that accorded any previous industrial accident.

Residents of the area were so upset by the stories that some, calling themselves "survivors," suffered long-term psychological difficulties much like the traumatized victims of catastrophes in which people were actually killed. When the time came for lawsuits, this psychological harm was one of the strongest claims that people made.

Eight months after the accident, while passionate arguments continued over the feasibility of evacuating people from the neighborhood of future reactor accidents, authorities in Ontario had to rush the evacuation of a quarter of a million people when a tank car leaked deadly chlorine gas.

The near disaster got only a few columns in the newspapers, quickly replaced by further talk about hypothetical reactor accidents. Still more instructive, a year after the Three Mile Island accident a chemical waste dump in New Jersey, next to New York Harbor, caught fire and sent a tower of noxious smoke high into the air. Schools on nearby Staten Island sent children home, and residents waited behind closed windows as chemical fog drifted past. If there had been a breeze blowing directly toward populated areas, millions of people would have been exposed to toxins—some of them as likely as any radioactive particle to cause cancer or other harm. In short, the waste dump fire came far closer than the Three Mile Island accident to a great public health catastrophe. The fire made headlines for a day or two and was then forgotten. A presidential commission was convened to investigate the Three Mile Island accident but not the New Jersey one; demonstrations brought 100,000 people to Washington, D.C., to protest reactors but not the far more ubiquitous and far less regulated chemical waste dumps. This was not a bias of the press alone. Nor was it an accurate response to the complex problems of the times. This was nuclear fear at work, single-minded and unappeasable.[28]

The reaction to the Three Mile Island accident guaranteed that no new reactors would be ordered in the United States, and few elsewhere. With less to fight about, the movement against reactors grew torpid, and the public began to lose interest. Besides, the press rarely sustains a fierce interest in any issue for more than a few years. Unless some new aspect or event renews the energy, there is only so much one can say about a topic before it becomes wearisome. In my counts of titles in the *Readers' Guide*, by 1984 the number of articles dealing with civilian nuclear energy had dropped to the level of a decade earlier, which was half the peak value. The same trend can be seen in another source that will be more useful from this point forward: the full text of news and periodical articles (mostly American) indexed in the Google News Archive. This source, too, shows, after the surge of interest in nuclear power in the early 1960s and the trough in the early 1970s, a climb to a peak in 1979 and then a steady decline into the early 1990s.[29]

The decline of interest did not come because feelings against reactors had mellowed. The proportion of *Readers' Guide* titles showing doubt or open hostility toward reactors continued to climb. In 1950 barely one-tenth of the titles had suggested that there was anything to worry about from the industry, but by 1986 the fraction had reached nine-tenths. The texts in the Google News Archive similarly become much more negative, concentrating on accidents. Wherever something new in nuclear energy was proposed, the opposition would be there, quietly waiting.

In summary, I find four main themes in the cluster of associations that influenced the way people thought about nuclear power. At the origin were (1) the technical realities of reactors, both the economic opportunities and the hazards, as seen by scientists and transmitted to the public. From these realities particular "facts," such as the hypothetical effects of widespread low-level radiation from a Maximum Credible Accident, were selected and stressed.

That special emphasis was largely because of (2) nuclear energy's social and political associations, especially ideas involving technology and authority. These associations explain what happened when reactors became a condensed symbol for all modern industrial society. They do not explain why nuclear power was singled out for this role. That was largely a result of (3) the old myths about contamination, cosmic secrets, mad scientists, and apocalypse that were historically associated with atomic power and radiation, myths with deep psychological resonances. No less important was (4) the threat of nuclear war, never for a moment forgotten.

20

WATERSHEDS

Immediately after the Hiroshima and Nagasaki bombings, daring Japanese newsreel photographers had filmed the consequences. The American occupation authorities had confiscated the films and hidden them away. Returned to the Japanese in 1968, the films were made into a documentary that showed doctors attempting to treat horrific wounds and vistas of streets paved by skulls. Excerpts appeared on television in a number of nations during the 1970s, while franker extracts played to audiences in colleges and churches. The audiences were stunned, scarcely able to speak after watching ten or fifteen minutes of film; some viewers fainted.[1]

For the first time, many people were starting to confront the dangers of nuclear war in a fully realistic way. For a decade and a half most people had shoved away the fearful ideas and fantasies, as if the problem could be left to their national authorities to handle. Since the 1963 treaty that banned bomb tests in the atmosphere, many had looked to diplomacy to gradually tame the arms race. Hopes rose with the 1968 Non-Proliferation Treaty, which established an international regime under which all but a few nations promised to allow inspection of their reactors, to verify that they were not making plutonium for bombs. Negotiations for a first Strategic Arms Limitation Treaty (SALT I) culminated in 1972, when President Richard Nixon and Soviet premier Leonid Brezhnev signed an agreement to restrict antiballistic missiles. In 1977 talks began toward a SALT II that was supposed to actually reduce the number of weapons.

None of these developments satisfied veteran opponents of nuclear weapons. They increasingly suspected that disarmament talks were merely a show to pacify the public. A renewed effort began where the disarmament movement of the 1950s had begun: Japan. Survivors of Hiroshima and Nagasaki, refusing to tolerate the world's acquiescence in an endless arms race, redoubled their efforts to spread their message. The documentary film about their suffering was only one of the tools they used to prod the world to think about weapons.

In the universities of Cambridge, Massachusetts, and a few other centers of dissenting thought, the disarmament advocates who had persisted almost unnoticed since the early 1960s began to find support among the opponents of nuclear power. Social psychologists argue that people have a "finite pool of worry"—if you spend less mental energy worrying about one thing, you have more available to worry about something else.[2] In the 1960s when people began to worry less about nuclear war, they had begun to notice various kinds of environmental risks. In addition, observers of the media have noted that there is a limited "news hole" that has to be filled with genuinely new topics. One can imagine an editor saying, "Not reactors again! We've had a dozen articles on that. Bring me something new." As the global reactor industry slowed down, the media found other interests, and reactor opponents found space to think of other problems.

Helen Caldicott concluded a polemic against reactors by warning, "When compared to the threat of nuclear war, the nuclear power controversy shrinks to paltry dimensions." Around the end of the 1970s Caldicott, and numerous other longtime reactor opponents such as the Union of Concerned Scientists and the Friends of the Earth, turned their main attention to armaments. Many of these people were moving back to their original concerns: striking evidence that anxiety about reactors had included a displaced component of nuclear bomb anxiety. Now this returned to its primary target.[3]

The antinuclear leaders were running hard to keep up with the public and the press as debate over bombs awoke from its enchanted sleep. Measures of the awakening were a steady rise, beginning in the late 1970s, in the number of magazine articles dealing with nuclear weapons (as seen in the *Readers' Guide* index of titles), and likewise of terms like "nuclear war" in newspaper text (as seen in the Google News Archive).

At first this seemed to be only a continuation of the debates, inspired by the reactor controversy, about proliferation and terrorists with atomic bombs. Novels on these subjects began to appear, and some of them were not the hack works of the preceding decade but thrillers by masters of the genre, each selling more than the one before. *Goodbye California* in 1978 had a mad terrorist threatening to destroy the state with bombs; *The Fifth Horseman* in 1980 had a mad dictator threatening New York City; *The*

Parsifal Mosaic, the number one best-seller of 1982, had a mad statesman threatening to trigger world war. The implicit message of all these stories was, as *The Parsifal Mosaic* put it, that "this world can be set on fire by a single brilliant mind." That sounded much like science-fiction thrillers ever since 1945, or indeed 1895.[4]

But something new was emerging. Close on the heels of the mad-bomber thrillers came articles and books that dealt with nuclear weapons directly and often realistically. By the early 1980s the number of factual American magazine articles and news items dealing with nuclear weapons had shot up to a level attained only once before, in the early 1960s. Meanwhile paperback racks in the United States and Europe, which a few years earlier had been loaded with books on reactor dangers, filled up with both sensational fiction and sober factual warnings about war. The most important was Jonathan Schell's 1982 essay *The Fate of the Earth*—the first nonfiction book about nuclear war to become a best-seller since Hersey's *Hiroshima* of 1946. Schell described a war that left behind "a republic of insects and grass." He argued eloquently that our species could be extinguished, an end not only to our personal lives but the entire human future. Hundreds of other fiction and nonfiction books in various languages appeared in a remarkable burst, well above the 1960s level. Even Dr. Seuss pitched in with a children's book about a perilous arms race between two groups of fanciful creatures.[5]

Why did anxiety about nuclear war emerge again around 1980? There may be a cycle to such things, as the exhaustion of fruitless anxiety wears off. But there were direct causes. Citizens usually responded rapidly to new facts about weapons, and there were dreadful new facts.

The United States had rapidly multiplied the number of warheads that it could hurl at an enemy, until the number leveled off around 1977 at roughly 7,000 nuclear bombs on intercontinental missiles along with 2,000 carried by bombers. The Soviet Union, as usual a half-decade behind, undertook a similar proliferation of missile warheads, passing the 7,000 mark around 1982. Meanwhile the number of shorter-range weapons climbed steadily into the tens of thousands.

Some people began to take notice of all this in 1977 during the first substantial weapons controversy since the antimissile fight of a decade earlier. The U.S. Congress was debating whether to allow production of neutron bombs, which would produce deadly direct radiation but little fallout. (They were called "clean" bombs, reflecting the well-established connection between fallout and filth.) Debate raged chiefly in Western Europe, the most likely place to see neutron bombs in use. Still more passionate protests erupted in Europe when the Soviet Union began deploying a new type of medium-range missile and NATO decided to reply with new missiles of its own. At first Americans scoffed at the Europeans' fears. But when the Soviet Union redoubled tensions by invading Afghanistan, and the U.S. Senate refused to ratify the proposed SALT II arms control treaty, most citizens realized that diplomacy was not even coming close to solving the arms race.

Now as in the 1950s, people who wanted an increased military budget joined their opponents in warning about nuclear peril. A group of conservative Americans formed a "Committee on the Present Danger," which launched a publicity effort to revive fear of a Russian attack. While there were economic interests that worked to hold down military spending (groups with other uses for the money), many powerful interests wanted more money spent. The committee had abundant links, such as interlocking boards of directors, with firms that lived off military contracts. Well funded and dedicated to its work, the committee strongly influenced American opinion.

A leading spokesman for the committee's viewpoint was Ronald Reagan. His election to the presidency in 1980 made many fear that the new administration would be willing to launch a nuclear war. And indeed talk of how to fight and "win" a nuclear war—ideas seldom heard since the mid-1960s—burgeoned among administration staff and outside supporters. In 1982 Reagan announced that the Soviet Union had taken the lead: there was a "window of vulnerability." This assertion was as false as the claims about bomber, missile, and antimissile gaps in earlier decades, but many Americans believed the president. Accelerating a trend begun under President Jimmy Carter, the Reagan administration worked to raise military spending by at least 50 percent. It fitted well with the Republicans' effort to divert as much money as possible from federal social programs to private industry.[6]

Was the Reagan administration actually preparing to launch nuclear war? Fears proliferated, along with passionate opposition. The 1970s battles for restrictions on air and water pollution, civil rights for blacks, and a halt to building reactors were at least partly won; the energies of social critics could return to the suddenly urgent topic of weapons. Many of the activists had won their spurs in other movements, especially the fight against reactors, and retained those movements' general distrust of modern society and authority. But unlike the reactor opposition and the 1960s anti-bomb campaigns, this movement did not rely almost entirely on relatively peripheral people such as housewives, students, and intellectuals. It drew equally on respected elites—mayors, elder statesmen, even some generals and admirals. An antiwar organization of medical professionals, Physicians for Social Responsibility, grew explosively in 1980 as it organized public conferences on "The Medical Consequences of Nuclear Weapons and Nuclear War." The Catholic bishops of the United States issued a 1983 pastoral letter condemning the arms race and the targeting of cities with their innocent civilians. This new movement did not focus on the supposed horrors of bomb tests and radioactive fallout but looked to the heart of the problem: the arms race itself.

In the fall of 1981, in ten major Northern European cities crowds of 100,000 or more demonstrated against deployment of the new intermediate-range missiles. Americans followed with demands for a Nuclear Freeze—a verifiable halt to *all* nuclear weapons deployment. They mobilized on a scale well beyond any other such outcry in the nation's history. In 1982 an enormous demonstration in New York City was followed by a series of votes on Nuclear Freeze resolutions in communities across the country. Demands for the freeze won, often by large majorities, in legislatures and referenda in hundreds of towns and cities and in seventeen states. A 1983 poll found that 81 percent of Americans favored a freeze.[7]

The leap in attention to nuclear war was not a reflection of gross changes in public beliefs. Polls around the world found that fear of imminent nuclear war, which had subsided somewhat in the 1960s, rose in the late 1970s and early 1980s, returning to the level of the 1950s—but these were shifts of no more than 10 percent of the population. Most people had known all along that they were in danger, whether or not they chose to talk about it.

The majority still preferred silence. A 1981 poll that gave Americans a choice for describing how they felt about future war found half of the respondents agreeing with the statement "While I am concerned about the chances of a nuclear war, I try to put it out of my mind."[8] When adults asked students what they thought of their future, they found the same disturbed responses as in the 1950s and early 1960s: strong fears of nuclear war (greater than in the late 1960s and early 1970s), tending to lead to denial. "We are the last generation," a young man told me—and went back to his pursuit of a career. The one clear difference was that this generation had heard far less than the children of 1945–1965 about weapons. Many American high school students in the 1980s did not even know that the United States had once dropped atomic bombs on Japan.[9]

Mainstream media producers mobilized to fight the apathy and ignorance. In the early 1980s television specials at last thrust realistic scenes of missile warfare into the public's face. One example was a 1981 CBS television news special that included a brief but disturbingly lifelike depiction of a missile exploding over Omaha, Nebraska. Nothing like that had ever appeared in a major television production. In 1983 still more explicit productions appeared in various media, showing the realistic effects of a war on ordinary people. Among these ABC television's "The Day After" became the most famous nuclear show ever. Heavily promoted as a news event, its television premier gave 100 million citizens their first look at what a real nuclear war might look like, complete with terrifying explosions and corpses.[10]

Alongside these the usual gaudy stories persisted, for example, *Damnation Alley*, an after-the-bombs movie screened in 1977. It could have been written in the 1950s, with its brutalized survivors battling swarms of giant insects and one another, ending with the hero and heroine reaching a land of blue skies and "a new life." The most popular of all, the 1984 *Terminator*, had heroic survivors battling killer robots with a less certain outcome. Such fantasies of tough-minded individual survival inspired many trashy paperback novels, detailed "survivalist" manuals, and a minor industry selling "survival" equipment with particular attention to guns.[11]

Even the most carefully researched books and films harbored traces of fantasy. From congressional studies to "The Day After," they showed war as a matter of a few thousand bombs all used within a few days. Historically, of course, most wars have lasted far longer than expected. Yet outside the military establishments, few people realized that once a nuclear war began it could continue for months or years, with tens of thousands of additional warheads hunting down the survivors. Such a scenario was even harder to contemplate than the vision of a day of apocalypse.

The best guess was that a fair fraction of the human race would survive even 50,000 bombs, but was that certain? When talk of actually fighting a nuclear war revived, a few scientists pointed out that the smoke from thousands of burning cities and forests could darken the skies for months. In 1983 a group of scientists launched a sophisticated publicity campaign warning against a "nuclear winter."

The idea of a Great Winter, an end of the world in ice, was a vision millennia old, but it now began to look as possible as missile warfare itself. (Later, more detailed studies showed that a nuclear war would not bring on an ice age; however, the smoke from even a limited war with a few hundred explosions could bring a "nuclear autumn" that would ruin agriculture around the world, threatening billions with starvation.)[12] Experts felt that the chance of outright race extinction was small. But some of them suspected that the chance did exist, growing year by year as warheads multiplied.

Was the war fear revival of the early 1980s no more than a pointless repeat of the temporary surge of the late 1950s and early 1960s? I think it brought something new: a better understanding of the real meaning of many of the old nuclear ideas and images. For example, Robert Lifton's term "numbing" came into widespread use; that pointed to a recognition of the psychological mechanisms of denial, learned helplessness, and so forth. The listing of scholarly research in *Psychological Abstracts* showed that the effect of nuclear war threats on people's thinking, a subject almost entirely neglected before the late 1970s, was now attracting numerous careful studies. Deepened insight also emerged in a burst of excellent poems, paintings, short stories, and novels that began where earlier artistic works on bombs had stopped.

Popular movies, too, were more subtle and complex, and more conscious of unconscious motives, than any nuclear movie of the 1950s. For example, the widely seen 1983 movie *Wargames* made the theme of the dangerous bad boy explicit. Here nuclear catastrophe was almost set loose by an irresponsible scientist, a childish computer—and a rebellious boy. In this and other productions the Men Who Started the War began to step into view. Now they were not evil madmen, but people wrenched by familiar personal and social strains. Meanwhile after-the-bombs genre films, reaching a peak in the popular *Mad Max: Beyond Thunderdome* of 1985, deliberately connected sadistic violence with technology. The savage battles of bomb survivors were no longer seen as the opposite of modern civilization, but as an exaggeration of trends already present in society.[13]

Beginning in the mid-1970s a few people took a closer look at some of the central ideas of nuclear strategy. Analysts agreed that once both sides had a few hundred missiles, possession of greater numbers or more modern types could scarcely make a nation come out ahead, in any meaningful way, in a nuclear war. Not winning the war but preventing it was the only rational goal. But if having more weapons than the adversary was meaningless to military practice, the way people perceived the difference might be crucial. Wouldn't having more weapons make it less likely that anyone would dare to use even one? The nuclear deterrence paradox—as Churchill had put it, "the worse things get the better"—was revealed to be not a military but a psychological conundrum.[14]

This "perception theory" had been accepted, if only tacitly, by many American military and political leaders since the early 1960s. They would act in accordance with the illusions of the Soviet leadership and the American public—or even foster those illusions, in hopes of weakening resolve on the one side and strengthening it on the other. They argued, for example, that there was urgent need to deploy modernized missiles in Europe, insisting that the missiles' presence would reassure allies and give the enemy pause; they neglected to mention that the deployment would have negligible practical influence on the military situation. In every such case the drive was to make armament "worse" as a physical destroyer in order to make it "better" as a mental deterrent. By the 1980s it was clear to all careful thinkers that nuclear policy had less to do with the physical weapons than with the images they aroused.

Since the 1950s there had been just one striking new image, the nuclear winter. Pictures of a smoke-shrouded planet dramatized the idea that at some point the stockpile of nuclear warheads would become—perhaps already was—the Doomsday Machine. Yet the idea had no effect on official strategic doctrine. After all, the Doomsday Machine, otherwise called Mutual Assured Destruction, or MAD, had long since been accepted as the keystone of the paradox of deterrence.

The most revealing use of traditional themes came from President Reagan in a March 1983 speech. He called on the nation to build a strategic defense system that would render nuclear weapons "impotent and obsolete." Like many other defense announcements, the motive seems to have been largely political. The secretary of state later said the speech came from a desire for "a big public relations splash." More precisely, the surprising strength of the Nuclear Freeze movement had showed that the public was not willing to keep endlessly spending money for nuclear weapons as deterrents. So Reagan changed the rules of the game.[15]

The speech was not only politically astute but also reflected an emotional commitment. After he became president, Reagan had kept putting off a briefing on nuclear war. Always attuned to movies, on viewing "The Day After" he was profoundly and uncharacteristically depressed. Now he finally attended a briefing on the nation's strategic plan in the event of an attack. It struck him as dreadfully like the movie. He took refuge in a belief that the nation's scientists could invent devices to counter anything the enemy threw at them.[16]

The speech took the government's official advisers by surprise; the impetus behind strategic defense had come from outside lobbying groups, supported by wealthy right-wingers and spearheaded by Edward Teller. In fact scarcely any other experts thought it likely that a system to defend the populace could be built in the foreseeable future. One scientist who participated in the project later admitted that "we worked on it for the money." A former secretary of defense publicly said that the only result of the scheme would be to "make lots of money for lots of groups."[17]

Yet important segments of the press supported the plan; a majority of citizens in the United States were immediately reassured, and Congress appropriated billions of dollars. The opponents had been far less effective than in the antiballistic missile controversy of the late 1960s. At that time they could battle specific weapons scheduled to be placed in designated neighborhoods. Now they were hitting out against nothing tangible but only a science-fiction vision of a future marvel. It was a striking demonstration that imagery could shape history in defiance of what the overwhelming majority of scientists and engineers held to be true.

For imagery was at the core of "Star Wars"—as various observers, spontaneously and independently, named the program within hours of Reagan's speech. They were referring to the space battles in the most popular of all Hollywood death-ray spectacles, but earlier precedents were also plain. As I showed in Chapter 3, marvelous rays had fascinated people for millennia.

The "Star Wars" genre further appealed to the public with another theme of ancient power: cosmic flight to battle in the heavens. That went all the way back to the magical ascents of shamans who wrestled with spirits in the skies, and the Revolt of the Angels in heaven. Since the early

1950s weaponized flying robots had been a staple of comics and television, starting with the Japanese atomic-powered robot Astro Boy (originally *Tetsuwan Atomu*, "Mighty Atom"). Among the young scientists at Livermore Laboratory whose work led directly to Reagan's speech, there was high interest in both space flight and the death rays of science fiction. As a historian of the events noted, "both antinuclear activists and missile defense enthusiasts were attracted to images of radical transformation of the world, presented as images of hope but certainly emerging out of deep despair."[18]

That the administration's Strategic Defense Initiative (as the program was officially named) was built on images was largely deliberate. Looking back from a quarter-century later, we can see that the tens of billions of dollars spent since the mid-1980s on ray weapons, defense missiles, and other schemes were entirely wasted if the aim was to produce a system that could defend against an intercontinental missile attack. But perception-theory strategists who knew that a missile defense system could not work still wanted to build one, for they thought the Star Wars imagery itself would instill caution in the enemy. It was not necessary to actually build anything. The idea that one might build it someday was enough to make a strong impression in diplomacy—and in domestic politics. If nothing else, the claims could divert the criticism of existing weapons into a sterile debate over weapons that were not yet even on the drawing boards.

Whether or not that outcome was intended, it happened. The Nuclear Freeze movement in the United States began to flag. After 1983 public attention to a freeze, as measured by attendance at demonstrations and by the number of American magazine article titles and newspaper references, dwindled swiftly. Meanwhile magazine articles on missile defense rose rapidly; by 1986 they exceeded the combined total of articles on Nuclear Winter, bomb testing, and other negative aspects of nuclear weapons.

The decline in articles directly concerned with bombs was due not only to the Star Wars diversion but to something that had also sapped the anti-bomb movement of the 1960s: opposition began to seem futile. The government, with the acquiescence of a majority of American citizens, remained determined to build more weapons. The European movement similarly withered when it became clear that its drive to block new missiles was a failure.

Moreover, now as in the 1960s politicians hastened to reassure the public by moving toward détente. Reagan's people suppressed their talk about actual warfighting; the president himself stopped exclaiming about the perils of Communist weapons. Congress slowed the American arms buildup. The Nuclear Freeze movement was not so futile after all.

Meanwhile Reagan was dismayed to learn that his anti-Soviet speeches and arms buildup had been taken in dead earnest in the Kremlin. A 1983 NATO war games exercise had particularly frightened Soviet premier Leonid Brezhnev and his colleagues, for it looked to them like preparation for a First Strike—which they were ill prepared to counter. They put their entire military on war alert. Some historians have suggested that (unknown to the public until much later) the world came nearly as close to nuclear war, purely by accident, as during the Cuban missile crisis. By 1984 both sides were frightened enough to seek an accommodation. This effort redoubled when a new, reform-minded leadership arose in the Soviet Union.

As the rhetoric turned mild, polls in the Western nations found overt fear of imminent war receding. Yet spy thrillers about nuclear war threats and survivalist fantasies remained common on the paperback racks. More careful polls found that latent public concern over the bombs remained high.

Reactors caught the public's attention again in 1986 when the meltdown of the Chernobyl reactor in the Ukraine threw huge quantities of radioactive materials into the atmosphere. Newspapers claimed that thousands or more had died and told of perilous fallout drifting on the winds across Europe. Thousands of miles from Chernobyl, mothers in England and Italy worried about whether they should let their children play outside, and temporary residents pulled up stakes to move even farther away. The abortion rate in Greece spiked in response to stories of dreadful risks for the unborn.

The press and public were making the best they could of scanty and confused information. In fact the immediate death toll was only thirty or so, chiefly reactor workers and firefighters. Experts calculated that the long-term damage would be at roughly the same level as the damage from bomb test fallout two decades earlier: a possibility of excess cancers spread among millions of people over the next half-century. Such illness could not be observed among the far more numerous cancers that would have happened anyway. The total number of Chernobyl fatalities would be somewhere between a few dozen and tens of thousands, depending on the still unknown effects of very low-level radiation.[19]

What mattered most was the imagery. Seeking a comparison with events outside the nuclear industry, the press most often settled on a 1984 disaster at Bhopal, India, where a chemical cloud had escaped from a pesticide plant. That cloud had killed outright not a few dozen people but over 2,000; the long-term damage to the health of another 10,000 or so was not hypothetical but immediately visible. Yet to the press and most of the public, the Chernobyl accident seemed more serious. That was largely because reactor worries centered on faint but widespread radioactivity; in Bhopal the hypothetical long-term effects of traces of chemicals dispersed across a large population were scarcely mentioned. It was not pesticides but nuclear power that *Newsweek* announced to be "a bargain with the Devil."[20]

Hardly any reporter described the most likely, but prosaic, direct health consequence at Chernobyl: an excess over the decades of thyroid abnormalities among nearby residents. (This prediction by experts turned out to be correct: the only visible long-term damage was several thousand thyroid cancers, requiring surgery but rarely fatal. These could have been avoided if Soviet authorities had warned citizens against drinking milk from the region.) The Western press gave far more attention to the plight of the Lapps in the far north, where government authorities ruled that radioactive fallout had made their reindeer unfit to eat. As innocents in a pristine environment, the pastoral Lapps and their animals could well stand for all victims of reactors—which in turn served as a symbolic representation of all that was feared from modern civilization.

Many went further to link Chernobyl with nuclear war, not omitting references to apocalypse. To *The New Yorker* the accident was "all that is given to us to know of the end of the world." Even Soviet Premier Mikhail Gorbachev called the accident "another sound of the tocsin, another grim warning" against the nuclear arms race. A reactor was not just a reactor.[21]

The catastrophe resolved controversies over nuclear power in many countries by the simple means of shutting down programs. In the United States, the growth of opposition and the 1979 Three Mile Island accident had already gone far to halt plans to build new reactors for generating electricity. Legal and regulatory obstacles were making the technology forbiddingly expensive. From the 1970s to the 1990s, American utilities cancelled seventy plants that had been on order and shut down twenty-two more. Environmentalist organizations that had led the fight against reactors now turned their attention to issues that seemed more urgent. The anti-nuclear movement dwindled into local watchdog groups, backed by a few national organizations like the Union of Concerned Scientists that kept reactors as a minor item on a long list of concerns.

Meanwhile a tide of opposition spurred nations from Sweden to Egypt to cancel their ambitious plans, or even pledge to phase out existing reactors. Some governments, particularly in France and Japan, continued to insist on building reactors and had political systems that prevented public outcries from reversing the decisions of their elites. Their antinuclear groups, resigned to impotence, grew as quiet as the rest. For even here, the exponential growth of the industry was ending. From the late 1980s on, construction starts around the world dawdled along at a constant low level.[22] There was not much for opponents to hurl frightening imagery against.

For Chernobyl also silenced the utopian missionaries of nuclear power. Already in the early 1980s, although the industry had continued to push for expansion, enthusiasm had been waning as the oil supply crises of the 1970s faded from memory. After 1986, nuclear proponents no longer fed the fires of controversy with grandiose plans. "Nuclear engineering as a profession went dark," sighed one veteran engineer, "as the power industry tried its best to stay out of the public mind . . . The indescribable sense of exclusive fun that we all felt . . . had disappeared . . . the edgy nuclear technology [was] reduced to crashing boredom."[23]

Perhaps reactor building would have languished even without the Chernobyl disaster, given the strength of the opposition and the weaknesses of the nuclear industry. Coal was readily available and cheap, scarcely restricted by pollution controls; oil, too, was cheap, helping people to forget their concern about national energy independence. For whatever reasons, the late 1980s marked a watershed. The effort to create a new age of abundant industrial nuclear energy, and the opposition it provoked, faded into silence.

If, as I have argued, fear of reactors was partly a displaced fear of nuclear war, it helped that steps were under way to reduce that fear. Gorbachev, as much as Reagan, had been frightened by knowledge of the horrors that his own mistakes might unleash. And Gorbachev, like Reagan, was under domestic pressure to do something about the nuclear risk. The Chernobyl accident put him under further pressure; some observers believe that by redoubling doubts that the Communist authorities were trustworthy, the disaster hastened the collapse of the Soviet Union. It was in the shadow of Chernobyl that Gorbachev met Reagan in Reykjavik, Iceland, in 1986.

To the consternation of their advisers, the two leaders came close to promising to radically reduce their nuclear arsenals. Only Reagan's devotion to the magic of missile defense prevented some kind of agreement. The world could see, however, that the pair were truly determined to avoid nuclear war. The following year they signed an agreement on intermediate-range nuclear missiles: for the first time ever, there would be actual reductions in the number of weapons. Although it did not become obvious until the disintegration of the Soviet bloc in 1989, another watershed was passed. The Cold War was ending.

The waning of hostilities revived a discussion that had engaged scholars, and not only scholars, for decades: Had nuclear weapons helped to keep the peace? The "Long Peace"—the absence of big wars among major powers ever since 1945—called for an explanation. To be sure, historians knew of even longer stretches without great wars. But the tension and ideological enmity between the two rival blocs had been extreme. On many occasions, observers up to the highest government levels had believed the world was lurching headlong toward war. Political scientists worked up various theories about how this was avoided, but most agreed that a main component, perhaps the crucial one, was nuclear deterrence—or, to be precise, nuclear fear.

The clever ploys of RAND strategists, advising that a few weapons could be exploded as a signal in an elaborate game, had never impressed the top leaders. Eisenhower, Khrushchev, Kennedy, Brezhnev, Reagan, Gorbachev: to each of them, and to many others in power, an international crisis evoked no thoughts so strong as the image of a world laid waste, if not the utter extinction of humanity. That imagery put matters in a wholly different realm from the traditional diplomacy that had produced two great wars within the century. As an eminent historian put it, nuclear weapons "forced national leaders, every day, to confront the reality of what war is really like, indeed to confront the prospect of their own mortality."[24]

To be sure, the bombs provoked not only caution but grave insecurities, which sometimes exacerbated the enmity. We can never know what history would have been like without nuclear fear. But history surely would have been different, and probably more violent.

21

THE SECOND NUCLEAR AGE

It was a new age for nuclear affairs, what some cultural historians have dubbed the "Second Nuclear Age"—radically separated in some ways (but not others) from all that went before. The détente at the 1986 Reykjavik summit meeting was confirmed by the fall of the Berlin Wall three years later. It was like coming into sunlight after four decades standing in the cold shadow of annihilation. Meanwhile the 1986 Chernobyl disaster settled many debates. Within a few years after the initial tempest of fears and protests, every nation where there was an effective opposition abandoned its ambitious programs. Deferring for the moment the history of weapons, in this chapter I will look into what the Second Nuclear Age meant for the civilian reactor industry.

The industry continued to flourish after a fashion. Although few new reactors were commissioned (and none in the United States), there was good money to make in maintaining and upgrading the existing power plants. This work went on largely out of the public eye. With no spectacular new disasters to command attention, news of the industry retreated to the business sections of newspapers.

The absence of disasters was not a matter of chance. Leaders of the industry understood that one more terrifying accident, anywhere, might force them all out of business. More prosaically, corporations and banks saw that they needed a better safety regime simply to protect their investments: at Three Mile Island, reactor operators with ordinary training had cost their company a billion dollars in half an hour. It was not enough just to obey all the government regulations (or, as had sometimes happened, obey most of them). For the industry to survive it would have to become truly reliable.

Already after the 1979 Three Mile Island accident the industry in the United States had created its own watchdog: the Institute of Nuclear Power Operators (INPO). Reverting to the self-reliance of the 1950s, each company's engineers set up their own rigorous procedures, sometimes even stricter than the increasingly draconian government regulations. In contrast to most other industry associations, INPO became an effective regulator. By the mid-1990s it employed some 400 people, comparable to a major federal agency. The first chief executive, Admiral Eugene Wilkinson (retired), had been groomed by Rickover, and he brought in many more "Navy nukes" with their submariners' tradition of fanatic precautions. When the institute tried to inculcate a new culture of responsibility, it turned out that executives and engineers could indeed be shamed into action by their peers.

INPO showed that it meant business in 1988, when executives of the Philadelphia Electric utility persisted in tolerating lax practices at their reactor. Working with the government's Nuclear Regulatory Commission, INPO convinced the board of directors to fire their top executives. Leaders elsewhere got the point.[1]

After Chernobyl the Europeans similarly reformed their practices, and in 1989 a World Association of Nuclear Operators was created, with more than 100 member nations. It attempted, with mixed results, to duplicate INPO's success by applying peer pressure. Occasional serious incidents continued to reach the public ear. For example, in 2002 it was revealed that corrosion in an American reactor (Davis-Besse near Toledo, Ohio), detected only by a lucky chance, had gone nearly far enough to threaten a meltdown. The incident attracted little attention and was soon forgotten. In Japan a 1999 accident killed two workers and exposed hundreds of people to some excess radiation; other incidents revealed that the Japanese nuclear industry regularly minimized the severity of problems and even falsified inspection reports.[2] Outside Japan these events were scarcely noticed.

Leading nuclear executives began again to warn that "the world nuclear power industry is in danger" from "negligence . . . overconfidence . . . arrogance and complacency." It was a continuation of the self-criticism that had prevented, at least so far, any more spectacular failures.[3]

In the early years of the twenty-first century interest in reactors began to revive. The industry had been quietly advancing even in the United States. Despite occasional problems that rarely attracted public attention, the 103 operating American reactors had steadily improved their efficiency and reliability, and were now providing some one-fifth of the nation's electricity. Worldwide there were more than 440 operating nuclear power plants, producing one-seventh of the world's electricity—including a third of Japan's electricity and nearly four-fifths of France's. Meanwhile engineers were developing a half-dozen new types of reactors that promised much greater efficiency and proof against accidents. Large majorities of scientists and other experts favored building more nuclear power plants.[4] Industrialists began to hope for a grand new wave of reactor construction.

Their optimism did not grow out of the old utopian fantasies, but from prosaic business plans. The world's demand for energy kept climbing, especially as economies in developing nations surged at unprecedented rates. It was now beyond doubt that oil could not meet this demand. There might be enough natural gas to fill in for a while, but even that prospect was far from certain. Solar and wind energy were expanding swiftly, but they were still too costly and small-scale to power civilization. There remained only two realistic options for the majority of the world's new electrical generating plants: nuclear or coal.

Ever since the price of oil spiked in the late 1970s, wherever people refused to build more reactors almost every new electrical plant had been a coal burner. But the hidden costs of the black stuff were becoming evident. Scientists had uncovered more and more health hazards of coal effluents, in particular microscopic soot particles that escaped pollution controls. In the early 2000s research confirmed the worst suspicions. Around the world, the burning of coal was causing at least 100,000 premature deaths every year.[5] But people were used to coal smoke, and regulation remained weak. It was a greater if more remote threat that came to the fore.

As early as 1965 a commission of climate experts had warned the American president of a new risk: future global warming due to the greenhouse effect of carbon dioxide gas emitted when oil and coal were burned. But not until the 1980s did a large number of scientists begin to worry about the danger they expected would emerge in the next century. In the summer of 1988 an international conference of climate scientists issued a statement to the world's governments: they should sharply restrict emissions of greenhouse gases, starting at once. Global warming, the scientists declared, was "a major threat to international security."[6] Meanwhile fierce heat, droughts, and forest fires struck the United States, and the public suddenly became aware of climate change as a risk. Environmental groups began to divert their attention from the fading issue of nuclear reactors. Nevertheless governments, under pressure from fossil fuel interests and antiregulation ideologues, did nothing but appoint panels to study the question. As decades passed the panels, with increasing urgency, repeated the call to curb fossil fuels.

By 2010 it was clear that the globe was in fact warming in the exact way that scientists had been predicting for decades, if not faster. An unprecedented heat wave in 2003 killed tens of thousands in Europe; the ice pack covering the Arctic Ocean was rapidly shrinking along with nearly all the world's glaciers; in 2010 another unprecedented heat wave in Eastern Europe joined with droughts elsewhere to send food prices soaring. The world's leading scientists and scientific academies issued grave warnings: before the century ended crop failures would bring deadly famines, refugees by the million would flee flooded coastlines, the seas would turn acidic, countless species would go extinct. During the two decades 1990–2009, American media as indexed by the Google News Archive printed more than four times more articles that combined the word "catastrophe" with "global warming" than ones that combined the word with either "nuclear reactor" or "nuclear war."[7]

To be sure, the warming would not become intolerable until the next generation's day; there was nothing like the urgency of the old nuclear war fears. Nevertheless one outspoken climate scientist (concerned for his grandchildren) declared that the problem was more dangerous than nuclear war. For somebody would have to deliberately precipitate a war crisis, whereas "the present threat to the planet and civilization . . . requires only inaction" to bring eventual catastrophe.[8]

People had many possible methods to maintain civilization without burning fossil fuels, ranging from solar power to energy conservation. But experts who studied the economics believed that global warming was advancing so fast that to prevent grave harm, the world might well need *all* of the methods together. That included nuclear reactors. Reactors emitted no greenhouse gases, and even including construction and fueling the industry emitted far less than fossil-fueled power plants. Starting around 2005, business magazines began to talk hopefully of a "nuclear renaissance."

Would the public tolerate a new wave of reactor construction? Opposition to nuclear power had begun in the 1960s as a set of local issues, and in the 1990s it had reverted to local issues. The only problem that regularly commanded widespread attention was finding a place to bury the industry's radioactive wastes. For the period 1990–2009 the Google News Archive recorded some 87,000 printed articles that included the terms "nuclear waste" or "radioactive waste," compared with only 12,000 that mentioned "nuclear accident" or "reactor accident." A 2007 survey found that Americans' beliefs about the likelihood of accidents made little difference in whether they would agree to building more reactors, but waste did matter; a large majority would accept reactors if the "waste problem" could be "solved."[9]

Waste politics in the United States centered upon a government effort to create a repository by tunneling into Yucca Mountain in the Nevada desert. Ham-handed attempts to impose a national problem on one locality alienated most Nevada voters. Poisons were unpopular even if, or perhaps especially if, they were hidden underground. Surveys found pervasive "dread, revulsion and anger—the raw materials of stigmatization and political opposition." Politicians got the message. After the Department of Energy had spent twenty years and some $10 billion on studies that convinced most experts (but not quite all) that the risk was negligible, the administration of President Barack Obama gave up attempts to certify Yucca Mountain for use. Although the matter now got tied up in lawsuits, it appeared that the search for a repository had to be started again from scratch.[10]

European nations, their plans for waste disposal likewise blocked by protests, sent their reactor products to a French plant for reprocessing to extract the plutonium. The French insisted that the residues be sent back to the original countries. In 2004 thousands of French and German protesters attempted to block a train returning radioactive materials to Germany for temporary storage: an entire transport route could be seen as a "locality" at risk.

Only Sweden and Finland made real progress. Their experts were confident that the wastes could be safely encapsulated in the masses of ancient crystalline rock that underlay Scandinavia. Instead of seeking a spot that was the very best for geology, they sought places that were safe enough while having little public opposition. In fact some localities were happy to consider a repository—mostly places that already had nuclear facilities. (The closer people lived to a nuclear plant, the more likely they were to accept another facility. The plants' neighbors had grown accustomed to their situation, and they had seen the industry bring jobs and money.)[11]

Everywhere wastes continued to accumulate—out of sight, out of mind— stacked in pools of water or sealed in giant canisters alongside the reactors that generated them. The curious result of the opposition to permanent repositories was materials scattered in hundreds of locations, incorporating tens of tons of plutonium. Yet virtually all experts agreed that no conceivable risk of burying the wastes was remotely as serious as the risk that ambitious governments or would-be terrorists would get their hands on some plutonium. "I've never come across any industry," sighed one official, "where the public perception of the problems is so totally different from the problems as seen by those of us in the industry."[12]

Meanwhile new reactors could be built where they did not arouse intense local opposition, or where the opposition could be ignored or repressed. China was the most ambitious, planning to build two reactors a year from the 2010s on. Almost as enthusiastic were India, South Korea, and several other emerging nations. The nuclear industry and its governmental allies in the United States and some European nations also laid plans to support a new wave of reactor construction as coal fell out of favor. In 2007 the U.S. Nuclear Regulatory Commission received its first full application for a new nuclear power plant since 1979.

———

Would the Second Nuclear Age eventually give way to a new golden age for the industry? The outcome would depend on the evolving record of safety or accidents, on economics, on how fast global warming became unmistakably dangerous, and, not least, on politics. "In most countries," a business journalist explained in 2006, "the future of nuclear power rests more on political considerations than on technological ones."[13] "Political considerations" would be driven by imagery. The Second Nuclear Age might have been less fearsome than the First, but that reality did not turn back the clock to 1960, when visions of a marvelous atomic-powered world had outweighed worries about reactors. The proponents of nuclear reactors no longer attempted to deploy millennial promises; most no longer believed them. Meanwhile the armory of images wielded by their opponents had expanded.

Consider the familiar pictures of a gleaming white containment dome and the soaring curves of cooling towers. The industry had originally promoted these images as a promise of the clean and efficient technological paradise to come. But the immense geometry and even the whiteness of the cooling towers could be (like Melville's white whale) ambiguous: an emblem of mysterious, overwhelming, inhuman powers. Media coverage of the Three Mile Island accident, seeking some visual icon for the invisible radioactivity hazard, had settled on the cooling towers, welding the image to visceral fears of poison. From the mid-1980s on, when a cartoonist or news photograph showed a cooling tower it was felt as ominous. Few remarked that many coal-fired plants used the same type of cooling structure.

More direct imagery came from the Chernobyl accident: the smoking wreckage of a reactor structure itself. An even more important image was the depopulated environs of Chernobyl. I described in Chapter 12 how old tales and paintings used ruins as an image of the downfall of civilization, usually connected with moral decay. In their emptiness the ruins also represented the individual as dead and forgotten, a symbol of physical and even spiritual death. Now there was Pripyat, an actual Empty City of hastily abandoned apartments and silent streets. Over the following decades the public could see it in television documentaries, art productions, and photographs posted on the Internet, a ruin gradually reverting to forest, the haunt of foxes and owls. It was another example of a singular trait of nuclear energy: its links to mythical images drew strength from actual facts.

In science-fiction stories of a world after nuclear war, the Empty City often stood within a deadly radioactive Forbidden Zone—another symbol of the utter negation of civilization and life itself. This, too, now seemed horribly real. To be sure, in the empty thousands of square miles around Chernobyl, wildlife flourished now that people were forbidden to stay. Opponents of reactors focused, however, on reports of malformed animals in the most severely polluted spots. Few remarked on the far larger industrial wastelands in Eastern Europe where wildlife was damaged by chemicals or even poisoned down to bare earth. What really impressed people was *radioactive* contamination, resonating so strongly with thoughts of monstrous mutant births and the forbidden secrets of overweening authorities. (After all, overweening Soviet authorities had indeed kept much about Chernobyl secret.)

Fear of radiation all by itself had severe consequences. The direct effects of the accident on Ukrainians were bad enough; estimates of the total deaths that would eventually result from the escaped radioactive substances ranged from hundreds to several thousands over the lifetimes of the survivors. But they did not call themselves by the positive term "survivors." They identified themselves, and were officially identified, as "victims." As such they were subject to all the problems found among the

Hiroshima and Nagasaki *hibakusha* and other traumatized and stigmatized groups. Hundreds of thousands of Ukrainians were deeply anxious about the possibility that radioactivity had permanently contaminated them. Many developed serious psychosomatic disorders. Still more became fatalistic, and grew careless about alcohol and other risks; some became suicidal. According to a definitive study, "the mental health impact of Chernobyl is the largest public health problem unleashed by the accident to date," a far worse cause of morbidity and economic damage than the radiation itself.[14] (The disproportionate mental impact applied still more to the Three Mile Island accident, where the actual damage from radiation was negligible but fears of contamination afflicted tens of thousands.)

Public fear of radioactivity naturally had an impact on plans for civilian nuclear energy. Unlike in the 1970s, this influence now worked less through direct political opposition than by raising doubts among industrialists and financiers. The biggest uncertainty in planning a nuclear plant was not predictable questions of technology; it was what changes and delays might be imposed by the changes of permits and regulation. The commitment of a billion dollars or so could not be taken lightly. The expense in interest charges alone could break a small private utility if there were long delays due to lawsuits or changes forced by new regulations.

That was the instructive fate of the Shoreham nuclear power plant on Long Island, New York. When the plant was proposed in 1966 as a futuristic marvel, it was expected to cost about $70 million. Mismanagement and engineering errors with the unproven design, plus regulators mandating costly retroactive design changes, delayed completion for decades. Meanwhile the reactor came under legal fire from local governments. In particular, they pointed out that they could not evacuate the population of Long Island speedily if there was a meltdown. In 1992, finally completed and ready to generate power, the plant was decommissioned without ever producing a single kilowatt. Altogether Shoreham cost a staggering $6 billion. No utility executive would care to risk duplicating that experience.[15]

Would public opposition be less intense in the Second Nuclear Age? When a researcher asked people in 2008 to list words they associated with the term "nuclear power," most of the responses were the same as for similar studies made three decades before—but not so overwhelmingly negative. The words "good" and "bad" showed up in about equal proportions; the words "electricity," "power," and "energy" were mentioned about as often as words like "death," "war," and "explosion." "Clean" was as common as variants of "waste." (An undercurrent of stranger associations remained in scattered responses of "mushroom," "end of the world," and even a few mentions of "love" and "sex.")[16]

The nuclear industry and polling organizations monitored public attitudes with care. At the end of the 1980s, in the wake of the news from Three Mile Island and Chernobyl, they found that slightly more Americans opposed getting electricity from nuclear energy than accepted it. The Second Nuclear Age brought a gradual shift; by 2010 polls were finding significant majorities of Americans in favor of reactors.[17] Meanwhile a poll of thirty nations found the world public on average almost evenly divided on their willingness to see nuclear reactors built to replace fossil-fueled plants. The support was strongest in some developing nations, weakest in Western Europe.[18]

There were many reasons for the gradually increasing acceptance of nuclear energy in the first two decades of the Second Nuclear Age. A new technology arouses more suspicion than a familiar one, and by now people were used to nuclear electricity. And it was a long time since they had been regularly pelted with frightening news items, movies, advertisements, and demonstrations deploring the hazards of reactors (it is a classic result of psychological experiments that recent experiences have a greater influence on attitudes than earlier ones). There may also have been a more subtle force at work. Most great images gain their potency from a creative tension, the endless and impossible effort of reconciling opposites. As the promises of a miraculous atomic-powered future waned, at some deep level the opposite side, too, was drained of energy. Nuclear power had become prosaic. To many people a reactor was just another industrial facility— hazardous, to be sure, but what industry was not?

I argued earlier that the twentieth-century protests against reactors fitted into a general revolt against technology and all rationalized hierarchical systems. This revolt had worn itself out by the end of the 1980s. Mistrust of scientists and technology did persist, showing up in opposition to genetic engineering and skepticism on topics ranging from vaccines to climate science. But in the new age, what remained of the vision of a society of unfettered self-expression attached itself less to rural communes and collective action than to personal immersion in the boundless realms of the Internet. In any case opposition to reactors had obviously failed as a strategy to halt the expansion of centralized corporate and government systems.

The new attitudes were visibly at work in the perennially popular television cartoon, *The Simpsons* (in the 2008 word-association survey, "Simpsons" was the only word that turned up repeatedly that had not appeared in similar surveys decades earlier). In the show's opening sequence and in a number of episodes that appeared over the years, a nuclear reactor was central to the fictional city of Springfield. At any moment a careless operator might set off a devastating meltdown; meanwhile emissions engendered a three-eyed fish and other monstrosities. But did anything strike those who watched these episodes with a visceral shock of fear or disgust? Far from it: the corrupt industrialist who owned the reactor, the careless operators, the goofy fish were only a few elements of the show's program of satirizing everything in sight with a wink and a chuckle. One episode even made fun of the apocalyptic vision of a chain reaction of reactor explosions, ending with cartoon mushroom clouds popping up around the globe. Next week Springfield was still there, reactor and all. In this characteristic product of the times, not only the nuclear industry but everything in modern society was dysfunctional . . . and amusing.[19]

A Hollywood attempt in 2010 to make a nuclear industry thriller failed to thrill anyone. The movie *Edge of Darkness* was based on a gripping 1985 BBC-TV miniseries involving a Karen Silkwood–like figure on the trail of corporate evildoers. Bored critics remarked of the 2010 product that "the original plot has not been altered much, although perhaps it should have been"; "stale and dated. It's a movie whose time has passed."[20]

So much of the emotional energy had leached out of nuclear imagery that expert media producers rethought some of their iconic characters. The transformation that came in 1945, when Superman's superpowers and weaknesses were all revealed to be connected with atomic rays, was now reversed. The owners of Spider-Man—who in his 1962 debut got his amazing powers from the bite of a radioactive spider—realigned him in their 2002 movie, deriving his powers from a genetically modified spider. The origins of other comic-book prodigies such as the Incredible Hulk were similarly reassigned from physics to biology.

In movies, a 2001 version of *The Time Machine* skipped the appalling nuclear war featured in the 1960 version. In the 2001 remake of the *Planet of the Apes,* humanity was felled by genetic engineering instead of the nuclear war implied in the 1968 original. The 2008 remake of *The Day the Earth Stood Still* replaced the 1951 prospect of nuclear warfare with the evil of environmental damage. The 1968 zombie movie *Night of the Living Dead* had inspired an entire genre with creatures that, it implied, resulted from a mysterious radiation. Late in the century, as zombies became increasingly popular, their origin shifted to biological experiments. As one zombie-movie director explained, "the connection with nuclear power and what it will do to us . . . Those fears aren't so relevant anymore."[21] In short, for media creators seeking an uncanny magic, whether to bestow amazing powers or to destroy civilization, nuclear physics was no longer scary enough.

This is not to say that the younger generations were unaware of the risks inherent in radiation and nuclear energy. If young people's exposure to the old imagery of monsters and explosions took place within a cartoon framework, it was exposure nonetheless. The few polls that reported responses by age group found young people *less* accepting of reactors than their seniors. But the deep forces of imagery were rarely visible in the daily entertainment or news media.

In the Google News Archive for the period 1990–2009, the titles of printed articles relating to "nuclear power" were largely positive or neutral (being largely concerned with the economics of the industry), by contrast to the mostly negative titles during the period of the Three Mile Island and Chernobyl accidents. Articles that included the term "nuclear accident" were outnumbered nine to one by articles that talked about "oil spill," while even more articles mentioned "genetically modified" something or other. "Water pollution" was mentioned in twice as many articles as "nuclear waste," "air pollution" in four times as many. "Global warming" was mentioned about as often as all the others added together. Once again we note that people can mobilize only a limited quantity of anxiety and activist effort: concern about one issue must dwindle when protests turn to another. After all, the new issues were becoming salient for good reasons. The exponential climb of population, industry, and many forms of pollution were visibly threatening entire ecosystems on which civilization depended.

When the scientist James Lovelock, a hero of radical environmentalists, declared in 2009 that our technology might "all but eliminate people from the Earth," he was referring not to nuclear war but what he called "global heating." Decrying the "false fears" of radiation, Lovelock called for more nuclear reactors.[22] So did Stewart Brand, whose *Whole Earth Catalog* had been a bible of the self-reliant, small-technology movement of the 1970s. So did Patrick Moore, a founder of Greenpeace, reversing his earlier fierce opposition to reactors. So did other veteran environmentalists, throwing the environmental movement back into the divided condition of the early 1970s. The nuclear industry found all this highly encouraging.

The economics were less encouraging. Nuclear reactors might be cheap to run, but they were forbiddingly expensive to build. And the technology kept shifting in search of greater safety. Coal was more familiar and cheaper. Meanwhile, a new technique to extract natural gas from shale promised another cheap energy source, although environmentalists pointed to pollution of water supplies by toxic materials released from the shale (as with coal wastes, that included radioactive elements). The cost of any fossil-fuel plant would rise sharply if environmentalists ever managed to impose controls on their direct damage to public health, their release of radioactive and other wastes, and their greenhouse-gas emissions. Forcing the fossil-fuel industry to pay for such problems could make reactors competitive. If the "nuclear renaissance" was going to happen, it would be driven more by environmental concerns than by traditional economics.

Yet even if all the reactors that people talked about really were built on schedule, they would barely suffice to replace the world's existing reactors as they came to the ends of their useful lives. The reactor construction industry, somnolent for so long, would need decades to get up to speed if it was to provide a much larger share of the world's electricity. The Second Nuclear Age would not soon be powered by fission.

Optimistic forecasts received a heavy blow in 2011 when a monstrous earthquake and tsunami wrecked a set of aging reactors in Fukushima, Japan. Only now did people realize that the reactors had not been protected against a tsunami of that scale, a kind that would happen barely once in 1,000 years (although an event of that probability was likely to hit somewhere among the world's hundreds of reactors every decade or so). Workers staved off a catastrophic release of radioactivity only through heroic improvisations. But enough material did escape to force authorities to evacuate nearly 100,000 citizens and to ban fish and farm products from the region. Half a world away, in America, television newscasters spoke ominously of drifting traces of radiation, and some families snatched up potassium iodide pills, which in fact would be worth taking only in case of massive radioactive fallout. In China people hoarded iodized salt in the mistaken belief it would protect them. Japanese food exports plunged. "The chances that any customer would eat radioactive fish or vegetables from Japan are negligible," a journalist remarked. "But the chance that they would think about radiation when eating Japanese produce is almost 100%." (Reporters might have noted a more serious potential health risk: countless tons of chemicals and toxic waste that the great tsunami had scattered across farmlands and swept into the ocean. But scarcely anyone mentioned that.)[23]

Officials in China, India, and some other developing nations announced that they would continue building reactors—albeit more cautiously—for they could not afford to buy oil and were literally sick from burning coal. But the governments of some advanced nations, including Germany, Italy, and Japan itself, announced they were abandoning their tentative plans for further reactor development and would close down existing reactors as soon as economically feasible. Elsewhere the lessons of Fukushima would make for even stricter regulation, and thus more costly nuclear power. In the United States, the pause had little immediate impact; for as one expert remarked, "The so-called nuclear renaissance wasn't going anywhere in the U.S. even before the Japanese earthquake."[24]

The Fukushima disaster inevitably brought a drop in public acceptance of nuclear power. Polls moved back toward the level of the early 1990s: in most places a small plurality again opposed building more reactors. Yet many environmentalists remained convinced that nuclear energy, however disagreeable, was preferable to the alternatives. If many Americans feared a reactor accident, roughly as many did not. Media discussions of the nuclear industry were rarely altogether negative, even while the Japanese crisis was under way. Rather than looking to the abolition of the industry, reporters asked whether, for the world as a whole, it would expand despite the news, or only stand in place? The Second Nuclear Age would remain, for a while at least, an age of gradual growth in some nations and stasis or retreat in others.

—————

Fukushima's imagery was similar to Chernobyl's, with its long weeks of crisis and uncertainty, its pictures of wrecked reactors, and a Forbidden

Zone of lingering radioactivity. Yet the public impact was less overwhelming. For one thing, Chernobyl had already put an end to claims of perfect safety and visions of reactors everywhere. For another, 2011 saw no large-scale radioactive contamination spread far away, such as had brought panic and agricultural restrictions all across Europe in 1987. The Japanese themselves, facing the stupendous damage that the tsunami had wrought across hundreds of miles of coastline, mostly met the nuclear problem with their customary stoicism; their anxieties were strongly felt but only occasionally displayed. Finally, as I remarked above, nuclear imagery in general may have evoked less visceral fear now than it had in the 1980s.

I have not yet mentioned one main reason for the public's partial acceptance of nuclear power in the Second Nuclear Age, perhaps the most profound reason of all. I have argued that nuclear reactors were lit by the reflected glare of nuclear weapons: that fear, disgust, and distrust of the industry stemmed in large part from its many intimate associations with the dreaded bombs. It follows that as fears of war dwindled with the end of the Cold War, so that explosions and radioactive fallout faded out of conscious reflection, anxiety about reactors would have dwindled in parallel. It was nuclear bombs, not nuclear power, that stood at the center of the complex of images. The crucial question is what changed in people's thinking about the bombs—and whether those changes were justified by the facts.

22

DECONSTRUCTING NUCLEAR WEAPONS

Whatever would they do with all that plutonium? By 2002 the United States and Russia had each agreed to dismantle nuclear warheads containing a total of thirty-four metric tons of the metal. Plutonium is so preposterously dense that thirty-four tons would fit in an ordinary closet—but there would be enough in each closet to make 13,000 bombs. What to do with it all? The stuff could serve as fuel for nuclear reactors, generating electricity as it turned into waste. The Americans designed a plant that would process the plutonium into an oxide reactor fuel that would be very hard to use for bombs. But by 2010 the plant, like many ambitious projects in the nuclear industry, was far behind schedule, far over budget, and had found no customer for the fuel. Meanwhile the two powers agreed to reduce their nuclear arsenals still more, scavenging another nine tons of plutonium apiece. It was like a gift from some malicious fairy that could neither be used nor thrown away.

The end of the Cold War had inaugurated the Second Nuclear Age, but it was only the first part of a great historical turn. Next came the astonishing disintegration of the Soviet Union and an accommodation between Western capitalists and Communist China, cancelling the fears of imminent war that had flared repeatedly from the 1950s into the 1980s. The number of nuclear warheads poised for use, which had peaked in 1986 at the fantastic level of 70,000, rapidly fell. Under new treaties, by 2012 barely 4,000 would be deployed on strategic delivery systems anywhere in the world. (Another 15,000 or so warheads were in storage, mostly in the United States and Russia.)

The public and the media found other issues to occupy their finite space for worry. The phenomenon was nicely demonstrated by a series of polls in the United Kingdom, which asked people what issues they thought were most important. Through the mid-1980s, between 15 and 30 percent of the public spontaneously brought up nuclear weapons. After 1991, there were never more than 4 percent who did so.[1] The decline of interest was paralleled by a decline in news coverage. Articles in the Google News Archive using the term "nuclear war" or "nuclear weapon," after peaking in 1982–1983, fell off steadily into the 1990s.

The Gallup organization probed American fears directly by asking citizens how likely they thought a nuclear war would be in the next ten years. From 1981 to 2001 the fraction who thought a war was "very likely" declined from 19 to 8 percent, while the share who thought it "very unlikely" rose from 23 to 33 percent. A significant change, but limited. Many adults would still acknowledge a fear of nuclear war if something drew the topic to their attention. But the news rarely did that.[2]

Yet the world was not safe. The United States and Russia each retained well over a thousand warheads poised to be launched on a few minutes' notice—plus tens of thousands more in storage. It was reliably reported that the Russians had even set up a sort of Doomsday Machine: in a crisis, if communications with Moscow were severed and nuclear blasts were detected, the device would automatically send orders to low-level commanders to launch all their missiles.[3] If the warheads on alert were no longer numerous enough to threaten the extinction of all advanced societies, they would suffice to throw civilization back many decades.

The other nuclear-armed nations were content to hold ready at most a couple of hundred warheads apiece—obviously enough to deter any sane adversary. That the two former Cold War enemies retained a hundred times more weapons than they could use for any rational purpose demonstrated that the old paradox remained in force: doom was still supposed to be staved off by the nightmare image of doom itself.

Was the cluster of nuclear war images acting as strongly as ever? There was only one new novel (Cormac McCarthy's *The Road*) on a par with iconic Cold War literary productions like *Alas, Babylon*, *A Canticle for Leibowitz*, or *On the Beach*. However, those and other works of the 1950s and 1960s remained in print into the twenty-first century; the paperbacks often held high positions in bookseller Amazon.com's sales rankings, well ahead of classics by esteemed authors like Saul Bellow and John Updike.[4] Meanwhile writers continued to publish pulp fiction featuring nuclear apocalypse. Particularly persistent were tales of the hero-survivor after a nuclear war. *The Survivalist*, a series of paperbacks by Jerry Ahern and ghost writers, which featured endless violence and self-righteous heroism with a right-wing flavor, persisted from its beginning in 1981 to a twenty-seventh volume issued in 1993. It won millions of readers—and thus plenty of imitators. For example, William Johnstone's *Ashes* series kept publishing new volumes from 1983 to 2002.[5]

Cultural historians typically look to movies to see what people worried about. Two scholars who exhaustively studied movies with nuclear themes, Jerome Shapiro and Mick Broderick, independently reported that such movies did not go into any noticeable decline after the late 1980s. Several hundred nuclear-related movies and other productions appeared in the first decade after the Cold War ended, the majority of them American.[6]

Some of these productions, like several sequels to the 1984 *Terminator* movie and the submarine thrillers *The Hunt for Red October* (1991) and *Crimson Tide* (1995), relied on the familiar threat that a nuclear war might be launched not by a nation's strategic choice but by an evil-minded clique or computer.[7] Other films, including *The Postman* (1997), *The Road* (2009), and *The Book of Eli* (2010), developed the old trope of postapocalypse social breakdown. Concentrating on the drama of survival, they only hinted that nuclear war had set the stage. But they exposed a new generation to emotionally telling images of measureless devastation.

Television, too, carried many of the familiar old plots. In 2001 the popular American television show *West Wing* built its drama around a crisis in a Russian missile silo. In the winter of 2006 a television viewer could follow no fewer than three major series on nuclear weapon themes that would have seemed overdone already in the 1970s. In *Jericho*, townspeople in mid-America dealt with criminal marauders in the days following the nuclear destruction of many of the nation's cities. In *24*, a secret agent was pitted against terrorists who exploded a nuclear device in Los Angeles. In *Heroes*, the plot centered on the efforts of its super-powered "heroes" to prevent the nuclear bombing of New York City.[8]

Meanwhile the rise of computer games opened a whole new way for people to experience imagery. For example, in the popular *Duke Nukem* series (1991–2002 plus later adaptations), you manipulated the eponymous hero, taking his viewpoint as your own. You battled, for example, killer robots created by Dr. Proton, a scientist driven mad by the radiation accident that altered his brain, and vicious creatures created with radioactive slime by another evil scientist. Dozens of other widely played games were set in an after-the-bombs wasteland where you battled outlaw gangs and/or mutant monsters.

If it all sounds familiar, that is because important mythical images are immortal. Since the spread of printed books, everything remains available down the generations. When witchcraft, astrology, and a hundred other systems of ancient imagination remain alive in many minds, we can hardly expect to lose nuclear imagery. After all, our subject is one of the most powerful complexes of images ever created outside of religions.

So had nothing changed since 1986? *Everything* had changed. Or at least it changed for a large and growing part of the public: the new generations. That included everyone born after about 1980. Back in the early 1980s, several researchers had asked American adolescents about nuclear war, and found that they thought about it often and seriously. A majority believed a nuclear war was likely in their lifetimes and doubted they would survive such a war.[9] After 1986 nobody seems to have repeated such surveys. Presumably those who dealt with adolescents did not see the kind of anxieties that were once so evident. In my own informal queries of young people, I have heard no echo of the nuclear anxieties that permeated the postwar generations.

The new cohorts of young people (not all so young as the twenty-first century progressed) had not grown up in a world where talk of nuclear war, radiation, reactors, and so forth showed up frequently in the news, and even in personal conversation, within a context drenched with anxiety. Often their first encounters with nuclear topics took place in the tedium of the schoolroom. As likely a place as any for a student to pick up some facts was Hersey's 1946 book *Hiroshima*, which teachers often assigned (the book kept a high position in Amazon.com's sales rankings, and a *Spark Notes* study guide was published in 2002). It was a fine work for showing what a 1945 fission bomb could do. It said nothing about missiles with multiple fusion-bomb warheads or the unfathomable paradox of deterrence. As for reactors, how many minutes would science and history teachers find to describe the industry, among all the other materials they were required to teach?

While there were a good many nuclear-bomb movies in the period 1990–2010, these mostly involved rogue individuals bent on destruction rather than the apocalyptic nuclear-war scenarios of the 1960s and 1980s. As one cultural historian noted, the nuclear threat had become "a cheesy plot device, not a viscerally felt reality." Movies that featured nuclear bombs were escapist fare, often veering into science fiction, representing "cultural avoidance and displacement."[10]

No longer were important social groups working to spread information or emotive imagery. The antibomb and antireactor songs and posters had gone away, along with the educational films and press releases of civil defense officials and the nuclear industry. The media no longer felt a strong public-service calling to inform citizens about nuclear energy (or much of anything else). For sensational stories, too, the media now had better choices. Aside from an occasional botched government nuclear waste initiative, the industry offered no news handle. As for military affairs, the exciting new weapons were missiles precisely targeted to kill enemies individually. What nuclear news items did come up mostly languished in the back pages of the newspapers . . . and many young people did not read even the front pages. Few sought out the specialized websites and blogs where nuclear information and opinions continued to swarm.

Surveys found that electronic entertainment expanded to take a far larger fraction of young people's time than reading or schoolwork, more so with each passing decade. The television screen and computer monitor were now the main devices presenting nuclear imagery. It was here that the new generations could see their choice of all the famous old nuclear movies, along with the countless new, derivative productions.

The small screen in a lighted living room notoriously gave a more detached experience than an enveloping movie viewed as part of a mass audience. Still more distancing from immediate emotional experience were recorded movies. You could freeze the action while you fetched a cola. You could view a sequence repeatedly to catch some nuance of the computer-generated background. And you could look at extra features—ludicrous outtakes, a documentary revealing the trickery behind the special effects. All this made for a less intimate, less impressive encounter. To be sure, the ancient brain core was still recording, unforgettably, the shocking blasts and dreadful monsters. These unconscious imprints were separated more than ever from the conscious experience of harmless entertainment.

That separation also held for a new medium, at first glance supremely immersive: computer games. These spread rapidly in the 1990s to take up many hours of emotional engagement. For example, more than half a million people bought *Fallout 3* in its first month of release in 2008. The award-winning game, sequel to 1997–1998 versions, featured battles with radioactive ghouls in the rusting and vine-covered ruins of ancient Washington, D.C. A young man who played such a game (and it was nearly always a male under forty years of age) felt his heart pumping, sweat sprang to his forehead, he knew terror and exhilaration. Yet at any point he could detach himself for a trip to the bathroom. The more feverishly you played, the less you were supposed to actually believe in the contents: even more than television, games occupied a realm intellectually separated from "real" life.

Arguably, playing a game in which you were blasting mutant horrors or degenerate savages would leave stronger unconscious emotional traces than watching a monster movie like *Them!* But unlike the audiences of *Them!*—many of whom had left the theater believing giant ants could truly be engendered—twenty-first-century gamers knew that their cartoonish adversaries were pure fantasy. It was certainly gripping to be Duke Nukem shooting mutant "Pig-cops," but when you paused to stretch you could hardly take the creatures seriously.

To be sure, there were a few nuclear strategy games. But cold chess-like strategy brought little engagement with the realities of nuclear war. The most popular such series, called simply *Nuclear War* (1989–1997), was called a "fun" game with a "light-hearted approach" full of "wacky humor."[11] The emotions such games evoked were far from what millions of people had felt, for example, at the conclusion of the 1959 movie *On the Beach* as the camera slowly scanned the empty streets of a dead city. In the strategy games as well as the shooting games you identified with a protagonist who was no victim, but a heroic warrior. This identification gives us a hint about a particular direction that nuclear-weapon imagery was taking, a shift I will discuss in depth in the next chapter.

The new approaches in games were part of a larger transformation. Most cultural critics agree that a new period began a decade or so before the end of the twentieth century, but they do not call it the Second Nuclear Age. They call it the age of Postmodernism—a word that first became common in the late 1980s.

The word "postmodern" covers a range of ideas. The meanings most important for our inquiry can be summarized in a single term: "referencing." From children's cartoons to rap music, postmodern productions bristle with references to previous works, appropriated to add sparkle to the production. Children were now less likely to have their first encounter with a mad scientist in an actual horror movie than in some passing sly reference in a comic book or Saturday morning cartoon. How many first met a nuclear reactor in the introductory sequence of *The Simpsons*, with its lovable but hilariously incompetent reactor operator? The result was *distancing*: you watch as from a distance, never fully engaged.

Referencing is especially distancing when, as often happens, it is *ironic*. To be sure, irony was on the rise long before the Cold War ended. As I will discuss later, some landmark works of the Cold War period were already built upon black humor and a disdain for mainstream social authority. In the postmodern period, however, irony was no longer a shocking position taken up by rebellious intellectuals: a sardonic attitude and outright mockery became the default position.

The mode was even marketed to children. Consider the Tyco toy company's popular Doctor Dreadful Radioactive Experiments (mid-1990s). Typical was the third in the series, the Nuclear Blob Mix, produced a bubbling yellow mess—"gross looking, great tasting!" Coincidentally, it resembled the mock-perilous yellow ooze in the back of the Mattel "Hot Wheels" Homer Simpson Nuclear Waste Truck ("Home delivery").[12]

Mockery does not entrain fear, but dispels it. The entertainer flatters the audience that they are "cool" (both coolheaded and up-to-date), sharing in a knowing and amused depreciation of the appropriated symbols. The nuclear ooze does not resemble real radioactive waste; it represents a quaint concept, something that might have seemed scary to our parents but not, heaven knows, to us.

The prototype for this approach to nuclear culture had appeared already in 1982: a documentary, *The Atomic Café*, regularly screened in classrooms ever since. It taught many to see "Duck and Cover" training as a strange aberration. In fact the training was a rational deployment of Second World War sheltering techniques, appropriate to the dangers of an attack with the few atomic bombs that existed in the early 1950s. But after hydrogen bombs were deployed by the thousand, the nuclear culture of the 1950s began to look farcical.

Many of the younger generation saw some of the older productions not just through references but in the original. For example, the cult television series *Mystery Science Theater 3000* (1989–1999) replayed the worst old science-fiction films, dubbed with a series of wisecracks. If young people happened to run across one of the old films elsewhere, or rent it, they could easily supply their own ridicule.

Yet if everything was fictional and amusing in conscious thinking, the exposure to movie, television, and game images continued to forge connections at the animal level of memory. Beginning with silly children's cartoons featuring evil scientists and their outlandish creatures, everyone was exposed to countless repetitions of all the associations developed in earlier centuries. Most young people would not have been aware of an autonomic reaction of disgust to the cartoon three-eyed fish featured on *The Simpsons*. The mushroom cloud rising over Los Angeles in the television thriller *24* was not likely to bring the teeth-clenching reaction of their seniors who had lived through the Cuban missile crisis. Yet associations were forged. To the player of *Fallout 3*, in the recesses of the brain where survival decisions are made, the horrid ghoul and radioactivity were indissolubly linked. For many young people nuclear weapons and reactors seemed horrid because, well, somehow anything nuclear was just bad, wasn't it?

The strange mating of visceral engagement with overt detachment was pushed to its limit in the *otaku* (fanboy) culture of young Japanese, whose influence was worldwide. The *otaku* withdrew into their rooms to work obsessively through *manga* (comics), *anime* (cartoons), computer games, and personal fantasies. These often featured radiation, annihilated cities, robots, and so forth. As one observer remarked, the young men's anxieties, nuclear and otherwise, might have been turned toward political action. But instead they were "transformed into the monstrous catastrophes and apocalyptic delusions depicted in the bizarre world of manga and anime . . . a profound psychological repression."[13]

Images of nuclear destruction are uniquely well fitted for this detached postmodern approach. For as I and others have pointed out, it is precisely as representation that nuclear weapons exerted their influence. Even the Hiroshima and Nagasaki bombings were conceived as rhetorical acts. Afterward, governments did not manufacture hydrogen bombs with an intention to explode them over cities; their purpose was to intimidate people or awe them. Geopolitical positions shifted when one or another nation simply hinted it might build nuclear weapons, without actually getting any. Arguably, the Bomb—not bombs, but "the Bomb"—was the first great postmodern object . . . or should I say image . . . but object and image are so entangled here that we can no longer make such a distinction.

Jacques Derrida, a leading explicator of postmodernism, explained in 1984 that this intermingling applied still more completely to the idea of "total nuclear war." Since that had not happened it was "fabulously textual," something "one can only talk and write about," not know through experience. Yet that did not mean it was *nothing but* a fable. For "total nuclear war" guided international relations and led to the creation of real stockpiles of weapons. The fiction and the reality, Derrida explained, were "not two separate things." The same can be said for the invisible, yet influential, "hundreds of thousands of deaths" some insisted must have been caused by the fallout from bomb tests or by a reactor accident like Chernobyl's.[14]

Another example was "the antiballistic missile." No actual hardware could be built capable of defending a nation against an enemy salvo. Governments spent their tens of billions of dollars only to make an impression on the enemy's leaders . . . and their own citizens. The cosmic magic of "the antiballistic missile" functioned as representation, like the Bomb itself—only more so, since working nuclear bombs did exist.

Sometimes the most powerful things in our heads are the ones we don't pay attention to. Ignoring or mocking a deep cultural symbol may just allow it to continue its work unobserved. The entire tangle of nuclear associations kept its old forms as it drifted in the abstract postmodern space. No matter what arms treaties were signed, powerful emotions associated with nuclear energy continued to resonate throughout society. A few events would be enough to mobilize the nuclear drama again in new forms.

23

TYRANTS AND TERRORISTS

It was only one bomb, small enough to fit in the trunk of a car. A band of fanatics stole it from the Israelis, smuggled it into the United States, and exploded it in a football stadium to kill tens of thousands. That was the centerpiece of a best-selling 1991 novel and popular 2002 movie, *The Sum of All Fears*.[1] Between the book and the movie the terrorist band changed from Palestinians to neo-Nazis while the stadium moved from Denver to Baltimore, but the details hardly mattered. In the many stories with a similar plot, bomb materials could be stolen from Americans or Russians; the catastrophe could be planned for Los Angeles or Miami. What did matter were two familiar themes: the proliferation of bombs in nations around the world, and evildoers intent on blowing things up. These themes were becoming inseparably entangled. The Second Nuclear Age had begun with a decade of release from the anxieties of the Cold War, but by the late 1990s nuclear fear was on the rise again.[2]

From 1945 through the 1980s, when people worried about the proliferation of nuclear weapons their main concern had been that nations would use the bombs in war. If Argentina or South Africa showed an interest in getting nuclear bombs, its aim would be to threaten, deter, or defeat neighboring states. These, it was presumed, would hasten to get their own bombs in turn. But it didn't happen. Proliferation, as one scholar pointed out in 2009, proceeded "at a far more leisurely pace than generations of alarmists have routinely and urgently anticipated." And careful study showed that aside from the United States and the Soviet Union, the few cases in which a nation did get its own bombs turned out "to have had remarkably limited, perhaps even imperceptible, consequences." Nobody was successfully threatened, deterred, or defeated by the bombs.[3]

Despite this reassuring history, many worried that one or another dictator must certainly want to possess nuclear weapons—and if he got them he would drop them on a neighbor. Could such a regional atrocity trigger a world war? Everyone understood that thanks to the horror of all things nuclear, objections to any use of nuclear weapons had become something like a sacred taboo. Once the line was crossed (for example, in the Near East, where the great powers had vital interests), would the conflict escalate into a general exchange of missiles that would annihilate everything? Fear of violating the nuclear taboo—a dread of everything "the Bomb" represented—was arguably a main reason that in all the tense crises and vicious regional conflicts since 1945, nobody came near to using even a single bomb.

But what if some evildoer set out on purpose to provoke a nuclear exchange? "You get Russia and America to fight each other," the *Sum of All Fears* villain declared as his aim, "and destroy each other!" The scenario of triggering mass missile attacks with a single explosion became less plausible with the end of the Cold War, yet it never vanished entirely from people's thinking. It was as if every bomb contained within itself the entire apocalypse.

Meanwhile events stimulated a new variety of fear. The world was shocked when United Nations inspectors, scouring Iraq in 1991 after the nation's defeat in the Gulf War, discovered that the Iraqis had secretly got well along toward building a few bombs. Worse, their bomb material was not plutonium but uranium-235 produced in a battery of centrifuges, easier to build and hide than a reactor. When the Iraqi program was forcibly dismantled, the issue faded from public consciousness. But not for long.

Proliferation worries began to revive in the late 1990s when Iraq obstructed the United Nations inspectors, raising suspicions it meant to try again. And in 1998, after twenty-four years of silence, India conducted its second bomb test, followed by more. Pakistan reacted swiftly with its own series of plutonium explosions. Worried talk about proliferation rose enough around 2002 to bring a third peak in the articles mentioning "nuclear war" in the Google News Archive, albeit much smaller than the peaks of the early 1960s and early 1980s. Adding to anxieties were reports of a massive Iranian effort to extract uranium-235 with centrifuges. In 2006 came an actual explosion of a North Korean device using plutonium from a reactor.

It was rare for more than 20 percent of Americans to be aware of a given news story (they notoriously do not pay much attention to foreign news), but since the late 1980s surveys recorded awareness well above that level for stories about nuclear weapon proliferation in Pakistan, Iraq, Iran, and North Korea. Significantly, after the U.S.-Soviet arms reduction agreement of 1991 the *only* nuclear-related news, civilian or military, that showed up in the surveys consisted of stories about weapons in third-world states.[4]

The nuclear ambitions that these regimes pursued were particularly worrisome because each was infected with some combination of irrationality, fanaticism, and tyranny. These elements featured in news accounts of the so-called "rogue states"—a term that rose to prominence during the 1990s, evoking the image of a vicious, solitary, uncontrollable beast—Libya, Iran, Iraq, and North Korea. Each of these favored certain terrorist organizations. If one of these states ever got a good stock of nuclear bombs, might it not give one to its protégés? Worse, each of these states was itself a brutish terrorist, as personified by its ruler. What would hold back Iraq's murderous tyrant Saddam Hussein from obliterating Tel Aviv if he had the means, or the bizarre North Korean dictator Kim Jong-Il from sneaking a bomb into Seoul or Washington?

Irrationality and fanaticism have been fundamental to images of terrorism for centuries. The third element, tyranny, is more modern. The anarchist bombmakers of the 1890s and the 1960s claimed to fight for democratic or socialist ideals. Some of the more recent murderers, like the right-wing extremist who blew up a federal government building in Oklahoma City in 1995, similarly imagined they were striking out on behalf of the suppressed individual. But by the start of the twenty-first century terrorists were mostly associated with repressive, authoritarian systems, such as Islamic fundamentalism or the fascism of the criminals in the movie *The Sum of All Fears*. It now seemed appropriate for dictators to sponsor terrorist groups.

As so often happened in nuclear affairs, the imagery got reinforcement from uncannily apt facts. Most important was a frightening revelation in Pakistan in 2004. The revered founder of the nation's weapons program, nuclear scientist Abdul Q. Khan, confessed that he had been busily transferring—indeed, selling—sophisticated warhead and centrifuge technology to Libya, Iran, and North Korea. It was like nothing so much as the plots of thriller novels in which a scientist, motivated by religious fervor or simple greed, stole secrets and gave them to terrorists.

Khan claimed he had not given secrets to terrorist gangs but to authoritarian nation-states. But how different were those? The lines were getting blurred. Many doubted that Khan could have done his nefarious work without the knowledge of Pakistan's government, or at least one of the military cliques within the regime. Here was a new idea: elements of a state might, so to speak, steal weapons technology from themselves, to place it in the hands of the enemies of their enemies. Not only was the line between dictator and terrorist getting blurred, but the image of the despotic leader was blending with the old image of the evil scientist.

The connection was far from new. The genealogy of well-organized criminal gangs or entire states with technologically adept masterminds runs back through Dr. No (1958–) and Ming the Merciless (1934–) to the pulp-novel master criminals Dr. Fu Manchu (1912–) and Sherlock Holmes's nemesis, Professor James Moriarty (1893–). These figures, while not exactly scientists themselves, wielded advanced technological powers, connecting us on back to Drs. Frankenstein and Faust. From here we can range forward through unbalanced scientists back to the modern terrorist. For when the fictional evildoer explained his scheme to trigger a world war, he sounded much like the mad scientist of earlier tales who plotted to blow up the planet.

In the real world, experts in international affairs had little fear that any leader of a nation would set off a bomb out of insane death-lust. The experts did worry that terrorists would somehow get a bomb—most likely by stealing nuclear materials. As we have seen, the idea formed a minor theme of concern from the 1950s through the 1980s. But stories of a fanatic who got hold of a terrible bomb had been mostly sidelined in implausible thrillers and science fiction. Events of the 1990s pushed the idea to the fore.

The disintegration of the Soviet Union brought a radical revision of thinking among nuclear experts. In the former Soviet lands, tons—not kilograms, tons—of bomb-grade material were unaccounted for. (It was not even possible to guess how much plutonium and enriched uranium was missing, for Soviet officials had lied when they claimed to precisely meet their production targets.) Worse, there were no good records for the tens of thousands of actual assembled weapons. Nobody could say how many "loose nukes" lay about (the term became popular in the mid-1990s). Meanwhile thousands of warheads sat behind chain-link fences with no more protection than a padlock and a sleepy guard.

Dozens of times newspapers reported incidents in which someone, usually a Russian, was caught attempting to peddle a small sample of weapons-grade uranium or plutonium. Nuclear facts were again reproducing earlier fictions. Perhaps the facts were imitating fiction: the thieves' hopes of finding a buyer for their wares may have been inspired by the old fantasies.

In the United States many policy experts and journalists called loudly for action. Leading senators demanded funds to help the former Soviet states remove or safeguard their nuclear materials. Articles and political advertisements warned of the deadly threat. It was another case of people deliberately spreading nuclear fear.

In 1993 Islamic extremists attempted to blow up the World Trade Center in New York City with conventional explosives, followed in 1995 by the blast that killed 168 people in Oklahoma City. Soon after, a poll of Americans found that nearly three-quarters of them believed there was a chance that terrorists could attack an American city with a "weapon of mass destruction." Another poll in 1998 found that half of all Americans believed that terrorists would explode a nuclear bomb in the United States within the next ten years. Experts wrote entire books discussing nuclear terrorism; columnists and politicians exclaimed that modern society faced a peril of vast dimensions. Nevertheless, the majority of people in the 1990s did not feel a deep personal concern. If they suspected a bombing might happen somewhere, unlike in the Cold War they thought the bomb was unlikely to strike just where they happened to live.[5]

The September 11, 2001, attacks on New York and Washington, D.C., did not initiate any new anxiety, then, so much as activate existing ones. The appalling pictures and stories that everyone saw and heard brought a visceral intensification of fears. What had been an abstract concern became almost a personal trauma. A series of Gallup polls found that the fraction of Americans who worried seriously about terrorism in general had been one in four in 2000. After the September 11 attacks the number leaped to six in ten, and after a few years leveled off at a relatively high four in ten. A more specific 2003 Gallup poll found four in ten Americans said they "often" worried that terrorists might attack the United States with nuclear weapons.

These were serious fears, though milder than the gripping dread of war that large majorities of Americans had felt during the tensest phases of the Cold War. In other developed countries, majorities usually rated terrorism as a second-rank concern like the environment and education, well below economic worries. People in most of the developing countries were even less likely to mention terrorism as a concern.[6]

In sum, in the early twenty-first century fear of terrorists with nuclear bombs stood on the shelf along with the other nuclear fears, but now it had moved to the front. Like fallout and reactor hazards in earlier decades, the threat satisfied the requirements for a risk whose likelihood people would tend to exaggerate by comparison with familiar risks like home accidents or fires. The image of a mushroom cloud rising over an American city was dreadful and memorable; the risk of harm was involuntary and unjust; the danger was novel and depended on unknowable secrets. Above all the idea was available in the mind, easily pictured by anyone raised on tales of nuclear weapons. Small wonder that the idea was popular with writers of thriller novels and television shows.

In November 2006 a former Russian intelligence officer living in London, Alexander Litvinenko, suffered an agonizing death through poisoning by radioactive polonium. *Time* magazine attributed "Litvinenko's excruciating and sinister death" to the effects of "a Chernobyl" inside him, and Litvinenko's father said his son "was killed by a little nuclear bomb."[7] It was the old cluster of associations among radiation, reactors, and bombs. But there was something else here, present in earlier decades but becoming more prominent in the Second Nuclear Age: a tight connection of radioactivity with poison and the traditional world of secret agents. After all, spies and poisons were old teammates. Not only history but physiology allied them: in both actual and vicarious experience, physical toxins and acts of treachery alike evoke visceral sensations of revulsion.

Ever since the 1940s some experts had warned that a well-organized group might manage to steal radioactive material, for example, from the wastes stored near reactors, and use it as a poison. They could then contaminate many city blocks with what was called a "dirty bomb"—we cannot escape references to the most primitive notion of pollution, dirt itself, again tying moral to physical disgust. To be sure, the terrorists could simply scatter the nuclear wastes about rather than blowing them into the air with a few tons of conventional explosives or a crudely constructed fission device of equivalent force. But the word "bomb" brought a more frightening, nuclear resonance.[8]

The "dirty bomb" was conventionally understood as vomiting death-dealing radioactivity. Of course, anyone determined to scatter substances that would poison a few hundred people right away and threaten many thousands of cancers over subsequent decades could find dozens of noxious chemicals that were far easier to get at than nuclear wastes. But nuclear materials were in fact the most likely choice for an intelligent terrorist. As the Chernobyl and other accidents had demonstrated, a dispersal of radioactive substances would be far more effective than any chemical in provoking terror and large-scale psychological and economic damage—to say nothing of media attention.

The increasing prominence of the connection between nuclear weapons and terrorists may have been related to the ending of the Cold War. One subtle but important consequence was that the old familiar enmities dissolved. Not only was the Soviet Union gone and China a trading partner, but when the world Communist movement faded, so did virulent hatred of capitalists. During the 1990s, the waning of opposition to the nuclear reactor industry was in parallel with a waning of hostility to corporations in general.

Why does all that matter? Because people *need* enemies. Social scientists report that a group can strengthen its cohesion by opposing an "outgroup," outsiders who draw us together in dislike and self-defense. And many psychologists believe that personal selfhood needs an "other," a type of person who stands for everything that we are determined not to be: we define ourselves in part by what we despise.

As the Cold War wound down, in many countries enmity came to rest on criminals. Russians for good reason became preoccupied with "corruption" (yet another word that covers both biological and social toxins). In the United States, starting in the mid-1980s passionate and in some aspects irrational campaigns targeted drug addicts, child molesters, and other especially nauseating criminals.[9] The incarceration rate, after lingering at around 2 per 1,000 from the 1930s through 1980, soared to 7 per 1,000 by the end of the 1990s. Citizens in many countries also increasingly targeted immigrants as outsiders and enemies. These foreigners had long been seen as a sort of intrusion polluting the purity of the nation, and moreover especially prone to crime. No wonder if there was new force in the image of criminal terrorists stealing in from abroad.

In the 1990s this imagery added its strength to the growing fears of tyrants/terrorists armed with nuclear bombs or radioactive poisons. Saddam Hussein, Kim Jong-Il, the ayatollahs of Iran, Abdul Khan and his clique, Osama bin Laden and his Al Qaeda network: all fitted the old archetypes with remarkable precision. The psychological association of criminal terrorist with evil mastermind with wicked despot looked like straightforward reality.

It did not matter that a rational leader could hardly find it expedient to set off a bomb as a terrorist act, let alone give a bomb to an independent terrorist group. The fear was of what an *irrational* leader might do. People have not found that the actual leaders of states are masters of calm reason. It became difficult to separate the sober warnings of government counter-terrorism experts from exclamations in television dramas.

Real leaders are complex, responsive to many pressures. Their counter-parts in novels and television shows were curiously one-dimensional, filling roles that could have been written a century earlier. The terrorists might threaten an explosion in order to extort political concessions, or they might set out to actually destroy a city, perhaps even to trigger a world war, acting out of avarice, revenge, political ideology, religious fanaticism, or simple insanity. Any motive would do. In *Tomorrow Never Dies* (1996), James Bond foiled a media mogul who tried to steal missiles and start a dreadful war in order to . . . improve his corporation's position in China. The only aim of the writers was to set up a hair-raising potential for villainy as a foil for the hero's efforts.

A fine description of the theme in a pure state (and a neat case of post-modern ironic referencing) is found in the 1996 satirical movie *Austin Powers: International Man of Mystery*. When Dr. Evil's ingenuity for con-cocting nefarious schemes fails him, he tells his cohorts: "Oh, hell, let's just do what we always do. Let's hijack some nuclear weapons and hold the world hostage." The cardboard terrorists of most fictions did not en-gage the public in symbolically deep ways. I have found little imagery in these tales that has the power and complexity of old nuclear themes like the ambiguous monster or the innocent child survivor. Actual terrorists often sought the apocalyptic transformation of society and the personal transmutation of martyrdom, but these grand themes were barely noticed in the popular culture of terrorism. To be sure, since the days of the anar-chists some of our best authors have written novels that attempted to get into the mind of an everyday terrorist.[10] But nuclear terrorism itself inspired no great novel, film, or other art in the way that nuclear war had done. The typical narratives of the Second Nuclear Age were simple fantasies of a terrorist or terrorist gang confronting a secret-agent hero.

And the secret agent? He (seldom she) was as symbolically shallow as his opponents. His history is easily traced back through the Cold War fables of James Bond and his colleagues to spy thrillers that proliferated after the First World War and earlier. Once the Cold War ended, the secret agent was usually pitted not against an entire foreign nation but against a clandestine group led by a crude villain. Hardly any author presented the pair as ambiguously related; the deceptions and violence of one were good, of the other evil, and that was that.

The secret agent did not attempt the apocalyptic passage through death with hopes for rebirth into full humanity. He remained stuck in a childish world of mayhem. In this regard the secret agent was cousin to the manly survivor-heroes who battled mutant creatures and savage gangs in the after-the-bombs fantasies that persisted in paperback thrillers and computer games. The secret agent, too, was often a survivor, triumphing over all sorts of threats. But none of these quick-shooting heroes were transformed by their experiences, unless to become coarser and more aggressive.

That absence of transformation fitted the postmodern culture, in which people looked skeptically on claims of genuine heroism, let alone transcendence. They asked only to be entertained by a sort of cartoon representation of a hero. Historian Paul Forman has pointed out that postmodern culture had little respect for methodical, selfless, socially responsible professionals, people like the monster-defeating scientists and spy-catching FBI agents featured in 1950s productions. Such figures had become more a target of satire than heroes to be admired. The new ideal was the self-motivated entrepreneur or rebel, transgressing all rules, using any means to capture his victory. You could not build a working society out of a population of terrorists, secret agents, or gun-wielding survivors, but you could make gripping stories.[11]

The terrorist stories did activate powerful emotions. Their unexpected explosions and hideous poisons created deeply embedded fears, not unlike the fears evoked by the plunging missiles and radioactive creatures of the First Nuclear Age. Meanwhile the image of the terrorist, in the usual manner of images, shed none of its old associations, all the way back to technological despots and nineteenth-century mad scientists. Adding nuclear bombs to the mix did not at first seem to bring the symbolic and psychological depth we have come to expect for anything caught up with nuclear affairs. Yet beneath the surface deeper feelings were stirring, driven once again by events in the news.

The secret heart of nuclear fear was not so much dread of a murderer as of being a murderer—and even of oneself, a suicide. In the first decades of the Second Nuclear Age, it was suicide that brought the most powerful new associations to the image of the terrorist. To be sure, the earlier terrorists and mad scientists might willingly die as they attacked their communities—after all, they were crazy. But they had not embraced suicide as a rational means to rational goals. That idea entered modern history in 1983, when militants suicide-bombed a U.S. Marines barracks in Lebanon. In the following decades the tactic became common, primarily in the Muslim cultural sphere but elsewhere, too, notably among the Hindu Tamils of Sri Lanka.

A chief problem of suicide bombing, as one observer pointed out, is that "by design, it unsettles the question of deterrence . . . To make the challenge to deterrence even more stark, a suicide bomber . . . is willing to kill innocent bystanders." That willingness could extend even to one's own people, as when a bomber welcomed the death of bystanders as fellow "martyrs." Even a threat of wholesale nuclear retaliation would not deter a group that believed the more deaths, the better.[12]

As I noted earlier (Chapter 12), the idea of suicide had been an important, if rarely acknowledged, component of nuclear weapons imagery since the 1950s. The concept of a nuclear warrior embracing death was perfected in a famous image: a cowboy-pilot riding a bomb to catastrophe in the movie *Dr. Strangelove*. By the time of the 1983 Nuclear Winter controversy if not earlier, most people understood that the strategy of Mutual Assured Destruction (MAD) worked precisely because it was as crazy as a suicide. Modern terrorism turned that strange logic on its head. Suicidal bombing was not only proposed as a rational political tactic but actually performed, time and again.

For many adults a mention of suicide (let alone suicide-murder) will evoke profoundly disturbing personal thoughts. But nobody could shut his or her eyes to the dreadful concept, for appalling pictures of carnage showed up in the news, year after year—striking as deep into the brain as the most monstrous creatures of past fictions. Now this pervasive dread of the terrorist could be activated by any thought of a nuclear explosion. That brought in further associations with older themes: Dr. Abdul Khan drags Dr. Frankenstein behind him, and in the distance we hear Godzilla's scream of desperate rage.

Traditionally people had a hard time imagining that any actual terrorist would wish to take lives not just a few at a time, but by thousands or millions: surely so dark a desire was not humanly possible? That hope was overthrown by events. The first serious blow came in 1995, when members of a large Japanese cult, Aum Shinrikyo, released poison gas in a coordinated attack on five subway trains in Tokyo. The gas killed a dozen people and harmed hundreds, but the cult had intended to kill far more; their ultimate aim was nothing less than global apocalypse. If anyone still doubted that a fanatic group could mobilize the will and means to kill on a very large scale, they were answered by Al Qaeda in the September 11 attacks.

I think the destruction of New York's World Trade Center had its deepest impact on people by showing that horrors of human intention that had seemed incredible must be taken as facts. As one nuclear authority put it shortly afterward, "The willingness of terrorists to commit suicide to achieve their evil aims makes the nuclear terrorist threat far more likely than it was before September 11." A writer reflecting back on events went further: "The reason 9/11 was so traumatizing for all of us, I believe, is that the vision we all had of the World Trade Center collapsing in a horrible cloud—for us it was effectively the mushroom cloud that we have been dreading for a generation."[13]

Many noted that the incessantly televised pictures of airplanes attacking buildings, billowing clouds of dust, and smoking wreckage resonated with familiar imagery of bombardment. Reporters immediately used the language long associated with nuclear apocalypse: "gates of hell," "like a nuclear winter."[14] Within a few days the site of the New York attack was universally called "Ground Zero." The phrase had originated in Los Alamos around 1945, reflecting the technical significance of the distance in thousands of feet from the point directly below an exploding atomic bomb. In New York it stood for a more mythic zero: an Empty Zone of total destruction, as in the familiar photographs of Hiroshima.[15]

Within a few months the term "nuclear 9/11" emerged as a common shorthand term for a kind of attack that many now feared more than before. A search of the American news media indexed in the Google News Archive for the combination "nuclear" and "terrorist" or "terrorism" shows a steady level of references from the 1980s up to September 2001, then a jump to a higher level maintained ever since.

To be sure, stealing and smuggling a nuclear bomb would be a formidable task; the only thing harder would be to attempt to make one yourself in a garage. Terrorists or despotic regimes seeking to kill people in large numbers could much more easily get their hands on chemical or biological agents. When governments talked about all these threats together, they began to use the phrase "weapons of mass destruction." The phrase was a product of the Second Nuclear Age, becoming common around 1990. It conventionally included chemical and biological weaponry—but in fact those are weapons of mass *mortality*. The word "destruction" pointed straight at nuclear weapons. The phrase was not entirely a euphemism for nuclear bombs, but it leaned deeply in that direction.[16]

During the national trauma Vice President Dick Cheney expressed what many were thinking when he remarked to an aide, "As unfathomable as this was, it could have been so much worse if they had weapons of mass destruction." Years later in a speech Cheney recalled the day in similar terms: "foremost in our minds was the prospect of the very worst coming to pass—a 9/11 with weapons of mass destruction." The prepared text handed out before the speech did not say "weapons of mass destruction;" it said "nuclear weapons."[17]

In the public mind after September 11, nuclear terrorism did trump all. The mailing of letters with deadly anthrax spores, which disturbed Americans soon after the September 11 attacks and demonstrated the potential for a biological assault of vastly greater magnitude, was largely forgotten within a year or two. The terms "biological 9/11" and "chemical 9/11" were almost entirely absent from the media. A search of pages on the World Wide Web that contained variations of the word "terrorist" plus the term "nuclear weapon" turned up about 6 million pages in 2006, whereas searches for variants of "terrorist" plus "biological weapon" (or "anthrax," "germ," etc.) got only 2 million and "chemical weapon" or "poison gas" fewer still.[18] In short, in the early twenty-first century the one truly great fear connected with terrorism, and thus near the forefront of Americans' anxieties in general, was the fear that terrorists might somehow get their hands on a nuclear weapon.

———————

Fear of nuclear terrorism, like other nuclear fears, had practical consequences. Since the 1950s, and accelerating in the 1990s, national authorities spent billions of dollars seeking to detect smuggled nuclear materials, safeguard radioactive wastes, dismantle warheads and reprocess the plutonium, and the like. It was far more than they spent on measures to contain possible chemical or biological attacks. No matter that many experts argued that it was very unlikely that a terrorist group would be able to procure or build a workable atomic bomb. And no matter that spending a fraction of the sum on the other prospective weapons would be valuable even if terrorists ignored them, by providing public health protection in case of an accidental chemical release or a natural epidemic. It was the dreaded nuclear risk that got money and attention, even before the September 11 attacks.

In 1999 an academic had predicted some consequences of such an attack (he imagined a one-kiloton atomic bomb that collapsed the Empire State Building). He thought citizens would demand "measures that would violate civil rights. Phones might be tapped, foreigners' movements monitored . . . Within days, the American way of life might change substantially."[19] His prediction was accurate. The unprecedented extension of presidential powers that began in the United States with the Manhattan Project, and expanded during the Cold War, was pushed further still.

A familiar social force was at work. Officials seeking to enlarge their sphere of authority (and what official does not?) would naturally emphasize the most horrifying of the threats from which they were defending the public. A chair of the Joint Chiefs of Staff said a single successful attack killing 10,000 civilians could "do away with our way of life"; a U.S. senator claimed that terrorism presented an "existential" risk to the nation.[20]

The intensification of surveillance and control measures extended even to torture of supposed terrorists. Many citizens swallowed the myth that torture was necessary if only to uncover what was commonly called a "ticking bomb"—often understood to be of the atomic variety. Only a minority of citizens with a knowledge of history worried that when surveillance and coercion are institutionalized, they are commonly turned against political dissidents and anyone else who threatens the position of authorities. Some feared a permanent loss of freedoms. The Cold War was over, yet nuclear fear continued to help political leaders and bureaucracies extend their powers.[21]

The connection between terrorism and nuclear fear also played a significant role in promoting the "war on terrorism" that Americans waged not only domestically but abroad. President George W. Bush repeatedly defined the threat as "the world's most destructive technologies" in the hands of "the world's most dangerous people." Cheney later summarized the views of the administration in a few succinct phrases: "You cannot keep just some nuclear-armed terrorists out of the United States. You must keep every nuclear-armed terrorist out of the United States."[22]

Such evocations of nuclear anxieties helped the Bush-Cheney administration mobilize public support for the invasion of Iraq in 2003. The president and other officials spoke of the Iraqi tyrant developing every sort of "weapon of mass destruction," but they got the greatest impact when they warned specifically of a "mushroom cloud." These warnings were based on flimsier reasons than assertions that Iraq might possess chemical or biological weapons. The most egregious case was Bush's State of the Union address preceding the invasion, in which the president said that Iraq had tried to procure uranium from Africa. It later emerged that American intelligence services had previously told the White House that this claim was baseless. The administration was willing to risk its reputation and the nation's by deliberately distorting facts in order to arouse nuclear fear.[23]

Concern about secret weapons was only one of several reasons the administration decided to invade Iraq, and for many officials it was not the most important reason. As a leading proponent of the invasion later explained, "we settled on one issue, weapons of mass destruction, because it was the one reason everyone could agree on." A former senior intelligence officer remarked that the claim that Iraq was preparing nuclear weapons "made sense politically but not substantively." It is an open question whether the administration could have won enough congressional and public support to invade Iraq if everyone had understood that in reality there was no prospect of an Iraqi atomic bomb.[24]

The tight-knit cluster of associations of the Second Nuclear Age that I have sketched in this chapter, with its deep resonances of dread and disgust— evil tyrants, secret cultists, criminal scientists, nuclear poisons, heroic secret agents, suicidal terrorists—was not found only in printed texts and on video screens. The complex of imagery altered world history.

24

THE MODERN ARCANUM

Another name for the philosophers' stone was the Arcanum, the great secret. Alchemists surrounded this central mystery with lesser symbols—naked kings coupling with queens, a green lion devouring the sun, a thousand pictures whose meanings were concealed from the profane. In the twentieth century nuclear energy became the modern Arcanum, no less surrounded with images that most people did not fully understand. I will now survey the entire period since Hiroshima, to see how nuclear energy was addressed, not by journalists or politicians or movie studios, but in even more subtle ways. From obscure sources among the public at large, mysteriously compelling symbols emerged.

In the early years of the First Nuclear Age, while many people who looked at the new technology saw dark devastation, still more saw a dazzling light. Some expressed feelings of religious awe. "Atomic power! Atomic power! It was given by the mighty hand of God!" Country-and-western singer Fred Kirby put that on the radio in 1945, making such a hit that he started a fad for atomic lyrics. Many agreed that the new power came from beyond the mortal sphere. From 1945 on, ministers and priests joined statesmen and pundits to speak of transcendental forces and the sacred duties of stewardship.[1] Watchful intellectuals warned that feelings of awe, insignificance, and hope for salvation, which properly belonged to holy things, were being directed instead toward atomic energy—what top scientists themselves called "the basic power source of the universe."[2]

Illustrators for magazines, comic books, and advertisements frequently represented nuclear energy by a blaze of light, for example, radiating from a nucleus. When observers at weapon tests told of being transfixed by the vision of the fireball, it was nothing new: incandescent balls or disks and haloes radiating light had stood for divine power in pictures everywhere from ancient Egypt to China. The solar fireball was also a traditional symbol for the goal of alchemists and mystics—gold, mystical wholeness, the

Arcanum. A luminous sphere was bound to evoke holy awe, suggesting godhead in its aspect of a tremendous otherworldly presence.[3]

Like any form of divine energy, nuclear energy could be seen not only as solar but as the chthonic opposite. Writers never tired of repeating how Oppenheimer at the Trinity test saw the bomb's cloud as divinity in its aspect of destroyer—an elegant version of the cartoonists' drawings of a bomb as brutish war-god. According to some psychologists, by the mid-1950s for many people "The Bomb" had become something even more: a prime symbolic representation, an apotheosis, of Death itself.

For some people the connection between bombs and apocalyptic power had a precise Christian meaning. A gospel song of 1950 warned that on the Day of Judgment, Jesus would "hit like an atom bomb." Some took that prediction literally. Preachers and religious tracts from the 1950s on-ward prophesied that the Second Coming of Christ would be heralded by nuclear missiles, fulfilling in plain fact the biblical prophecy of falling stars, scorching heat, rivers of blood, and so forth. Of course the preach-ers' followers hoped to be among the remnant of the faithful who would be saved in the Last Judgment. Some millenarian sects moved to remote areas or built fallout shelters, just to make sure.[4]

This was no minor trend. In the United States the most popular nonfic-tion book of the 1970s, reportedly selling some 15 million copies (at least three times as many as any other book I have mentioned except the Bible) was Hal Lindsey's *The Late Great Planet Earth*. It predicted a nuclear Ar-mageddon, miracles and all. The book spun off a movie and four sequels, inspiring many other paperback books prophesying a nuclear Doomsday that would fulfill Christian hopes and fears. In the 1980s an estimated 5 to 10 million Americans, including President Reagan, believed there was a good chance that the Armageddon of the Bible was imminent in nuclear form. These ideas persisted into the twenty-first century. A series of twelve novels (1996–2004), beginning with *Left Behind: A Novel of the Earth's Last Days*, sold 60 million copies and many spinoffs by 2005.[5] Fundamen-talist Christians eagerly waited for the Jews of Israel to build the Third Temple as a precondition for the final battle, typically assumed to feature nuclear devastation (after all, Armageddon is an actual location in the Holy Land).

The aftermath, as vividly described in after-the-bombs tales, would serve admirably as the prophesied time of tribulation, punishment, and cleansing. The trial by fire envisioned by preachers since antiquity would burn the wicked and purify the virtuous, a process like the refining of gold in the alchemical furnace. But in one scenario that was increasingly discussed starting in the 1950s, before the bombs came the purest Christians would be lifted up "to meet the Lord in the air" (1 Thessalonians 4:16–17).

Or would God spare us all? A substantial minority of people took refuge in trusting that God would never permit the world to be destroyed. That was less than a complete answer to the problem of bombs even for people of firm religious faith, and still less comfort for everyone else. In a secular age, symbols of salvation would be more impressive if they were wrapped in an aura of science.

⸻

In 1951 Arthur C. Clarke published a story about a colony on the moon where the human race survived after nuclear war had left the earth a phosphorescent ruin. Survival of a remnant rocketed into space became a familiar idea—a secular version of the saintly lifted into the air to avoid the end times. After the Soviet Union launched its Sputnik in 1957, many took to spotting satellites; the lights cruising among the stars stimulated thoughts of space flight, but also of missiles. All this talk of danger or salvation by way of technology in the skies activated the deep old myths of magical flight and battle in the heavens.[6]

Thoughts turned to denizens of outer space. It became a science-fiction cliché that if space aliens attacked the earth, nations would forget their differences and unite, solving the problem of human warfare. Another solution appeared in Clarke's famous 1953 novel, *Childhood's End*: the aliens became a Wellsian corps of virtuous technocrats, hovering in great ships over the world's cities to forbid war.[7] The novel ended in a terrestrial apocalypse and a reborn race rising into the sky, described with barely concealed religiosity. In the popular 1951 film *The Day the Earth Stood Still*, a messenger from the stars who demanded that humans make peace with one another was recognizable as a miracle-working Christ figure.

This last messenger had arrived in a glowing, silvery disk propelled by atomic energy: a "flying saucer." Since 1947 an avalanche of reports about strange objects in the skies, first in the United States and then around the world, roused intense public attention. Some observers called it an "atomic psychosis" brought on by fear of the bombs, and the idea is worth inspecting.

Flying saucers were no trivial phenomenon. Reports of sightings of un-identified flying objects (UFOs), as the things were named in the 1960s, continued at a rate of hundreds each year. Sightings and media attention rose to higher peaks at various times in the 1970s, 1980s, and 2000s. Numerous highly successful movies featured visitors from outer space. Polls in the 1960s found that about half the population believed that UFOs were real, and by the 1970s more than one-tenth of Americans said they had personally seen one. In the early twenty-first century a third of Americans still thought it likely that UFOs existed.

This was another belief indifferent to boundaries of class and education; the sightings could come from sober businessmen or engineers as much as anyone. Leaving aside a few cases that were difficult to explain, millions of ordinary people looked at planets, weather balloons, and other ordinary things and thought they were seeing something uncanny. It was a singular example of the mass projection of cultural images onto external stimuli.

The projected material, especially in the first decade of sightings, in-cluded much about nuclear energy. This was commonly suggested as the UFOs' mysterious power source, and rumors circulated that radiation monitors had gone wild when the objects were nearby. People noticed that sightings were reported especially often near AEC installations. By the 1950s the saucer pilots were widely assumed to be alien beings, surely far advanced not only in technology but in wisdom—the Martian engi-neers. Had nuclear bomb tests drawn their attention to our planet? Had they come to save us from war?

The psychologist Carl Jung concluded that the sightings had much to do with the Cold War. He suggested that in the anxiety caused by the di-vision of the world into two hostile camps, people would like to pray for a divine answer, but many found such a miracle implausible. They looked instead for salvation through superhuman technology and compassion.[8]

Such an answer was openly promised in tracts and personal proselytiz-ing that inspired the growth of more than 100 American cult groups. Major movies and widely read books resembled schizophrenics' fantasies in their preoccupation with secrets and with superior beings at once menacing and benign, who conveyed redemptive messages of cosmic importance. The UFO theme addressed anxiety about all modern technology. By the

1970s some men and women were reporting messages from space that called for reforms to prevent not only nuclear war but also environmental perils (notably nuclear reactors).

The UFO phenomenon was a rare historical event: the emergence of a major popular symbol. A host of seemingly ordinary people had worked a grand creative act, inventing a new representation for the dangers and hopes of personal and social transmutation, a new myth peculiarly appropriate to the modern age. This myth had originated in close connection with nuclear fear, and carried an implicit response. Awesome modern technology, said the UFO stories, must be accompanied by a full-scale transformation of civilization and everyone within it.

Nuclear energy's remarkable affinity for themes of transmutation and the sacred was sometimes better concealed. The most impressive of all nuclear symbols, found in the great majority of nuclear productions, seemed straightforward: a towering white cloud. In the late 1970s the psychologist Michael Carey found that photographs of these clouds had made an unforgettable impression on nearly everyone he interviewed, imparting a vision of overwhelming and numinous power. As *Pravda* remarked, the "mushroom-shaped cloud" seemed to hang "suspended over the future of mankind."[9]

Why mushroom-shaped? Other words could have been chosen. Observers of the first Trinity test wrote of a "dome-shaped" column, the "parasol," a "great funnel," a "geyser," a "convoluting brain," and even a "raspberry." One Japanese witness of the Hiroshima explosion thought it looked like a "jellyfish"; a Bikini test in 1946 was accurately described as a "cauliflower" cloud. Yet at Bikini a reporter also spoke of "the mushroom, now the common symbol of the atomic age," and the term became almost synonymous with nuclear bombs. Already at the Trinity test more observers had mentioned mushrooms than anything else. Somehow it was hard to imagine people feeling the same awe for a "cauliflower cloud."[10]

The mushroom cloud was a folk symbol, created by nobody in particular for reasons that nobody explained. But a symbol of what, exactly?

In the 1950s the respected scholar R. G. Wasson noted that in Western culture mushrooms were usually associated with dank, dark places, rot and poison—in short, with death. But if that was all people had in mind at bomb tests, they could have used the English term for a specifically poisonous fungus, namely "toadstool." Wasson pointed out that mushrooms were also associated with food, and therefore with life. The mysteriously swift growth of these organisms was part of their folklore, and indeed descriptions of the Trinity test spoke of the cloud not only as a static shape but equally often as something that "mushroomed" up. The mushroom, even if just a knob growing on a rotting log, could represent life opposing death—perhaps even life arising from within death, that is, transmutation.[11]

Mushrooms had traditional associations with thunderbolts, witches, and fairies, that is, with magical power. Wasson suggested that these associations were connected with a particular mushroom, the fly amanita, well known among folklorists for its poisonous and magical associations, and used as a hallucinogenic drug by Siberian shamans to help them into ecstatic trances in quest of cures and mystical knowledge. "Magic mushrooms" of an unidentified species also figured prominently in ancient Hindu texts and the lore of medieval Chinese Taoism, connected with both religion and the elixir of immortality.[12]

In 1955 Wasson discovered isolated Mexican villages where another mushroom, the white psilocybin, was traditionally used to induce religious trances. The drug culture of the 1960s adopted the "magic mushroom" as a vehicle of mystical transports that they expected would lead toward a rebirth of individuals and all society.

The atomic bomb mushroom likewise could be used optimistically. General Electric's 1952 animated film *A Is for Atom* opened with a bomb explosion, and the cloud then transformed into a faceless giant with muscular arms crossed—an ominous warrior or powerful servant. The 1957 Walt Disney movie *Our Friend the Atom* made it all explicit with its tale of a fisherman who uncorked a mushroom-cloud genie. The image of the atomic genie became familiar—but already in 1908 Anatole France had compared the cloud rising from a city blasted by an atomic device to the cloud from the *Arabian Nights* fisherman's bottle.[13]

The atomic genie was a cousin of the golem, of course. With his rebirth into the world after woeful cramped imprisonment, with his irresistible rage that could be converted to benevolent power, the genie excelled as a symbol of the peril and promise of psychological transmutation.

Another image takes us even deeper. At one point in *A Is for Atom* the genie faded into a representation of an atom itself. The picture was utterly conventional—a little ball ringed by ellipses, standing for a nucleus with electrons whirling around it. This ringed atom was the most universal of all symbols of nuclear energy, seen in countless presentations from the 1920s on. Anything so hackneyed merits close scrutiny.

Early in the century physicists had imagined electrons circling the nucleus like planets orbiting a sun. They abandoned that model in the late 1920s when they found that electrons actually oscillate through a pattern of probable locations, nothing like a solar system. Anyway the electrons had nothing to do with the energy within the nucleus. A ringed atom was scarcely more accurate than a genie as a representation of nuclear energy.

Cut off from real physics, the ringed atom evolved in a significant direction. The original solar-system model had not been symmetrical, yet popularizers began to draw symmetrical pictures, most often with a fourfold symmetry, which had no scientific justification whatsoever. The definitive version was promulgated by the Atomic Energy Commission on its official seal. From a number of designs the commission chose a ringed atom with four ellipses around a central nucleus, the whole enclosed in a circle. "I consulted some of our scientists," the designer later recalled, and "they told me that the design . . . did not mean anything. It is only symbolic."[14]

Any student of symbols could instantly identify a design characterized by a central object, symmetry and especially fourfold symmetry, perhaps contained in a circle. It is a mandala. This class of patterns is the most important of all human symbols.

Jung pointed out the importance of mandalas after he found such patterns in the recondite alchemical texts he was studying and independently in some of his patients' dreams. Subsequent work by many scholars found mandalas in the earliest scribblings of children around the world; in plans for sacred cities from Middle America to China; in therapeutic drawings of schizophrenics, including tribesmen who had never held a pencil before; in high religious art, such as the rose windows of cathedrals; in the psychedelic visions of mushroom eaters—almost anywhere one looked. The mandala proved to be a universal human archetype.[15]

The mandala typically appears in a context of spiritual and psychological questing. The most thorough students of mandalas, the Indo-Tibetan sages who gave them their name, used them as maps of a mystical state embracing everything from sexual congress to the entire cosmos. As Jung put it, mandalas stood for the potent union of every opposite, male and female and all the rest—that is, for the philosophers' stone. In sum, the ringed atom, the ubiquitous symbol of nuclear energy, followed a pattern universally associated with transmutation.[16]

A related symbol, displayed in every American city and town in the 1960s, stood in black on bold yellow to point to a fallout shelter. Three triangles arranged in a triangle within a circle, its threefold symmetry closely resembled Indo-Tibetan mandala patterns. The designer probably took the device from a handbook in which the three triangles were described as "an ancient symbol for the Godhead."[17]

Fireball, flying saucer, mushroom cloud, genie, ringed atom, fallout shelter symbol—in their various ways all said the same thing. If one knew how to read them, these pictures were posted across the public image of nuclear energy like a banner: "CAUTION—HIGH SYMBOLIC POWER!" Nuclear energy had become a full symbolic representation for the entire bundle of themes involving personal, social, and cosmic destruction and rebirth. By symbolic representation I mean a widespread cluster of associations, condensing a large number of meanings, all loaded with symbolic power, into a single thing. Nuclear energy had become precisely the new Arcanum, permanently connected with the most terrifying, fascinating, and sacred of all human themes.[18]

25

ARTISTIC TRANSMUTATIONS

How could we deal with this unimaginable force, nuclear energy? The symbols developed in popular thought scarcely answered that question, pointing only toward hazy and grandiose transmutations. Some writers and filmmakers rose to the challenge, creating works of high art that called for moral courage. But the message ran into difficulties that were only gradually understood.

In 1947 the poet W. H. Auden characterized postwar times as the "Age of Anxiety." A bleak mood had fallen upon the artistic and literary elites. Novelists, painters, social critics, theologians, all spoke of an anguished hollowness. Historians and psychologists reported that anxiety had been growing steadily among thoughtful people throughout the twentieth century; a hundred influences had created an existential dilemma. But many sophisticated thinkers connected the problem most particularly with nuclear bombs.[1]

Hardly anyone claimed that bombs were the only source of modern anxiety, but what could stand as a better symbol of our general problem? No discussion about contemporary anguish and futility seemed complete unless it mentioned the bombs. Typically such mentions were brief and stereotyped, as if a phrase or two said it all. Like reactors only more thoroughly, the bombs—or nuclear energy in general—served as a condensed symbol for the worst of modernity.

With the news from Hiroshima sensitive thinkers quickly realized that doomsday was no longer just a religious or science-fiction myth, but as real a part of the possible future as tomorrow's breakfast. Worse, the future might lead into blank nothingness. Some declared that death itself lost meaning when a nuclear apocalypse could rob us not only of our lives but of our progeny and even the world's memory of what we had been. What use was anything when a foolish accident might annihilate everything that mattered? This became a recurrent theme in poetry and even music and dance: the bombs had placed at the base of things an unspeakable void.[2]

Such despair often led to denial. People looking in the direction of modern war might perceive a great void simply because they dared not open their eyes. For example, Kurt Vonnegut, whose best novels dealt with nihilism, personally struggled with the classic denial mechanism. As a captured soldier he had been in Dresden on the night in 1945 when incendiary bombs converted the city to a desert of ash; afterward he simply could not remember details of the experience. His innocent faith in science and humanity was destroyed, not just at Dresden but still more when he heard the news from Hiroshima. In his novels of the 1950s and 1960s Vonnegut kept groping for images of meaninglessness and annihilated civilization without ever quite facing bombs directly.[3]

Particularly frank was the British novelist John Braine, who admitted that he could not write a book except when he forgot about nuclear weapons. "Whenever I think about the H-bomb, I can only give way to despair, which is to say that I can only stop thinking of it." Filmmakers were equally baffled. In 1962 a magazine surveyed the world's greatest directors of artistic films, asking them how they would treat hydrogen bombs in a movie. Louis Malle replied, "For nothing in the world would I make a film about the H-bomb." Elia Kazan: "It's too much for me." John Cassavetes: "Frankly, the very thought of atomic warfare . . . threatens to throw me into a panic."[4]

Those who did manage to get into the subject tended to produce works of despair. Pat Frank, for example, confessed that he forced himself to write nuclear war novels from a sense of duty. Each of Frank's three novels on nuclear war was more pessimistic than the one before. Philip Wylie followed a similar downward path. Characteristic of sophisticated nuclear fiction was a novel that won critical respect and millions of readers around the world, Mordecai Roshwald's *Level 7* of 1960. Its characters were depersonalized soldiers, locked deep in sterile tunnels under the command of unseen authorities. They inadvertently began a missile war that destroyed all life on the surface of the earth, then perished themselves one by one, obedient and unthinking. In the end no life remained. Such works found only emptiness in the modern situation.[5]

Psychologist Robert Jay Lifton and others spoke of a "numbing" induced by the specter of nuclear destruction: a condition that took denial almost to the extreme of blindness. The refusal to face up to dreadful reality was infectious, Lifton argued, a numbness that spread into entire regions of public thinking, from international politics to thoughts of death itself. To be sure, the accusation of numbness was often aimed at people who simply did not choose to take up political activism as the accusers wished. Yet there is plenty of evidence that many did ignore the bombs not out of conscious choice but because they could not stand to gaze into that fearsome glare. In the following pages I will discuss people who thought deeply about the dilemmas of nuclear war, but for most people, outside a few periods of heightened anxiety, the chief response was avoidance.[6]

In the early years there were some optimistic responses, reflecting the "Atoms for Peace" hopes. Futuristic essays by intellectuals were joined by paintings that tried to express the vitality of atoms in a positive way. These petered out once missiles and hydrogen bombs multiplied. As a literary historian remarked of another medium, science fiction, in the 1950s the best writings on nuclear war themes typically had happy endings, but in the 1960s the "dominant mood . . . could only be described as nihilistic."[7]

From 1945 into the 1980s a sense of anxiety and disintegration ran through all the arts from music to sculpture, although it was impossible to say how much that trend had to do specifically with nuclear fears. For example, Jackson Pollock's explosively disorganized "drip" paintings looked like a response to the new sense of fragmentation and uncertainty—but neither Pollack nor his critics could say how much of that was related specifically to nuclear energy. Although retrospectively one scholar would later claim that "the awe or aura of the atomic bomb was pervasive among leading Abstract Expressionist painters," this was an outsider's perspective.[8] Aside from occasional polemical drawings and writings that were more political messages than art, most creative people did not attempt to deal explicitly with the bombs.

To be sure, a few individuals and entire groups claimed to represent the "atomic age." Surrealist Salvador Dalí said that the Hiroshima explosion "filled me with terror" by crushing his fantasy of personal immortality; it initiated a "nuclear" period of his art with disjointed, increasingly mystical paintings. A "Nuclear Art" movement founded in Milan in 1952 linked mindless paint gestures and images of deformed humans to Hiroshima. And it seemed clear enough what a follower of the movement, Yves Klein, meant in his notorious 1958 Paris exhibit *Le Vide* (The Void), consisting of an empty gallery painted white. Klein explained, "we are living in the atomic age, where everything material and physical could disappear from one day to another." Going a step further was Klein's friend Jean Tinguely, whose ironic 1960 contraption of motors, noise, and smoke performed its sole function—self-destruction—before an elite audience at New York's Museum of Modern Art. However, it was only a small minority of artists, musicians, playwrights, and so forth who pointed even indirectly at the bombs as a main inspiration for their works of disintegration.[9]

The weapons were not really beyond the reach of the arts. Beginning around 1980 as war fears reawakened, hundreds of young artists began to treat nuclear weapons in paintings and sculptures of high quality. For example, a 1983 show by ninety artists in New York portrayed in every conceivable way the aggressive impulses embodied in nuclear war and, still more, their hideous consequences. A traveling exhibition of forty-four "nuclear" works included Alex Grey's 1980 *Nuclear Crucifixion*, a horrific image of Christ impaled on a nuclear bomb cloud—making at last a connection with Western culture's greatest image of the transmutational passage through death. Other old symbols brought explicitly to the fore included the vine-covered ruined city and the Horsemen of the Apocalypse, presented as skeletons riding phallic missiles. Some artists found new ways to create moving statements, for example, invoking the bomb shelter's grave and womb symbolism. Many of these works remained (as Grey confessed) "just elaborate despair." Yet working out the despair explicitly was an important step, rarely made openly before 1980. It was the same for poets, a group that had been silent about nuclear issues in the 1950s aside from a few occasional pieces. A 1982 reading in New York City by a large group of "Poets against the End of the World" was followed in the early 1980s by the publication of at least five more anthologies of American nuclear-weapons poetry.[10] This surge in the arts was short-lived, ending along with the Cold War.

In the Second Nuclear Age that began in the 1990s, postmodern modes of referencing and simulation came to the fore. Artists looked back at the seductive technology of the Manhattan Project, gathering or reproducing artifacts of the early bombs or photographing project sites. Especially famous was the Chinese-American artist Cai Guo-Qiang, whose stay in Japan in the early 1990s inspired him to do a series of "mushroom cloud" performances at American sites. "The figures of mushroom clouds," said the artist, "are the visual creation that symbolizes this century, overwhelming all other artistic creation of the time." Cai's subsequent work frequently featured explosions, referencing everything from terrorist bombings to the nuclear apocalypse. Among other objects he displayed a dried *lingzhi*, what Chinese Taoists for many centuries called the mushroom of immortality.[11]

Some who would not approach the bombs directly looked toward them through a screen of irony. Most critics who discussed postmodernism agreed that nuclear fears played a central role in the rise of the ironic mode, and especially its distancing tendencies. A combination of mordant social criticism with pessimism and sheer absurdity in art could be traced back to the Dada movement during the First World War, but it was in the 1960s that "black humor" became a major movement. An example of a leading work in the genre was *Cat's Cradle*, in which Vonnegut joked about everything from religion to third-world dictators, but primarily about an atomic scientist whose invention accidentally and ridiculously destroyed the world. Most influential of all was Joseph Heller's *Catch-22*. Beneath jeers about military insanity, Heller slipped a mordant story of a man who, with classic denial, refused to face the fact of death. "The Cold War is what I was truly talking about," said Heller. Nuclear concerns were still more obvious in the most successful black humor movie, *Dr. Strangelove*. And the greatest painting done in the ironic mode, James Rosenquist's *F-111* of 1965, featured handsome images of consumer goods pasted over images of a warplane and an atomic bomb cloud.[12]

In the twenty-first century the ironic mode flourished more than ever, though in a mood more playful than foreboding. The noted Japanese artist Takashi Murakami created a major traveling exhibit and book, *Little Boy*, around the theme of the nuclear influence on postmodern art. Murakami himself contributed a series of paintings of a mushroom cloud in the shape of a skull, at first glance a ghastly personification of Death. In fact the design was appropriated from a children's cartoon show, where the skull-cloud signaled the death of the villain (but it's okay to laugh, kids, the villain will return in next week's show, resurrected).[13]

In sum, while no doubt many conditions fed the new taste for combining criticism of modern society with images of absurd death, the most obvious influence was nuclear fear. Macabre irony was a step upward from plain denial or despair: most black-humor productions implicitly compared our current society with more rational and kindly possibilities. But these works offered no advice except to grin bitterly and scramble for personal escape. It seemed that, as Vonnegut insisted, "there is nothing intelligent to say about a massacre."[14]

The problem of nuclear weapons was that humans sometimes wished to cause harm—hardly a new problem, but one whose scope had expanded immeasurably. The old answers needed modification to deal with the new scale of danger. Thus the author of one major artistic production that preached a solution, written before the Second World War, revised it extensively in response to the news from Hiroshima. In its prewar version, Bertolt Brecht's play *Galileo* portrayed the scientist as a flawed hero, blamed only for failing to persist as a champion of truth. In Brecht's postwar version the scientist became a complete intellectual traitor, guilty of allowing evil authorities to abuse his discoveries.[15]

It became common to fasten the problem upon scientists. Another widely seen play, Heinar Kipphardt's *In the Matter of J. Robert Oppenheimer*, directly convicted a physicist of betraying society by heedlessly turning over his knowledge to authorities. It was the old warning against Faust's sin of prideful power divorced from moral responsibility. Some serious fictional works portrayed atomic scientists as not only Faustian but warped, for example, unable to approach women normally, as in mad-scientist movies.[16]

Tales of scientists were one of the few modes of literary production relating to nuclear energy that not only continued but expanded after the Cold War's end. Histories and biographies of the American "atomic scientists" of the 1940s flourished; no less than seven significant biographical studies of Oppenheimer appeared in the years 2005–2008. The deepest work came in another medium. The director of the San Francisco Opera asked composer John Adams to prepare a modern version of *Doctor Faustus* in which Oppenheimer would make "a pact with the Devil." Adams took up the idea, he said, because the unleashing of atomic energy "was *the* great mythological tale of our time." Moreover, atomic bombs "dominated the psychic activity of my childhood . . . a source of existential terror" that seemed to be "the ultimate annihilator of any positive emotions or hopes." He was now ready to confront that nihilism. Adams's opera *Doctor Atomic*, premiered in 2005, presented a subtle combination of dozens of nuclear-related themes, interwoven with an exploration of Oppenheimer's enigmatic personality. It was a major work of artistic integration, if a bit too complex and enigmatic for some critics.[17]

Histories and biographies of the German and Soviet atomic scientists also proliferated, and inspired another highly acclaimed and surprisingly popular work of art. Michael Frayn's play *Copenhagen* sensitively investigated Werner Heisenberg's moral role in the Nazis' attempt to build an atomic bomb. Again the work was as complex and ambiguous as the subject itself.[18]

Portraying a scientist as a conflicted Faust evaded the main moral problem. The scholarly and artistic works showed scientists as exceptional people, driven by exotic aspirations. In fact, large majorities of the citizens of every nation that built nuclear weapons were delighted to get them.

Some writers did address general human responsibility. The issue was central, for example, in the only explicitly bomb-related work of high literary quality to achieve huge popular success: *Lord of the Flies.* In William Golding's 1954 novel, schoolboys fleeing atomic bombardment were marooned on an island, where most of them descended into murderous savagery—the primal evil lurking behind all warfare. Such works did not quite despair. For all their bleakness they upheld morality as a conceivable solution, seeing limits to human evil.[19] Was it really plausible that normal adults would conspire to deliberately massacre whole cities of innocents?

There were words to answer that question: Dresden, Tokyo, Leningrad, Hiroshima. And if you saw those as excusable by-products of war there remained a word still harder to dismiss: Auschwitz. Images of nuclear energy, like all modern thought, felt a tidal pull from the systematic extermination of Europe's Jews. It became common for intellectuals to name Hiroshima and Auschwitz in the same sentence. Even the term most often used for the genocide, "Holocaust," was also widely used to describe a future nuclear war. It was not only that the slaughter at Auschwitz was ten times greater than at Hiroshima, more slaughter than in any other single location in history. The real lesson of the Nazi camps was that intelligent and educated adults, looking their innocent victims in the face, could organize torments that until then had been literally inconceivable. Jung rightly called this outbreak of collective sadism an "unmitigated psychic disaster" with far worse repercussions than the destruction of Hiroshima. And the Nazi atrocity was only one of the twentieth century's genocidal horrors.[20]

A few psychologists who studied nuclear fear warned that the world public was refusing to comprehend the human forces that could destroy it. People who tried to understand the situation hesitated when they found that the investigation led them on a frightening journey into themselves. The urge to mayhem was of course a familiar theme as the bond joining the mad scientist with his monster, a theme that hundreds of works associated with "control" of nuclear energy. Most works projected the dangerous feelings outward onto someone else: if not atomic scientists then Cold War generals, corporate evildoers, terrorists, or faraway Russians (or Americans).

A better symbol stood right at hand, for explosions were a traditional symbol of inner destructiveness. In the 1930s Melanie Klein noticed that small children sometimes took literally the idea of "bursting with rage." Such explosion analogies could fit especially well with nuclear energy, yet to my knowledge nobody explored this until some young painters took it up in the 1980s. Back in 1946 the cliché expert Mr. Arbuthnot had hinted at the problem. "The atomic energy in your thumbnail," he told his interviewer, "could, if unleashed, destroy a city." The interviewer exclaimed, "For God's sake, stop, Mr. Arbuthnot! You make me feel like a menace to world security in dire need of control by international authority." The only thing to add would be what Karel Čapek's fictional atomic explosives expert was told back in 1924: if it had not been in you, it would not have been in your invention.[21]

Some works of nuclear art did address the enormity of human evil head-on without losing hope. Though few, they were among the most powerful works of the age. One reason was that these works spoke of redemption through individual love, achieved not by inborn goodwill but through painful effort. A second reason for their power was that they used transmutation imagery.

The imagery appeared clearly in Russell Hoban's critically acclaimed 1980 novel *Riddley Walker*. Young Riddley had to survive as a victim in a brutalized world long after a nuclear war; meanwhile he had to work out his angry and loving relationship to dangerous father figures—which could easily remind readers of nuclear authorities. A similar after-the-bombs novel published in 1985, Denis Johnson's *Fiskadoro*, also took up a boy's relations with father figures alongside broader problems of the loss of reason in the individual and in a degenerated postwar society.

A few sentences can say little about such themes, which entire novels were barely able to unfold, but I can address the symbolism. Both boys embodied the old trope of the innocent child as both victim and survivor. Riddley Walker was preoccupied with rebirth of the spirit, seen as torn asunder by a conflict between scientific "cleverness" and intuition. In his pilgrimage Riddley encountered such symbols as a mutant telepathic community, a mandala laid out across the ruins of England, and even an overweening technologist's personal explosion. The boy in *Fiskadoro* got caught up in still more explicit transmutation symbols: a primitive rite of passage and an apocalyptic religious movement.[22]

The symbolism was buried deeper in a comparable work of the Second Nuclear Age, Cormac McCarthy's 2006 novel *The Road*. McCarthy's after-the-bombs world was one of the darkest ever described, a shattered gray land that lacked even birds—the very seas were covered with ash. The presentation of human nature was equally disastrous. Yet the central figure of the book, another innocent boy deeply engaged with his father, represented the fragile survival of an unquenchable quest for goodness. In the brief final pages the boy, subtly presented as messiah-like, grew up within a reviving family, community, and natural world. It was the old story of rebirth by the grim passage through darkness. Such works showed that many of the transmutation themes associated with nuclear energy could be woven together with serious artistic insight.[23]

Although such works seemed at first to be about world catastrophe, they really centered on individual rebirth through the hard-won ability to love. This move from exterior social problem to interior spiritual problem was structural in one of the best-known artworks with a nuclear theme, Alain Resnais's 1959 film *Hiroshima mon amour*. It opened with grisly reenactments of the city's desolation. Scene by scene the film moved forward until the conclusion seemed like an ordinary story of a sexual liaison amid the bright lights of the rebuilt city. The plot centered on a woman who, during the war in France, had been locked in a dark cellar and descended into madness. Eventually she emerged from that symbolic grave, but it was only when she saw how her lover from Hiroshima had also survived that she could put the past behind her and truly live again.

In transmutation imagery the way to a saving union leads through chaos, the anguish of temporary dissolution. That passage was the subject of the film, which maintained that destruction was essential for spiritual salvation. Three times the woman said to her lover, who forced her to face her memories, "You destroy me. You are good for me."

All these fine novels and films did help people to explore the motives that supported nuclear weapons, and most of them pointed out a path to personal redemption through the great task of love. However, these works went astray when they drew artistic strength from the analogy between individual and social transmutation. Comparing the personal passage through destruction with an entire society's travails was a false analogy: the descent of civilization into chaos is not a desirable step to anything. To find a way forward, people would have to explore the specific structures of technological society.

Art on nuclear themes rarely gave a social analysis as penetrating as its psychological analysis. A few fine early novels, notably *Level 7* with its military hierarchy and *Lord of the Flies* with its tribal savagery, did explore the social roots of organized violence. And they implied a solution: a society structured to respect individuals. But these and similar works were allegories that had no plain connection to actual social organization. Not until the late 1960s did some people look in detail into the relations between nuclear war and the structures of modern civilization, and their attention soon shifted to debates over the environment and nuclear reactors. Would the reactor debate and the other vigorous environmentalist questioning have something new to offer?

Reactors were seldom addressed in the arts, but the 1970s did bring a few poems by significant authors. Like most poets, they believed that people should use more feelings and less abstract logic—the antinuclear stance. Along with a few painters, they drew art out of the antinuclear movement's ideal of spontaneous intuition opposed to mechanical authority.[24] Nuclear power advocates produced nothing like that. But wherever a reactor was built as intended, clean-shaped and efficient, it could be a statement as impressive as any poem.

In works of antinuclear art the cluster of symbolic polarities of the reactor controversy—nature versus culture, feelings versus logic, and all the rest—was usually connected with the cluster of themes that dominated works on nuclear war. The most obvious shared concern involved authority. That concern had the potential for putting problems of social structure, the central question in reactor debates, together with problems of human aggression, the central question in nuclear war. However, most writers on nuclear issues spoke only vaguely about how modern society's organization might be encouraging and organizing destructive forces. Many settled for depicting authority in the person of a wrongheaded scientist or corrupt corporate executive, sidestepping the key societal problems.

The real question was, what sort of transformed society should we seek? The alternatives became clear during the reactor controversy as antinuclear thinkers rejected the vision of a rationally organized White City in favor of Arcadian individualism. That vision was in accord with their goals for personal renewal—the private aspect of transmutation, the victory of human feelings over mechanical logic.

The entire argument drifted into silence in the Second Nuclear Age as people turned to other concerns. The visions on each side seemed impossible to realize, let alone reconcile. Public interest in all environmental questions had declined since the 1980s, even as futuristic technological tales became mere food for satire. Yet a new issue was emerging that would force all the issues back into public view. If the old anxieties about nuclear catastrophe no longer stimulated much worry, the space could be filled by a new anxiety: climate catastrophe.[25]

Novelist Ian McEwan neatly summarized the views of a typical intelligent citizen in the early twenty-first century in his 2010 novel *Solar*. The protagonist "vaguely deplored" global warming, but "he was unimpressed by some of the wild commentary that suggested the world was in peril . . . drought, floods, famine, tempests, unceasing wars for diminishing resources. There was an Old Testament ring to the forewarnings . . . that suggested a deep and constant inclination." He supposed that when the threat of nuclear war had faded away, "the apocalyptic tendency had conjured yet another beast"—nothing that he needed to take very seriously.[26]

In fact climate had not taken the place in the public mind once held by nuclear war, let alone nuclear terrorism. Most Americans ranked global warming near the bottom of their list of worries (people elsewhere ranked it a bit higher). A large section of the public took comfort in a belief that there was no serious risk at all, a claim originally advanced in public relations campaigns funded by fossil-fuel and conservative ideological interests. Moreover, unlike nuclear imagery, climate change did not have many hooks sunk in ancient mythical and psychological themes. Besides, the risk would not become acute for decades. Climate change inspired nothing like the twentieth century's frantic response to nuclear energy, the many hundreds of novels and movies and television productions that had commanded the world's attention.

The world's image makers had failed to come up with convincing pictures of what climate change might realistically mean. The media illustrated news about greenhouse gas emissions with television clips of storm floods or photos of drought-stricken farmland, while political cartoonists sketched a withered desert landscape, whirling tornadoes, or buildings half underwater. These images were limited by their familiarity in connection with ordinary weather problems. More-specific images appeared as actual climate changes began to show up. Since warming was fastest in the Arctic, pictures of ice and glaciers became, perversely, a sign of a climate change story. The images only reinforced a notion that global warming was distant in space and time, a problem for foreigners and future generations. As one critic remarked when reviewing a show of mediocre artistic paintings on climate change, "a far more compelling case" was made by the scientists' plain graph of the rise of global temperature.[27]

As late as 2010 nobody had produced a widely known novel or movie that showed, in a realistic and personalized story, the travails that climate change was predicted to bring, and in some places had already brought: the squalid ruin of forests and coral reefs; deadly heat waves and droughts; torrential downpours; the press of refugees from inundated regions. The first outstanding novelist to attempt a work dealing directly with global warming, Ian McEwan, had to admit that "The best way to tell people about climate change is through non-fiction."[28]

Although fiction and art had done little to propose how society might be reconstructed to meet either the nuclear or the environmental challenge, there were plenty of nonfiction articles and books making the attempt. For both problems, the solutions proposed tended toward the two extremes we have seen in the debate over nuclear reactors. The debate resurfaced in the twenty-first century. Should we hold back global warming by building thousands of nuclear reactors (or yet more visionary devices, say, artificial clouds to block sunlight)? Or was it time to turn our back on all large-scale technology?

Many writers proposed returning to a simpler and presumably more humane way of life. There was much talk of recycling, conserving resources, using local products. Technology, insofar as it was absolutely necessary, would be small-scale, subordinated to local control and genuine human needs. Individuals and communities were exhorted to transform into a less rationalized and regimented state, one less concerned with material goods, more in harmony with the natural world, more spiritual.

An opposing vision aimed to achieve individual and social rebirth not by pulling back but by pushing ahead harder. In writings, speeches, and advertisements, engineers and executives insisted that the advance of technology would bring a more satisfactory life. Liberated from superstition by scientific knowledge, brought to a deeper sympathy with foreign peoples by electronic communications, saved from the degradation and violence that come with material poverty, each of us would become a better person. If you did not notice these statements, you could not overlook the ideology built into the sweeping highways, the glass-fiber web of the Internet woven around the globe, the urban spaces where shimmering towers rose above plazas lined with neat rows of trees.

I find both views of social transmutation valuable but incomplete. In our times, too many writers and artists have weighted the balance heavily toward mystical communion with nature while rejecting rational organization; too many technologists, in their own structures of metal, have weighted the balance heavily in the opposite direction. Our arts have not yet given us an entirely convincing image of a society that will merge Arcadia with the White City—a wealthy society that will not encourage the pursuit of wealth regardless of harm to the ecosystems that sustain us, a society that will not support our cruel desire to identify enemies and destroy them. Yet work toward such a condition has been pursued by many, and not without partial successes.

Of particular interest are some writings of the early twenty-first century about climate change.[29] These authors insisted that a simpler way of life, less preoccupied with the consumption of material goods, was inevitable: if our society did not embrace it, the laws of geophysics would force it upon us. But they were not calling for a retreat from civilization, technology, and rational organization. On the contrary, they explained, we were challenged to engage in a great scientific-technological undertaking. Even as we found satisfaction in orienting our personal lives more toward local communities and genuine human needs, we could join to construct an advanced civilization that did not require ever larger amounts of the dwindling stock of fossil fuels.

Scientific understanding was increasing at a pace even swifter than the rise of population and pollution. Every year brought more efficient and environmentally safer designs for wind turbines, solar power stations, and, yes, nuclear reactors. These devices could be under local surveillance, yet connected in continent-spanning networks to exchange power efficiently. Meanwhile every year brought new devices to enhance work and leisure. If some activities based on fossil fuels or other diminishing resources must be curtailed, visionaries spoke of more satisfying activities. We could walk to visit a neighbor, and return home to chat with a colleague on the far side of the world.

Products of the arts, whether paintings or turbines, could only vaguely point a way forward. There was another medium in which people expressed their feelings and ideas more specifically: political action. There was no work of the emotions and intellect so vast and effective as the interlocking organizations that millions of citizens formed at every level from their neighborhoods to the international stage. The pattern was set in the 1960s through 1980s in the protests against nuclear devices. Decades later global warming attracted the main attention, with protesters standing in front of coal-fired power plants instead of reactors. For they found it increasingly plausible that our profligate use of fossil fuels would turn the Earth into what a leading climate scientist called "a different planet" . . . and a harsher one.[30] Would that risk teach governments to advance sustainable technologies, and to cooperate with one another so they could assure their citizens a good life? Or would the deterioration of the environment drive us back upon the brutish proliferation of nuclear weapons, ever more desperate terrorism, and a renewed danger of the catastrophe of war?

A PERSONAL NOTE

This book has been about the forces of imagery and their pressure upon policies. Every work of history speaks implicitly to current problems, and I feel a responsibility to make plain my personal opinions about what policies we should follow.

In all that I have read I am most impressed by ideas that David Lilienthal offered in the 1960s and later. As head of the AEC and afterward he struggled with nuclear energy fantasies, finally realizing that he and others had abandoned the normal way of dealing with problems. Instead of making piecemeal compromises, "Our obsession with the Atom drove us to seek a Grand Solution." Some people dreamed of doing away with every nuclear bomb under a system of world control, while others wanted to dismay their enemies by building bombs without limit. As for reactors, some people said the solution was to tear them all down regardless of what the alternatives might be, while others said civilization could survive only by basing its energy supply on fission. Lilienthal saw that these extraordinary ideas rested on myth, and he rejected them all in favor of more modest approaches. We will not get where we want to be with one giant step, but with a thousand little ones.[1]

First, energy supply. Much more electricity will be needed before the entire world reaches minimal prosperity. None of the ways to generate this electricity is fully satisfactory. In terms of both my family's health and the health of the environment, I would personally rather live near any existing nuclear reactor than near a plant fired by fossil fuels such as coal. And the new generation of reactors is much safer still, invulnerable to the types of accidents that befell Chernobyl and Fukushima. As for nuclear wastes, for all their problems they are easier to keep from the public than coal wastes and many other dangerous industrial products. The only truly serious problem with a new generation of reactors is that they would offer an opportunity for evil-minded groups or nations to steal or make fearful weapons. But the risk is nowhere near as great as thriller novels pretend it is, and building more reactors will not greatly change the level of the risk.

All that would hold even if global warming did not pose an enormous threat . . . but it does. All of the world's leading science academies and science societies tell us that our current trajectory of burning fossil fuels is very likely to bring major climate disruptions. How can we improve world prosperity while burning less fuel? There are many partial solutions: more-efficient transport and housing, using solar and wind power, capturing carbon emissions, and so forth. But if the future begins to look seriously dangerous—which will most likely happen within the next few decades— we will need "all of the above." That would include building many nuclear power plants. It takes so long to ramp up a power system that the advanced economies should begin *now* to revive their reactor industry, in case we need it to carry a share of the great burden.

A good first step would be to provide a level playing field among the various possible energy sources. To be sure, reactors have large government subsidies, hidden and overt—but fossil fuels enjoy far higher levels of support. Instead of subsidizing fossil-fuel production and use, governments should charge these industries their real costs to society and environment. The many tens of thousands of deaths and millions of health problems caused annually around the world by coal smoke, for example, should be paid for by the plants that emit the smoke.

Better still if all industries could be forced to handle their wastes better. Should not one gram of cancer-causing substances produced from fossil fuels be sequestered as carefully as one gram produced by a nuclear reactor? In practice, it would suffice if governments were less hesitant about establishing long-term nuclear waste repositories and more attentive to halting leaks from coal-ash dumps, oil drilling, natural-gas production, and so forth—so that nuclear wastes were handled, say, only ten times more carefully than their equivalents. Of course such a sensible policy would have to push past the fallacies of nuclear fear, seeing each form of energy clearly in its own right.

Second, war. We must never forget that however thrilling the thought of a terrorist attack may be, there is only one thing aside from climate change that could wreck our entire civilization, and that is full-scale warfare. Tens of thousands of nuclear bombs remain in existence, waiting to fulfill their purpose. Nevertheless these weapons, like reactors, deserve less obsessive and exclusive attention than they have often received. Today a war between advanced nations would be intolerable even if they used only their "conventional" weapons, whose capacity for mayhem has advanced immeasurably. In our age of vulnerable computer systems and precisely targeted missiles, nuclear war is not the most efficient way to threaten an enemy. (It is not even clear that nuclear weapons were ever of any real use except as a talisman in domestic politics.) Whether the world ever eliminates nuclear weapons totally is thus not crucial. But certainly there is no need for any nation to have more than a couple of hundred bombs as a deterrent; the United States and Russia have a long way to go in arms reduction.

Our main business, then, is to understand and uproot the domestic and international causes of war itself. So this book has not been about our real problems at all: it has been about what distracts us from them. It is in non-military areas, as Lilienthal stressed, that we will build up true security, and it is also there that we may find, instead of hostile calculations, the pleasure of productive work.

Our work over the long term should begin with the fact that democratically governed nations do not try to destroy one another.[2] We can be safe not just for a while, but forever, if the current total lack of military threats among democracies can be gradually extended to the rest of the world. Thus the survival of civilization is tied up with the struggle against institutions whose authority rests on force rather than free consent.

That struggle will not be won with nuclear weapons. In fact the more we rely on such weapons, the further we ourselves move from democracy, as we give unelected officials an authority to censor information, to suppress dissent, and even to hold the actual keys to survival. More truly democratic government is also necessary to overcome the distorted distribution of resources promoted by lobbyists for wealthy corporations (notably in the fossil-fuel and armaments industries). The way to safety and prosperity is to work actively toward a more democratic politics, domestically and internationally.

In the meantime we must do something soon about our tendency to possess more weaponry than any rational use could require, while emitting more greenhouse gases than the planet's ecosystems can tolerate. Let me suggest just one step. Let us move a few tens of billions of dollars, a small fraction, from the annual budget for the development and maintenance of weapons and from subsidies to fossil fuels, and devote them instead to improving our system of supplying and using energy. (That research budget has been woefully underfunded for decades.) This proposal may seem too modest, but history shows that research and development do much to determine what eventually comes to pass. Anyone who tries to shift this little lever will learn what tremendous forces rest upon it.

Finally, what of nuclear fear itself? And nuclear hope, which can also go to harmful extremes? I wish we could have a moratorium on the exaggerated images that incite such emotions. Exclaiming that reactors contain astonishing power terrifies the citizen; exclaiming that bombs can lay waste whole nations tempts the citizen to possess more. If any group has been like children playing with matches, it has been those who thought they could evoke fantastic images to good purpose.

The best way to affect imagery is to alter reality. For example, shifting an important fraction of scientific and engineering talent from producing weapons and fossil fuels toward research on better energy sources would bring hope immediately as a gesture, and more hope over decades as the research effort changed our actual situation. We can also work with imagery directly if we approach the task with full respect for truth. Let us tell each new generation in new ways the fact that someday any town in the world might look like central Hiroshima in August 1945. More important, let us spread the word that any town can look like the finest quarters of Paris without destroying the health of the countryside, with nature and civilization, order and freedom, feelings and logic, all working in cooperation. Nuclear energy and its imagery have made familiar the dramatic possibilities for gain and loss that our own human energy can release. In the long run we may find that they have aided us toward the true transmutations we need.

NUCLEAR HISTORY TIMELINE

NOTES

FURTHER READING

INDEX

NUCLEAR HISTORY TIMELINE

1895   Discovery of X-rays.

1898   Discovery of radioactivity.

1902   Discovery of nuclear transmutation.

1934   Artificial radioactivity created.

1936   *The Invisible Ray* (film).

1938   Uranium fission discovered.

1939   Studies show how to build atomic bomb. Second World War begins.

1942   First uranium chain reaction, Chicago. Manhattan Project established.

1945   Trinity atomic bomb test; bombing of Hiroshima and Nagasaki. World War II ends.

1946   Atomic Energy Commission (AEC) created in the United States. *Hiroshima* (book).

1949   Soviet Union tests atomic bomb.

1950   Korean War begins.

1952   United Kingdom tests atomic bomb. United States tests hydrogen fusion device.

1953   USSR tests hydrogen fusion device. Eisenhower launches Atoms for Peace.

1954   United States tests hydrogen bomb; radiation harms *Lucky Dragon* crew; fallout controversy begins. Oppenheimer disgraced.

1955   Peaceful Uses of Atomic Energy International Conference held in Geneva. Soviet Union tests hydrogen bomb.

1956 World's first full-scale nuclear power plant goes online at Calder Hill, UK.

1957 Windscale reactor (UK) catches fire, releases radiation. Soviet Union launches Sputnik artificial satellite. SANE antibomb group founded in United States. Shippingport, first US commercial power reactor, goes online.

1958 Campaign for Nuclear Disarmament launched (UK).

1959 US deploys its first Polaris nuclear missile submarine and first intercontinental ballistic missile. *On the Beach* (film).

1960 France tests atomic bomb.

1962 Kennedy asks Americans to build fallout shelters. Cuban missile crisis.

1963 Limited Test Ban Treaty bans atmospheric tests (US, UK, USSR). Demonstrators oppose proposed Bodega Bay nuclear power plant.

1964 China tests atomic bomb. *Dr. Strangelove* (film).

1968 Non-Proliferation Treaty establishes inspection of peaceful reactors.

1969 Strategic Arms Limitation Treaty (SALT) talks inaugurated.

1970 US deploys Multiple Independently-targetable Reentry Vehicle (MIRV) multiwarhead missiles. Earth Day marks coming-of-age of environmental movement.

1972 US, USSR sign Anti-Ballistic Missile Treaty and SALT I.

1974 India tests atomic bomb. AEC broken up into an energy agency and Nuclear Regulatory Commission (NRC).

1975 *We Almost Lost Detroit* (book).

1977 Protesters occupy site of Seabrook nuclear power plant.

1979 *The China Syndrome* (film). Reactor meltdown at Three Mile Island, Pennsylvania. SALT II treaty signed. Vela satellite detects explosion that points to possession of atomic bombs by Israel and South Africa. Soviet Union invades Afghanistan.

1981 Reagan administration begins arms buildup, provoking Nuclear Freeze campaign. *The Fate of the Earth* (book).

1983   Reagan announces Strategic Defense Initiative to destroy incoming missiles ("Star Wars"). *The Day After* (television film).

1986   Reactor catastrophe at Chernobyl, Ukraine, with widespread contamination. Reykjavik meeting of US president and Soviet premier begins to wind down Cold War.

1987   US, USSR sign Intermediate-Range Nuclear Forces Treaty restricting midrange missiles. Yucca Mountain is chosen for study as sole US nuclear waste repository.

1989   Berlin wall falls, end of Cold War. Attempts to put Shoreham nuclear power plant into service are abandoned.

1991   Gulf War; inspectors find that Iraq had made surprising progress toward atomic bombs. South Africa says it dismantled its weapons. US, USSR sign Strategic Arms Reduction Treaty (START).

1993   Former Soviet Union states Kazakhstan, Ukraine, and Belarus agree to relinquish weapons and nuclear materials in their possession. US, Russia sign START II treaty to further reduce arsenals.

1994   Report of attempt to smuggle weapons-grade uranium out of Russia.

1996   US, UK, Russia, France, China sign Comprehensive Test Ban Treaty.

1998   Second Indian atomic bomb test, quickly followed by Pakistani tests.

2001   Attacks of 9/11 heighten concern about terrorists with nuclear devices.

2003   UN inspectors report suspicions that Iran is developing atomic bombs. US invades Iraq in futile search for "weapons of mass destruction."

2004   Pakistan's Abdul Khan confesses to selling nuclear weapons technology to Iran, Libya, and North Korea.

2006   North Korea tests atomic device.

2007   Intergovernmental panel reports that global warming is very likely. NRC receives first full application for a nuclear power plant since 1979.

2010   Attempt to certify Yucca Mountain as a waste repository is abandoned by US administration, disputed in law courts.

2011   Earthquake and tsunami wreck reactors at Fukushima, Japan, releasing radioactivity; some advanced nations abandon reactor programs.

NOTES

Abbreviations

The following abbreviations are used in the notes.

BAS   *Bulletin of the Atomic Scientists*
DDE   Dwight D. Eisenhower Library, Abilene, Kans.
JCAE Records of the U.S. Congress Joint Committee on Atomic Energy, RG 128, National Archives, Gaithersburg, Md.
MB   Museum of Broadcasting, New York City
NYT   *New York Times*

1. Radioactive Hopes

1. Muriel Howorth, *Pioneer Research on the Atom: Rutherford and Soddy in a Glorious Chapter of Science; The Life Story of Frederick Soddy* (London: New World, 1958), 83–84.

2. "Inexhaustible": Frederick Soddy, "Some Recent Advances in Radioactivity," *Contemporary Review* 83 (May 1903): 708–720. "Dragonfly": Soddy, "The Energy of Radium," *Harper's Monthly* 120 (Dec. 1909): 52–59. "Race which could transmute": Soddy, *The Interpretation of Radium*, 3d ed. (London: Murray, 1912), 251.

3. "World's demand": Soddy, "The Energy of Radium," 58; "Coming struggle": Soddy, "Transmutation, the Vital Problem of the Future," *Scientia* 11 (1912): 199, as quoted in Thaddeus J. Trenn, "The Central Role of Energy in Soddy's Holistic and Critical Approach to Nuclear Science, Economics, and Social Responsibility," *British Journal for the History of Science* 12 (1979): 261–276.

4. W. Kaempffert, "Science Presses On Toward New Goals," NYT *Magazine*, 28 Jan. 1934, pp. 6–7.

5. "Famous problem": John A. Eldridge, *The Physical Basis of Things* (New York: McGraw-Hill, 1934), 330, 333. Ernest Rutherford, *The Newer Alchemy* (Cambridge: Cambridge University Press, 1937).

6. One introduction to the enormous literature on alchemy is Betty Jo Teeter Dobbs, *The Foundations of Newton's Alchemy, or "The Hunting of the Greene*

Lyon" (Cambridge: Cambridge University Press, 1975). See also Carl G. Jung, *Collected Works*, vols. 12–14, trans. R. F. C. Hull (Princeton: Princeton University Press, 1968–1972).

7. Mircea Eliade, *The Forge and the Crucible: The Origins and Structure of Alchemy*, trans. Stephen Corrin (New York: Harper & Row, 1962), 169. "Divine furnace": Evelyn Underhill, *Mysticism: A Study in the Nature and Development of Man's Spiritual Consciousness*, 12th ed. (1930; reprint, New York: Dutton, 1961), 140–148, 221 and chap. 9.

2. Radioactive Fears

1. As reported in *New York Press*, 8 Feb. 1903, and other clippings, box 42, William Hammer Collection, Museum of American History, Smithsonian Institution, Washington, D.C.

2. Soddy to Rutherford, 19 Feb. 1903, Rutherford Papers, Cambridge, microfilm copy at Niels Bohr Library, American Institute of Physics, College Park, Md.

3. Wyn Wachhorst, *Thomas Alva Edison: An American Myth* (Cambridge, Mass.: MIT Press, 1981), 102–103.

4. H. G. Wells, *The World Set Free* (1913; New York: Dutton, 1914), 222.

5. W. Churchill, *Pall Mall*, 24 Sept. 1924, as quoted in Raymond B. Fosdick, *The Old Savage in the New Civilization* (Garden City, N.Y.: Doubleday, 1931), 24–25. Sigmund Freud, *Civilization and Its Discontents*, trans. J. Rivière (London: Hogarth, 1930), 144.

6. *The Times* (London), 11 Nov. 1932, p. 7.

7. George H. Quester, *Deterrence before Hiroshima: The Airpower Background of Modern Strategy* (New York: John Wiley, 1966). For fiction of the period, Bruce H. Franklin, *War Stars: The Superweapon and the American Imagination* (New York: Oxford University Press, 1988).

8. "Lay down their arms": F. W. Parsons, "Stupendous Possibilities of the Atom," *World's Work* 42 (May 1921): 35.

9. Soddy, "Some Recent Advances in Radioactivity," *Contemporary Review* 83 (May 1903): 708–720. W. C. D. Whetham, "Matter and Electricity," *Quarterly Review* no. 397 (Jan. 1904): 126. "Some fool": Whetham to Rutherford, 26 July 1903, Rutherford Papers.

10. Jean-Baptiste Cousin de Grainville, *Le Dernier Homme* (1805; reprint, Geneva: Slatkine, 1976); Mary Shelley, *The Last Man* (1826; reprint, Lincoln: University of Nebraska Press, 1965).

11. Mary Shelley, *Frankenstein; Or, the Modern Prometheus* (London, 1818), chap. 19.

12. E. M. Butler, *The Myth of the Magus* (Cambridge: Cambridge University Press, 1948).

13. Jules Verne, *For the Flag* (1896; reprint, Westport, Conn.: Associated Booksellers, 1961).

14. H. A. Kramers and Helge Holst, *The Atom and the Bohr Theory of Its Structure*, trans. R. B. Lindsay and R. T. Lindsay (New York: Knopf, 1926), 103.

15. "Shambles": Joseph Conrad, *The Secret Agent* (1907; reprint, Garden City, N.Y.: Doubleday, 1958), 303. Anatole France, *Penguin Island*, trans. A. W. Evans (New York: Dodd, Mead, 1909), book 8.

16. Robert Nichols and Maurice Browne, *Wings over Europe: A Dramatic Extravaganza on a Pressing Theme* (1928; reprint, New York: S. French, 1935).

17. Waldemar Kaempffert, *Science Today and Tomorrow*, 2d ser. (New York: Viking, 1945), 266.

18. Ibid., 90–91; see also 73.

19. Daniel Kevles, *The Physicists: The History of a Scientific Community in Modern America* (New York: Knopf, 1983), 180–183.

20. For a review of the vast literature see Daniel Lawrence O'Keefe, *Stolen Lightning: The Social Theory of Magic* (New York: Random House, 1982). Note also Mary Douglas, *Purity and Danger: An Analysis of Concepts of Pollution and Taboo* (Harmondsworth, England: Penguin, 1966).

21. Raymond B. Fosdick, *The Old Savage in the New Civilization* (Garden City, N.Y.: Doubleday, 1931), 23–24.

22. Robert Millikan, *Science and the New Civilization* (New York: Scribner's, 1930), 58–59; see also 94–96, 111–113.

23. "Moonshine": *NYT*, 12 Sept. 1933, p. 1.

24. George Wise, "Predictions of the Future of Technology: 1890–1940," Ph.D. diss., Boston University, 1976, chap. 6.

3. Radium: Elixir or Poison?

1. "Old Age": *Salt Lake City Telegraph*, 6 Nov. 1903; "Secret of Life": *Los Angeles Herald*, 6 Oct. 1903. These and more in William Hammer Collection, Museum of American History, Smithsonian Institution, Washington, D.C., boxes 42–44. Robert A. Millikan, *Science and Life* (Boston: Pilgrim, 1924), 27.

2. Frederick Soddy, *The Interpretation of Radium*, 3d ed. (London: Murray, 1912), 250.

3. Matthew P. Shiel, *The Purple Cloud* (London: Chatto & Windus, 1901).

4. "Secret of Sex": *New York Evening Journal*, 28 Jan. 1904, Hammer Collection, box 43. Autry: *The Phantom Empire*, episode 5 (Mascot, 1935).

5. "Penetrating": quoted in Edmund Morris, *The Rise of Theodore Roosevelt* (New York: Ballantine, 1979), 547.

6. Radium tube: Géza Róheim, *Magic and Schizophrenia* (New York: International Universities Press, 1955), 95–96; see also 110–113.

7. Otto Glasser, *Dr. W. C. Röntgen* (Springfield, Ill.: C. Thomas, 1945), 60.

8. *Adventures of Captain Marvel* (Republic, 1941) from *Whiz Comics* (New York: Fawcett Comics, 1940–1941).

9. Garry Wills, *Reagan's America: Innocents at Home* (Garden City, N.Y.: Doubleday, 1987), 361, 447.

10. For this and much following see Barton C. Hacker, *Elements of Controversy: The Atomic Energy Commission and Radiation Safety in Nuclear Weapons Testing, 1947–1974* (Berkeley: University of California Press, 1994).

11. H. J. Muller, "Artificial Transmutation of the Gene," *Science* 66 (1927): 84–87.

12. *Gehes Codex der Bezeichnungen von Arzneimitteln . . .* , 5th ed. (Dresden: Schwarzeck, 1929), s.v. "Radi-."

13. William D. Sharpe, "The New Jersey Radium Dial Painters: A Classic in Occupational Carcinogenesis," *Bulletin of the History of Medicine* 52 (1979): 560–570.

4. The Secret, the Master, and the Monster

1. Ernest Rutherford, "The Transmutation of the Atom," 13th BBC National Lecture (London: British Broadcasting Corporation, 1933), 25. Waldemar Kaempffert, *NYT Magazine*, 24 May 1936, pp. 6 ff.

2. Carolyn Merchant, *The Death of Nature: Women, Ecology, and the Scientific Revolution* (San Francisco: Harper & Row, 1980). "Interrogated nature:" Jean-Baptiste Cousin de Grainville, *Le Dernier Homme* (1805; Geneva: Slatkine, 1976), 141. Melanie Klein, "On the Theory of Anxiety and Guilt," in Klein et al., *Developments in Psychoanalysis*, ed. Joan Riviere (London: Hogarth, 1952), 271–291.

3. "Les propriétés les plus intimes de la matière": Poincaré et al. to Nobel Prize for Physics Committee, Jan. 1903, Protokoll vol. 3, Swedish Academy of Sciences, Stockholm, p. 134; I thank John Heilbron for this quote and for much else. "Satisfaction": Robert Millikan, *Science and the New Civilization* (New York: Scribner's, 1930), 60.

4. Waldemar Kaempffert, "Science Launches New Attack on the Atom's Citadel," *NYT*, 15 Nov. 1931, sec. 9, p. 4. Kaempffert, "Atomic Energy—Is It Nearer?," *Scientific American* 147 (Aug. 1932): 79–81.

5. Autry: *The Phantom Empire* (Mascot, 1935). Corrigan: *The Undersea Kingdom* (Republic, 1936), episode 1. Atom furnace: *Spaceship to the Unknown* (Universal, 1936). See Douglas Menville and R. Reginald, *Things to Come: An Illustrated History of the Science Fiction Film* (New York: New York Times Books, 1977), 68–72.

6. "Penetrate": Mary Shelley, *Frankenstein* (1818; reprint, New York: Bantam, 1981), 33.

7. "Docile servant": Michael Mok, "Radium: Life-giving Element Deals Death in Hands of Quacks," *Popular Science Monthly* 121, no. 1 (July 1932): 9 ff. Raymond B. Fosdick, *The Old Savage in the New Civilization* (Garden City, N.Y.: Doubleday, 1931), 23–24.

8. *The Master Mystery* (Octagon, 1918), as described in Douglas Menville and R. Reginald, *Things to Come: An Illustrated History of the Science Fiction Film* (New York: New York Times Books, 1977), see 68–72.

9. Karel Čapek, *R.U.R. (Rossum's Universal Robots)*, trans. Paul Selver (Garden City, N.Y.: Doubleday, Page, 1923).

10. A brief introduction is Claude Lévi-Strauss, "The Structural Study of Myth," in Thomas A. Seboek, ed., *Myth: A Symposium* (Bloomington: Indiana University Press, 1958), 81–106.

11. I am grateful to Canaday for an unpublished essay. See also John Canaday, *The Nuclear Muse: Literature, Physics, and the First Atomic Bombs* (Madison: University of Wisconsin Press, 2000); Shelly Chaiken and Yaacov Trope, eds., *Dual-Process Theories in Social Psychology* (New York: Guilford Press, 1999).

12. *The Invisible Ray* (Universal, 1936). Universal Studios, *Advance Publicity* (pressbook), 1936, New York Public Library, Theater Collection, Lincoln Center, New York, n.c. 240.

13. Dario De Martis, "Note sui deliri di negazione," *Rivista sperimentale di Freniatria* 91 (1967): 1119–1143.

14. Karel Čapek, *An Atomic Phantasy: Krakatit*, trans. Lawrence Hyde (London: Allen & Unwin, 1948), 287.

15. *Madame Curie* (Metro-Goldwyn-Mayer, 1944).

16. For an introduction to the history of nuclear physics see Daniel Kevles, *The Physicists: The History of a Scientific Community in Modern America* (New York: Knopf, 1978).

5. The Destroyer of Worlds

1. NYT, 3 Feb. 1939, p. 14; *Scientific American* 161 (1939): 2, 214–216.

2. Leo Szilard, *Leo Szilard: His Version of the Facts*, ed. Spencer Weart and Gertrud Weiss Szilard (Cambridge, Mass.: MIT Press, 1978), 3.

3. Bruce Bliven, "The World-Shaking Promise of Atomic Research," *Reader's Digest* (July 1941): 103–106, from *New Republic* (16 June 1941).

4. Robert Heinlein, "Blowups Happen," *Astounding Science-Fiction* 26, no. 1 (Sept. 1940): 51–85; see John W. Campbell Jr., editorial on pp. 5–6.

5. "Angus MacDonald" [Robert Heinlein], "Solution Unsatisfactory," *Astounding Science-Fiction* 27, no. 3 (May 1941): 56–86.

6. "V-3?," *Time* 44 (27 Nov. 1944): 88. Ken Tachikawa, "San Francisco Tukitobu," *Shin-Seinen*, July 1944, pp. 52–64, as described by Maika Nakao, "The Image

of the Atomic Bomb in Japan before Hiroshima," *Historia Scientiarum* 19 (2009): 119–131, quotation on 128.

7. Barton C. Hacker, *The Dragon's Tail: Radiation Safety in the Manhattan Project, 1942–1946* (Berkeley: University of California Press, 1987).

8. Alice Kimball Smith and Charles Weiner, *Robert Oppenheimer: Letters and Recollections* (Cambridge, Mass.: Harvard University Press, 1980), 250.

9. "Vital war plant": Interim Committee Minutes, 31 May 1945, reprinted with other useful documents in Robert C. Williams and Philip L. Cantelon, eds., *The American Atom: A Documentary History of Nuclear Policies from the Discovery of Fission to the Present* (Philadelphia: University of Pennsylvania Press, 1984), 62. See Michael J. Hogan, ed., *Hiroshima in History and Memory* (New York: Cambridge University Press, 1996), for discussion of historiography. Historians have reached a consensus that the "Truman administration used it [the bomb] primarily for military reasons but also hoped that an additional result would be increased diplomatic power" (ibid., p. 17). Also J. Samuel Walker, *Prompt and Utter Destruction: President Truman and the Use of Atomic Bombs against Japan* (Chapel Hill: University of North Carolina Press, 1997); Walker, "History, Collective Memory, and the Decision to Use the Bomb," in *Hiroshima in History and Memory*, ed. Michael J. Hogan (New York: Cambridge University Press, 1996), 187–199.

10. A. H. Compton to James B. Conant, 15 Aug. 1944; for this and what follows see Alice Kimball Smith, *A Peril and a Hope: The Scientists' Movement in America, 1945–47*, rev. ed. (Cambridge, Mass.: MIT Press, 1970).

11. K. K. Darrow to R. S. Mullikan, 4 June 1945, "Nucleonics" folder, Metallurgical Laboratory records, Argonne National Laboratory, Argonne, Ill.

12. For this and the following see William L. Laurence, oral history interviews by Louis M. Starr, 1956–1957, and by Scott Bruns, 1964, Columbia University Library, New York.

13. William L. Laurence, *Dawn over Zero: The Story of the Atomic Bomb*, 2d ed. (Westport, Conn.: Greenwood, 1972), xiii, 116.

14. Lansing Lamont, *Day of Trinity* (New York: Atheneum, 1965), 235–236. For another version see Laurence, *Dawn over Zero*, 191.

15. Farrell's report is in Martin Sherwin, *A World Destroyed: The Atomic Bomb and the Grand Alliance* (New York: Knopf, 1975), app. P, 312–313.

16. W. Laurence, *Men and Atoms: The Discovery, the Uses, and the Future of Atomic Energy* (New York: Simon & Schuster, 1959), 117–120; Bhagavad-Gita, XI.

17. W. Laurence, oral history interview, 319.

18. "Husky": George L. Harrison, quoted in Richard G. Hewlett and Oscar E. Anderson Jr., *A History of the United States Atomic Energy Commission*. Vol. 1, *The New World 1939/1946* (Washington, D.C.: U.S. Atomic Energy Commission, 1962), 386; Truman Journal, 25 July 1945, reported in *NYT*, 2 June 1980,

p. A14; Churchill recollected by Harvey H. Bundy, "Remembered Words," *The Atlantic* 199, March 1957, p. 57.

19. Specifically, equivalent destruction would have required 210 sorties, and 120 for Nagasaki: John Mueller, *Atomic Obsession: Nuclear Alarmism from Hiroshima to Al-Qaeda* (New York: Oxford University Press, 2009), 10.

6. The News from Hiroshima

1. Leslie R. Groves, *Now It Can Be Told: The Story of the Manhattan Project* (New York: Harper & Row, 1962), 327–328.

2. "Atom Bomb" (Paramount newsreel, Aug. 1945), text courtesy of Public Information Office, Argonne National Laboratory, Argonne, Ill.

3. "Darkening heavens": War Dept. release in Henry D. Smyth, *Atomic Energy for Military Purposes: The Official Report . . .* (Princeton: Princeton University Press, 1946), app. 6.

4. H. V. Kaltenborn, NBC radio, 6 Aug. 1945, R78:0345, MB.

5. Frank Sullivan, "The Cliché Expert Testifies on the Atom," in Sullivan, *A Rock in Every Snowball* (Boston: Little, Brown, 1946), 28–36.

6. William L. Laurence in "The Quick and the Dead" (NBC radio, 1950), on RCA Victor records, copy at Niels Bohr Library & Archives, American Institute of Physics, College Park, Md. Pea: Sullivan, "Cliché Expert," 31.

7. Edward A. Shils, *The Torment of Secrecy: The Background and Consequences of American Security Policies* (Glencoe, Ill.: Free Press, 1956), 71.

8. See Paul Boyer, *By the Bomb's Early Light: American Thought and Culture at the Dawn of the Atomic Age* (New York: Pantheon, 1985). Norman Cousins, "Modern Man Is Obsolete," *Saturday Review* 28 (18 Aug. 1945): 1; book version (New York: Viking, 1945).

9. Robert Jay Lifton, *Death in Life: Survivors of Hiroshima* (New York: Simon & Schuster, 1967), 486. Comparable experiences elsewhere: Kai T. Erikson, *A New Species of Trouble: Explorations in Disaster, Trauma, and Community* (New York: Norton, 1994), 226–242 and passim.

10. "Atomic plague": Peter Burchett, *London Daily Express*, 5 Sept. 1945; see Wilfrid Burchett, *Passport: An Autobiography* (Melbourne: T. Nelson, 1969), 120, 162–176.

11. Jessica Stern, *The Ultimate Terrorists* (Cambridge, Mass.: Harvard University Press, 1999), 32, 35–37; Eddie Harmon-Jones and Piotr Winkielman, eds., *Social Neuroscience: Integrating Biological and Psychological Explanations* (New York: Guilford Press, 2007).

12. Dan Jones, "The Depths of Disgust," *Nature* 447 (2007): 768–771; Paul Rozin et al., "From Oral to Moral," *Science* 323 (2009): 1179–1180.

13. Hersey, *Hiroshima* (1946; reprint, New York: Bantam, 1959), 89. Aldous Huxley, *Ape and Essence* (New York: Harper, 1948).

14. "Supernatural": Harry S. Hall, "Scientists and Politicians," *BAS* (Feb. 1956), reprinted in *The Sociology of Science*, ed. Bernard Barber and Walter Hirsch (New York: Free Press of Glencoe, 1962), 269–287.

15. "Guilty men": *Time* 46 (5 Nov. 1945): 27. "Touched very deeply": J. Robert Oppenheimer, "Atomic physics in civilization," manuscript, box 29, Bulletin of Atomic Scientists Collection, University of Chicago Library. "Known sin": Oppenheimer, *The Open Mind* (New York: Simon & Schuster, 1955), 88.

16. "I'm a Frightened Man," *Collier's* 117 (5 Jan. 1946): 18 ff.

17. "Action-goading fear": *NYT*, 26 May 1946, p. 7.

18. Kai Bird and Martin Sherwin, *American Prometheus: The Triumph and Tragedy of J. Robert Oppenheimer* (New York: Vintage, 2006), 349.

19. Bertrand Russell in British Broadcasting Corp., *Atomic Challenge: A Symposium* (London: Winchester, 1947), 155. Sullivan, "Cliché Expert," 32.

20. "Oppie's plan": Herbert Marks in Daniel Lang, *Early Tales of the Atomic Age* (Garden City, N.Y.: Doubleday, 1948), 102.

21. Fonda: SANE Education Fund, "Shadows of the Nuclear Age: American Culture and the Bomb" (WGBH-FM broadcast and cassettes, 1980), cassette 1.

22. Soviet quotes: *Current Digest of the Soviet Press* 2, no. 30 (1950): 27; 2, no. 29 (1950): 39; 2, no. 28 (1950): 465.

23. Joseph Alsop and Stewart Alsop, "Your Flesh *Should* Creep," *Saturday Evening Post* 219 (13 July 1946): 49.

24. Leslie R. Groves, *Now It Can Be Told: The Story of the Manhattan Project* (New York: Harper & Row, 1962), 415, 438–439; Stephane Groueff, *Manhattan Project: The Untold Story of the Making of the Atomic Bomb* (Boston: Little, Brown, 1967), 3, 31–32.

25. *NYT* index, Oct. 1945. Radio: Executive Office of the President, Division of Press Intelligence, "Atomic Energy," 22 April–25 July, 1947, folder "Radio & Press References," box 7, JCAE.

26. Scott Shane, "Cold War Nuclear Fears Now Apply to Terrorists," *NYT*, 15 April 2010; Robert E. Hunter, "Expecting the Unexpected: Nuclear Terrorism in 1950s Hollywood Films," in *The Atomic Bomb and American Society: New Perspectives*, ed. Rosemary B. Mariner and G. Kurt Piehler (Knoxville: University of Tennessee Press, 2009), 211–237.

27. David Caute, *The Great Fear: The Anti-Communist Purge under Truman and Eisenhower* (New York: Simon & Schuster, 1978), chap. 5 and p. 541.

28. *Seven Days to Noon* (British Lion, 1950): Sullivan, "Cliché Expert," 36.

29. Stewart Alsop, introduction to Ralph Lapp, *The New Force: The Story of Atoms and People* (New York: Harper, 1953), ix. Parliament: Margaret M. Gowing

with Lorna Arnold, *Independence and Deterrence: Britain and Atomic Energy, 1945–1952*, 2 vols. (New York: St. Martin's Press, 1974), 1: 52.

30. "Torn from nature": "The Story of Five Bombs" (U.S. Department of Defense, 1946?), text in Public Information Office, Argonne National Laboratory, Argonne, Ill. "Probe": *Congressional Record* 83:1, vol. 1C (1953): 239, as quoted in Gerald J. Ringer, "The Bomb as a Living Symbol: An Interpretation," Ph.D. diss., Florida State University, 1966, 116–117.

31. See Robert Jay Lifton, *The Broken Connection: On Death and the Continuity of Life* (New York: Simon & Schuster, 1979), 354–357.

32. Gary Wills, *Bomb Power: The Modern Presidency and the National Security State* (New York: Penguin, 2010), 1.

7. National Defenses

1. Frank Sullivan, "The Cliché Expert Testifies on the Atom," in Sullivan, *A Rock in Every Snowball* (Boston: Little, Brown, 1946), 34. For a general history 1945–2005 see Dee Garrison, *Bracing for Armageddon: Why Civil Defense Never Worked* (New York: Oxford University Press, 2006).

2. Philip Wylie, *Tomorrow!* (New York: Rinehart, 1954). "We have taught": Wylie in Reginald Bretnor, ed., *Modern Science Fiction: Its Meaning and Its Future* (New York: Coward-McCann, 1953), 240.

3. "Criminally stupid": Ralph Lapp, "An Interview with Governor Val Peterson," *BAS* 9, no. 7 (Sept. 1953): 241.

4. "Dangerous reductions": Lambie to Sherman Adams, 9 July 1953, folder "Candor (1)," box 12, White House Central Files, Confidential File, DDE.

5. "War games": for example, Minutes of Cabinet Meeting, 10 June 1955, box 5, Ann Whitman Cabinet Files, DDE. See the cabinet minutes for spring and summer of 1956; Wilson quote in 13 July 1956, box 7.

6. *Survival under Atomic Attack* (Washington, D.C.: U.S. Government Printing Office, 1950), 30; film: (Castle Films, 1951).

7. "Please don't let them": SANE Education Fund, "Shadows of the Nuclear Age: American Culture and the Bomb" (WGBH-FM broadcast and cassettes, 1980), cassette 4.

8. "Calf crop": "Atomic Bomb—Operation Crossroads" (CBS radio, 28 May 1946), transcript S76:0502, p. 12, MB.

9. Irving L. Janis, *Air War and Emotional Stress: Psychological Studies of Bombing and Civilian Defense* (New York: McGraw-Hill, 1951), 239.

10. "Supernatural . . . heads in the sand": Rensus Lickert on "You and the Atom" (CBS radio, 30–31 July 1946), R76:0223–0224, MB.

11. Bernard Brodie, ed., *The Absolute Weapon* (New York: Harcourt, Brace, 1946), 74. On strategy debates (omitted here) see Fred Kaplan, *The Wizards of*

Armageddon (New York: Simon & Schuster, 1983); Lawrence Freedman, *The Evolution of Nuclear Strategy*, 3d ed. (New York: Palgrave Macmillan, 2003).

12. Jessica Stern, *The Ultimate Terrorists* (Cambridge, Mass.: Harvard University Press, 1999), 43.

13. P. M. S. Blackett, *Fear, War, and the Bomb* (New York: McGraw-Hill, 1949). "Blow a hole"; "not a military weapon": David E. Lilienthal, *The Journals of David Lilienthal*, 4 vols. (New York: Harper & Row, 1964–1969), 2: 473–474, 391.

14. "Really delightful": Stanley A. Blumberg and Gwinn Owens, *Energy and Conflict: The Life and Times of Edward Teller* (New York: Putnam's, 1976), 119.

15. Enrico Fermi and I. I. Rabi, in Herbert F. York, *The Advisors: Oppenheimer, Teller and the Superbomb* (San Francisco: W. H. Freeman, 1976), app.

16. "Frankenstein": Lilienthal, *Journals*, 2: 581.

8. Atoms for Peace

1. Drew Pearson, ABC radio, 1 Jan. 1950, in folder "Broadcasts—Pearson," box 106, JCAE.

2. C. D. Jackson to Walter B. Smith, 10 Nov. 1953, folder "OCB—Misc. Memos (2)," box 1, Jackson Records, DDE.

3. "Scare the country": probably a paraphrase of Jackson's recollection, John Lear, "Ike and the Peaceful Atom," *The Reporter* 14, no. 1 (12 Jan. 1956): 11. "Bang-bang": Jackson to Lewis Strauss, folder "Atoms for Peace," box 5, Ann Whitman Administration Files, DDE.

4. William Laurence, "Paradise or Doomsday?," *Woman's Home Companion* 75 (May 1948): 33.

5. David E. Lilienthal, *The Journals of David Lilienthal*, 4 vols. (New York: Harper & Row, 1964–1969), 2: 16–17, 635; Lilienthal, *Change, Hope, and the Bomb* (Princeton: Princeton University Press, 1963), 23.

6. Richard G. Hewlett and Francis Duncan, *Atomic Shield, 1947/1952* (University Park: Pennsylvania State University Press, 1969), 435–438.

7. Report by Atomic Energy Commission, 6 March 1953, folder "NSC 145," box 4, White House Office of the Special Assistant for National Security Affairs, NSC Series, Policy Papers, DDE.

8. "To the general public": Leonard S. Cottrell Jr. and Sylvia Eberhart, *American Opinion on World Affairs in the Atomic Age* (Princeton: Princeton University Press, 1948), 36. "Restricted to the upper": Elizabeth Douvan and Stephen Withey, "Public Reaction to Nonmilitary Aspects of Atomic Energy," *Science* 119 (1954): 1–3.

9. Stefan Possony, "The Atoms for Peace Program," in F. L. Anderson Panel, "Psychological Aspects of United States Strategy: Source Book . . .," Nov. 1955,

folder "Rockefeller (5)," box 61, White House Central Files, Confidential Files, DDE, p. 203.

10. "Aladdins Wunderlampe": Karl Winnacker, *Nie den Mut verlieren: Erinnerungen* . . . (Düsseldorf: Econ, 1971), 311–312; "priest": Laura Fermi, *Atoms for the World: United States Participation in the Conference on the Peaceful Uses of Atomic Energy* (Chicago: University of Chicago Press, 1957), 64.

11. Lilienthal, *Change, Hope*, 111–112.

12. Memo of conference, 14 Jan. 1955, folder "AEC 1955–56 (8)," box 5, Ann Whitman Administration File, DDE.

13. Lewis L. Strauss, *Men and Decisions* (Garden City, N.Y.: Doubleday, 1952), 4, 429. "Beneficent use": Strauss, remarks at Rockhurst College, Mo., 24 May 1955, fiche SPCH-1, AEC speeches, Public Document Room, Nuclear Regulatory Commission, Washington, D.C., p. 30.

14. Strauss, remarks for National Association of Science Writers, New York City, 16 Sept. 1954; my thanks to George Mazuzan for a copy of this AEC press release.

15. Victor Cohn in Sharon Friedman, *Science in the Newspaper*, no. 1 (Washington, D.C.: American Association for the Advancement of Science, 1974), 21.

16. Francis K. McCune, "Atomic Power—A Challenge to U.S. Leadership," *General Electric Review*, Nov. 1955, p. 10.

17. L. W. Cronkhite in Atomic Industrial Forum, *Atomic Energy, a Realistic Appraisal. Proceedings of a Meeting* . . . (New York: AIF, 1955), 1.

18. James M. Lambie Jr. to Sherman Adams, 6 Oct. 1954, folder "Atomic Industrial Forum," box 11, Lambie Papers, DDE.

19. *Our Friend the Atom* (Walt Disney, 1956). Heinz Haber, *The Walt Disney Story of Our Friend the Atom* (New York: Simon & Schuster, 1956).

20. "Good Atoms": George L. Glasheen, "What Schools Are Doing in Atomic Energy Education," *School Life* 35 suppl. (Sept. 1953): 153.

9. Good and Bad Atoms

1. William Laurence, *Dawn over Zero: The Story of the Atomic Bomb*, 2d ed. (1946; reprint, Westport, Conn.: Greenwood, 1972), 254.

2. Ruth Ashton, "The Sunny Side of the Atom" (CBS radio, 30 June 1947), transcript in box 22, Federation of Atomic Scientists Collection, University of Chicago Library.

3. *The Atom Comes to Town* (U.S. Chamber of Commerce, 1957).

4. Piston rings: AEC Press Release no. 153, 28 Jan. 1949, Records of the AEC, Germantown, Md.

5. Claude Lévi-Strauss, "The Structural Study of Myth," in *Myth: A Symposium*, ed. Thomas A. Seboek (Bloomington: Indiana University Press, 1958), 81–106. "Good & Bad Atoms," *Time* 49 (31 March 1947): 81.

6. "Hard core": Lebaron Foster, "Public Thinking on the Peacetime Atom," in Atomic Industrial Forum, *Public Relations for the Atomic Industry: Proceedings of a Meeting . . .* (New York: AIF, 1956), 85.

7. "Humanité de plus en plus mécanisé": Charles-Noël Martin, *Promesses et menaces de l'énergie nucléaire* (Paris: Presses Universitaires de France, 1960), 250.

8. Eisenhower's message to 1955 Geneva Atoms for Peace Conference, repeated by Richard Nixon at the 1971 conference, in International Conference on the Peaceful Uses of Atomic Energy, Fourth, *Peaceful Uses of Atomic Energy; Proceedings,* 3 vols. (New York: United Nations, 1972), 1: 86.

9. "Dangerous to touch": Burton R. Fisher, C. A. Metzner, and B. J. Darsky, *Peacetime Uses of Atomic Energy,* 2 vols. (Ann Arbor: Survey Research Center, University of Michigan, 1951), 2: 12–16; see also 25–28. Workers: Joseph Blank, "Atomic Tragedy in Texas," *Look* 21 (3 Sept. 1957): 25–29; George T. Mazuzan and J. Samuel Walker, *Controlling the Atom: The Beginnings of Nuclear Regulation, 1946–1962* (Berkeley: University of California Press, 1984), 327–332. Mazuzan and Walker generously shared the drafts of this book with me. Glowing man: for example, *The Atomic Kid* (Mickey Rooney Productions, 1954).

10. Wouter Poortinga and Nick F. Pidgeon, "Exploring the Dimensionality of Trust in Risk Regulation," *Risk Analysis* 23 (2003): 961–972; William R. Freudenberg, "Risky Thinking: Facts, Values and Blind Spots in Societal Decisions About Risks," *Reliability Engineering and System Safety* 72 (2001): 125–130.

11. Heinar Kipphardt, *In the Matter of J. Robert Oppenheimer,* trans. Ruth Speirs (1964; reprint, New York: Hill and Wang, 1968), 127.

10. The New Blasphemy

1. Here and in the following see Robert A. Divine, *Blowing on the Wind: The Nuclear Test Ban Debate, 1954–1960* (New York: Oxford University Press, 1978).

2. Great Britain, Commons, *Debates* 315 (1 March 1955): 1895.

3. "The U.N. in Action" (CBS-TV, 17 March 1953), T77:0329, MB.

4. *Asahi,* 17 March 1954, trans. in folder "Weapons Tests 1954," box 712, JCAE.

5. Estes Kefauver, *NYT,* 17 Oct. 1956, p. 1; Nikita Khrushchev, *NYT,* 31 May 1957, p. 8. *The Day the Earth Caught Fire* (British Lion; scripted 1954, produced 1961).

6. Mary Douglas and Aaron Wildavsky, *Risk and Culture: An Essay on the Selection of Technical and Environmental Dangers* (Berkeley: University of California Press, 1982).

7. A handy compendium is Paul R. Baker, ed., *The Atomic Bomb: The Great Decision*, rev. ed. (Hinsdale, Ill.: Dryden, 1976). See Robert Lifton and Greg Mitchell, *Hiroshima in America: A Half Century of Denial* (New York: Avon, 1996); J. Samuel Walker, "History, Collective Memory, and the Decision to Use the Bomb," in *Hiroshima in History and Memory*, ed. Michael J. Hogan (New York: Cambridge University Press, 1996), 187–199.

8. Joseph Alsop and Stewart Alsop, 18 Jan. 1950, quoted in Norman Moss, *Men Who Play God: The Story of the H-Bomb and How the World Came to Live with It* (New York: Harper & Row, 1968), 33. William Randolph Hearst, *Los Angeles Herald Express*, 16 March 1954. Pius XII in *NYT*, 19 April 1954, p. 12.

9. Robert Jay Lifton, *Death in Life: Survivors of Hiroshima* (New York: Simon & Schuster, 1967), 110 and passim.

10. Stuart Galbraith IV, *Monsters Are Attacking Tokyo: The Incredible World of Japanese Fantasy Films* (Venice, Calif.: Feral House, 1998), 49–50.

11. "Guard dog": Lawrence R. Tancredi, *Hardwired Behavior: What Neuroscience Reveals about Morality* (New York: Cambridge University Press, 2005), 34.

12. Antonio R. Damasio, *Descartes' Error: Emotion, Reason, and the Human Brain* (New York: Avon, 1994).

13. Paul Slovik, *The Perception of Risk* (London: Earthscan, 2000), xxxii; see also chap. 26.

14. Greg J. Stephens et al., "Speaker-Listener Neural Coupling Underlies Successful Communication," *Proceedings of the National Academy of Sciences* 107 (2010): 14425–14430.

15. Karl K. Szpunar et al., "Neural Substrates of Envisioning the Future," *Proceedings of the National Academy of Sciences* 104 (2007): 642–647.

16. Eliezer Yudkowsky, "Cognitive Biases Potentially Affecting Judgement of Global Risks," in *Global Catastrophic Risks*, ed. Nick Bostrom and Milan M. Cirkovic (New York: Oxford University Press, 2008), 91–119, quotation on 103. Reading: Deborah A. Prentice and Richard J. Gerrig, "Exploring the Boundary between Fiction and Reality," in *Dual-Process Theories in Social Psychology*, ed. Shelly Chaiken and Yaacov Trope (New York: Guilford Press, 1999), 529–546.

17. Anyone doubtful of this should at once read Sigmund Freud, *The Psychopathology of Everyday Life* (1901). A recent review of the experiments is Ruud Custers and Henk Aarts, "The Unconscious Will: How the Pursuit of Goals Operates Outside of Conscious Awareness," *Science* 329 (2010): 47–50; for a popularization see Dan Ariely, *Predictably Irrational: The Hidden Forces That Shape Our Decisions*, rev. ed. (New York: Harper, 2009.), esp. chap. 1.

18. Susan Sontag, "The Imagination of Disaster," in Sontag, *Against Interpretation and Other Essays* (New York: Delta, 1966), 208–225.

19. Yuki Tanaka, "Godzilla and the Bravo Shot: Who Created and Killed the Monster?," *Japan Focus* (2005), online at http://www.japanfocus.org/products /topdf/1652.

20. John Brosnan, *Future Tense: The Cinema of Science Fiction* (New York: St. Martin's Press, 1978), 95.

21. "Prone to terror": review of *The Magnetic Monster, New York Herald Tribune*, 14 May 1953; like other reviews I found this in the New York Public Library, Theater Collection, Lincoln Center, New York. "It's radioactive!": for example, *The Crawling Eye*, alternate title *The Trollenberg Terror* (Eros, 1958).

11. Death Dust

1. Robert Gilpin, *American Scientists and Nuclear Weapons Policy* (Princeton: Princeton University Press, 1962).

2. James J. Orr, *The Victim as Hero: Ideologies of Peace and National Identity in Postwar Japan* (Honolulu: University of Hawaii Press, 2001). "Guinea pigs": Robert Jay Lifton, *Death in Life: Survivors of Hiroshima* (New York: Simon & Schuster, 1967), 512.

3. See George T. Mazuzan and J. Samuel Walker, *Controlling the Atom: The Beginnings of Nuclear Regulation 1946–1962* (Berkeley: University of California Press, 1984), chap. 2.

4. *U.S. News & World Report* 38 (25 March 1955): 21–26.

5. "Face the Nation" (CBS-TV and radio, 19 June 1955), transcript in folder "Broadcasts—general," box 106, JCAE.

6. Paul Slovik, *The Perception of Risk* (London: Earthscan, 2000); Jonathan Haidt, "The New Synthesis in Moral Psychology," *Science* 316 (2007): 998–1002; Dan M. Kahan et al., "Cultural Cognition of Scientific Consensus," *Journal of Risk Research* 14 (2011): 147–174. See also Chapter 19.

7. John B. Martin, *Adlai Stevenson and the World* (Garden City, N.Y.: Doubleday, 1977), 373.

8. Letters in folders "155-B Sept. 1956 (1,2)," box 1215, White House Central Files, General File, DDE.

9. David Lilienthal to Carroll L. Wilson, 26 May 1958, folder "Lilienthal," Wilson Papers, Massachusetts Institute of Technology Archives, Cambridge, Mass.

10. *NYT*, 18 May 1957, p. 2.

11. Paul Slovik, "Perception of Risk," *Science* 236 (1987): 280–285, reprinted in Slovik, *Perception of Risk*, chap. 13; Jon Palfreman, "A Tale of Two Fears: Exploring Media Depictions of Nuclear Power and Global Warming," *Review of Policy Research* 23 (2006): 23–43.

12. Slovik, *Perception of Risk*, 323, referring to J. Lichtenberg and D. MacLean (1992), "Is Good News No News?," *Geneva Papers on Risk and Insurance* 17: 362–365.

13. For a summary of this and similar public relations matters see J. Flynn, "Nuclear Stigma," in *The Social Amplification of Risk*, ed. Nick Pidgeon et al. (Cambridge: Cambridge University Press, 2003), 326–354. On reactions to testing into the 1970s see A. Constandina Titus, *Bombs in the Backyard: Atomic Testing and American Politics*, 2d ed. (Reno: University of Nevada Press, 2001); Barton C. Hacker, *Elements of Controversy: The Atomic Energy Commission and Radiation Safety in Nuclear Weapons Testing, 1947–1974* (Berkeley: University of California Press, 1994).

14. Edward Teller with Allen Brown, *The Legacy of Hiroshima* (Garden City, N.Y.: Doubleday, 1962), 81–91; Teller, "We're Going to Work Miracles," *Popular Mechanics* 113 (March 1960): 97 ff.

15. Teller and Brown, *Legacy of Hiroshima*, 56. *Crack in the World* (Security-Paramount, 1965). "Digging too deep": C. M. Kornbluth, "Gomez," in *A Treasury of Great Science Fiction*, 2 vols., ed. Anthony Boucher (Garden City, N.Y.: Doubleday, 1959), 1: 305.

16. Walter R. Guild in John M. Fowler, ed., *Fallout: A Study of Superbombs, Strontium-90, and Survival* (New York: Basic Books, 1960), 91.

17. John Bowlby, *Attachment and Loss*, vol. 3: *Loss: Sadness and Depression* (New York: Basic Books, 1980), chap. 4; Paul Kline, *Fact and Fantasy in Freudian Theory* (London: Methuen, 1972), 181–182, 355.

18. "The Contaminators," *Playboy* 6, no. 10 (Oct. 1959): 38. See advertisements in *NYT*, 5 July 1962, p. 54; 18 April 1962, p. 26.

19. Herblock [Herbert Block], *Washington Post*, 24 Oct. 1961, p. A14. Snow: Benjamin Spock, "Do Your Children Worry about War?," *Ladies' Home Journal* 79, no. 8 (Sept. 1962): 48.

12. The Imagination of Survival

1. *Public Papers of the Presidents of the United States: John F. Kennedy, 1961* (Washington, D.C.: Government Printing Office, 1962), 625.

2. "Vienna appeal" quoted in Committee for the Compilation of Materials on Damage Caused by the Atomic Bombs, *Hiroshima and Nagasaki: The Physical, Medical, and Social Effects of the Atomic Bombings*, trans. Eisei Ishikawa and David L. Swan (New York: Basic Books, 1981), 577.

3. "Ten tons": White House press release, 21 Jan. 1964.

4. Nevil Shute [Nevil Shute Norway], *On the Beach* (New York: William Morrow, 1957).

5. Edward Teller with Allen Brown, *The Legacy of Hiroshima* (Garden City, N.Y.: Doubleday, 1962), 239.

6. Ibid., 241.

7. James J. Hughes, "Millennial Tendencies in Response to Apocalyptic Threats," in *Global Catastrophic Risks*, ed. Nick Bostrom and Milan M. Cirkovic (New

York: Oxford University Press 2008), 73, 84; Eliezer Yudkowsky, "Cognitive Biases Potentially Affecting Judgement of Global Risks," ibid., 114.

8. John M. McCullough, *Atomic Energy: Utopia or Oblivion?* (Philadelphia: The Inquirer, 1947), 27.

9. "Good habits": quoted in Mikiso Hane, *Peasants, Rebels, and Outcastes: The Underside of Modern Japan* (New York: Pantheon, 1982), 72; see also 36.

10. Richard Rafael in *Astounding Science-Fiction* 27 (May 1941), quoted in Paul A. Carter, *The Creation of Tomorrow: Fifty Years of Magazine Science Fiction* (New York: Columbia University Press, 1977), 242, see also 231–233, 241–244.

11. Anna Freud, *The Ego and the Mechanisms of Defense*, rev. ed., in *The Writings of Anna Freud*, vol. 2 (New York: International Universities Press, 1966), 170–171.

12. *You Can Beat the A Bomb* (RKO, 1950).

13. Pat Frank, *Alas, Babylon* (1959; reprint, New York: Bantam, 1980).

14. Walter M. Miller Jr., *A Canticle for Leibowitz* (Philadelphia: Lippincott, 1959).

15. Michael Ortiz Hill, *Dreaming the End of the World: Apocalypse as a Rite of Passage* (Dallas: Spring Publications, 1994).

16. Mick Broderick, "Rebels *with* a Cause: Children versus the Military Industrial Complex," in *Youth Culture in Global Cinema*, ed. Timothy Shary and Alexandra Seibel (Austin: University of Texas Press, 2006), chap. 4; Hill, *Dreaming*, 59. On all these matters see also Jerome F. Shapiro, *Atomic Bomb Cinema: The Apocalyptic Imagination on Film* (New York: Routledge, 2002).

17. Karl Menninger, *Man against Himself* (1938; reprint, New York: Harcourt Brace Jovanovich, 1966), 180; see also pt. 2, passim.

18. Donald N. Michael, "The Psychopathology of Nuclear War," *BAS* 18, no. 5 (May 1962): 28–29.

19. Philip Wylie, *Tomorrow!* (New York: Rinehart, 1954); Wylie, *Triumph* (Garden City, N.Y.: Doubleday, 1963).

20. Philip Wylie, "Blunder: A Story of the End of the World," *Collier's* 117 (12 Jan. 1946): 11–12, 63–64.

21. Truman F. Keefer, *Philip Wylie* (Boston: Twayne, 1977), 19, 127, and passim.

22. SANE Education Fund, "Shadows of the Nuclear Age: American Culture and the Bomb" (WGBH-FM broadcast and cassettes), 1980, cassette 8.

23. Stephen B. Withey, *4th Survey of Public Knowledge and Attitudes Concerning Civil Defense* (Ann Arbor: Survey Research Center, University of Michigan, 1954), 72.

24. SANE Education Fund, "Shadows."

13. The Politics of Survival

1. Christopher Driver, *The Disarmers: A Study in Protest* (London: Hodder & Stoughton, 1964); Lawrence S. Wittner, *Rebels against War: The American Peace Movement, 1941–1960* (New York: Columbia University Press, 1969).

2. Campaign for Nuclear Disarmament, *The Bomb and You*, pamphlet, n.d. "Act Now or Perish!" quoted in Norman Moss, *Men Who Play God: The Story of the H-Bomb and How the World Came to Live with It* (New York: Harper & Row, 1968), 182. See David Boulton, ed., *Voices from the Crowd: Against the H-Bomb* (Philadelphia: Dufour, 1964).

3. Bertrand Russell, *The Autobiography of Bertrand Russell: 1872–1914* (Boston: Little, Brown, 1967), 3–4, 220–221, and passim; Ronald W. Clark, *The Life of Bertrand Russell* (New York: Knopf, 1976), 84–86, 264.

4. Frank Parkin, *Middle Class Radicalism: The Social Bases of the British Campaign for Nuclear Disarmament* (Manchester: Manchester University Press, 1968).

5. Midge Decter, "The Peace Ladies," *Harper's* 226 (March 1963): 48–53. Whether women are inherently more fearful than men is moot, but their greater social tendency to display fearfulness is well documented.

6. Parkin, *Middle Class Radicalism*, 58–59.

7. J. B. Priestley, *Instead of the Trees* (London: Heinemann, 1977), 85–87.

8. George Clark, quoted in Driver, *Disarmers*, 126; see also 128.

9. One entry to this large and important topic is Frederic J. Baumgartner, *Longing for the End: A History of Millennialism in Western Civilization* (New York: St Martin's Press, 1999).

10. See Gerald J. Ringer, "The Bomb as a Living Symbol: An Interpretation," Ph.D. diss., Florida State University, 1966.

11. After publishing this in 1988 I learned that unknown to historians, a few international relations experts had independently made this discovery. See Spencer Weart, *Never at War: Why Democracies Will Not Fight One Another* (New Haven: Yale University Press, 1998).

12. Herman Kahn, *On Thermonuclear War*, 2d ed. (Princeton: Princeton University Press, 1961); see 145 ff.

13. Moss, *Men Who Play God*, 198. Susan T. Fiske, Felicia Pratto, and Mark A. Pavelchak, "Citizens' Images of Nuclear War: Content and Consequences," *Journal of Social Issues* 19 (1983): 41–65.

14. J. B. Priestley, "Sir Nuclear Fission," *BAS* 11, no. 8 (Oct. 1955): 293–294. Priestley, *The Doomsday Men: An Adventure* (London: Heinemann, 1938).

15. Robert Wallace, "A Deluge of Honors for an Exasperating Admiral," *Life* 45 (8 Sept. 1958), 104 ff.; "Talk of sex": J. Robert Moskin, "Polaris," *Look* 25 (29 Aug. 1961): 17–31.

16. "More machine than man": W. B. Huie, "A-Bomb General of Our Air Force," *Coronet* 28 (Oct. 1950): 89. "Irrefutable logic": Ernest Havemann, "Toughest Cop of the Western World," *Life* 36 (14 June 1954): 136.

17. Margot W. Henriksen, *Dr. Strangelove's America: Society and Culture in the Atomic Age* (Berkeley: University of California Press, 1997).

18. Richard B. Stolley, "How It Feels to Hold the Nuclear Trigger," *Life* 57 (6 Nov. 1964): 34–41. Max Born, "What Is Left to Hope For?," *BAS* 20 (April 1964): 4.

14. Seeking Shelter

1. "Discussion . . . footing": Guy Oakes, *The Imaginary War: Civil Defense and American Cold War Culture* (New York: Oxford University Press, 1994), 160. "Awaken . . . business": Charles Haskins to McGeorge Bundy, 21 Feb. 1961, folder "Civil Defense," box 295, National Security Files, John F. Kennedy Library, Boston (hereafter JFK). In general see Kenneth D. Rose, *One Nation Underground: The Fallout Shelter in American Culture* (New York: New York University Press, 2001).

2. Drafts, 24–25 July 1961, folder "Berlin Speech," box 60, Theodore Sorenson Papers, JFK.

3. Copy of bank advertisement in folder "ND 2–3, 9–10/61," box 598, Central Subject Files, JFK.

4. On the shelter debate see Arthur I. Waskow and Stanley L. Newman, *America in Hiding* (New York: Ballantine, 1962).

5. Twilight Zone, "The Shelter" (CBS-TV, 29 Sept. 1961). Every man for himself ("*Sauve-qui-peut*"): Arthur Schlesinger Jr., "Reflections on Civil Defense," folder "Civil Defense 12/61," box 295, National Security Files, JFK.

6. *Time* 78 (20 Oct. 1961): 25.

7. Alice L. George, *Awaiting Armageddon: How Americans Faced the Cuban Missile Crisis* (Chapel Hill: University of North Carolina Press, 2003), 153.

8. Robert F. Kennedy, *Thirteen Days* (New York: W. W. Norton, 1969), 79, 87–90, 98, 180.

9. McNamara in *The Fog of War: Eleven Lessons from the Life of Robert S. McNamara* (film directed by Errol Morris, 2003).

10. George, *Awaiting Armageddon*, 160–165.

11. Michael Scheibach, *Atomic Narratives and American Youth: Coming of Age with the Atom, 1945–1955* (Jefferson, N.C.: McFarland, 2003).

12. Sibylle K. Escalona, "Growing Up with the Threat of Nuclear War: Some Indirect Effects of Personality Development," *American Journal of Orthopsychiatry* 52 (1982): 600–607; see John Mack, "The Perception of US-Soviet Intentions and Other Psychological Dimensions of the Nuclear Arms Race," pp. 590–599 of the same issue.

13. Robert Liebert, *Radical and Militant Youth: A Psychoanalytic Inquiry* (New York: Praeger, 1971), 234.

14. "The Face of the Future," *Look* 29, no. 1 (12 Jan. 1965): 73.

15. William Abbott Scott, "The Avoidance of Threatening Material in Imaginative Behavior," in *Motives in Fantasy, Action, and Society*, ed. John W. Atkinson (Princeton: Van Nostrand, 1958), 572–585. Robert Jay Lifton and Richard Falk, *Indefensible Weapons: The Political and Psychological Case against Nuclearism* (New York: Basic Books, 1982).

16. B. Ashem, "The Treatment of a Disaster Phobia by Systematic Desensitization," *Behavior Research and Therapy* 1 (1963): 81–84.

17. Bertrand Russell, "My View of the Cold War," reprinted in David Boulton, ed., *Voices from the Crowd: Against the H-Bomb* (Philadelphia: Dufour, 1964), 142–145, quote on p. 143.

15. Fail/Safe

1. On all these matters see Paul Bracken, *The Command and Control of Nuclear Forces* (New Haven: Yale University Press, 1983).

2. Charles Yulish to Arthur Sylvester, 4 Feb. 1968, folder "Airplane Crash in Spain (Correspondence)," JCAE.

3. Eugene Burdick and Harvey Wheeler, *Fail-Safe* (New York: McGraw-Hill, 1962).

4. Peter Bryant [Peter Bryan George], *Two Hours to Doom* (London: Clark Boardman, 1957), published in the United States as *Red Alert* (New York: Ace, 1958).

5. Robert Brustein, *New York Review of Books*, 6 Feb. 1964, pp. 3–4.

6. Michael Ortiz Hill, *Dreaming the End of the World: Apocalypse as a Rite of Passage* (1994; reprint, Putnam, Conn.: Spring Publications, 2004).

7. "Toy": John Brosnan, *Future Tense: The Cinema of Science Fiction* (New York: St. Martin's Press, 1978), 163.

8. Vannevar Bush, *Endless Horizons* (Washington, D.C.: Public Affairs Press, 1946), 105.

9. C. Rogers McCullough, Mark M. Mills, and Edward Teller, "The Safety of Nuclear Reactors," International Conference on the Peaceful Uses of Nuclear Energy A/CONF.8/P/853, July 1955; *NYT*, 11 Aug. 1955, p. 11.

10. Edward Teller, "Reactor Hazards Predictable," *Nucleonics* 11, no. 11 (Nov. 1953): 80.

11. Interview with Henry Hurwitz, 1980.

12. Westinghouse Company, *Infinite Energy*, pamphlet, quoted in Stephen Hilgartner, Richard C. Bell, and Rory O'Connor, *Nukespeak: Nuclear Language, Visions, and Mindset* (San Francisco: Sierra Club, 1982), 190.

13. Steven L. Del Sesto, *Science, Politics, and Controversy: Civilian Nuclear Power in the United States, 1946–1974* (Boulder, Colo.: Westview, 1979), 122–135.

14. Atomic Industrial Forum statement, 14 Nov. 1955, in U.S. Congress, 84:2, Joint Committee on Atomic Energy, *Peaceful Uses of Atomic Energy: Background Material for the Report of the Panel on the Impact of the Peaceful Uses . . .*, vol. 2 of McKinney Panel report (Washington, D.C.: Government Printing Office, 1956), 599.

15. Edward Teller to Clifford Beck, 9 Jan. 1957, folder "Reactor Safety Project—AEC," Leland Haworth Papers, Brookhaven National Laboratory, microfilm in Niels Bohr Library, American Institute of Physics, College Park, Md.

16. "Fallible": David Lilienthal, *Change, Hope, and the Bomb* (Princeton: Princeton University Press, 1963), 101–103, 106–107. Lilienthal, *The Journals of David E. Lilienthal*, 4 vols. (New York: Harper & Row, 1964–1969), 3: 21; "orgasms": 4: 165, 491.

17. David E. Lilienthal, *Atomic Energy: A New Start* (New York: Harper & Row, 1980), 39.

16. Reactor Promises and Poisons

1. Rodney Southwick to Joseph Fouchard, 10 March 1962, folder "IR&A6, Reg. PG&E—Bodega Bay," box 8330, AEC Secretariat files, Nuclear Regulatory Commission archives, Washington, D.C.

2. Jessie V. Coles to the Dairymen of Sonoma and Marin Counties, 25 March 1963, folder "IR&A6, Reg. PG&E—Bodega Bay," box 8330, AEC Secretariat files, NRC archives.

3. For all this see George T. Mazuzan and J. Samuel Walker, *Controlling the Atom: The Beginnings of Nuclear Regulation, 1946–1962* (Berkeley: University of California Press, 1984), 358.

4. Pare Lorentz, "The Fight for Survival," *McCall's* 84, no. 4 (Jan. 1957): 29, 73–74. Robert Rienow and Leona Train Rienow, *Our New Life with the Atom* (New York: Crowell, 1958), 35–36, 100–102, 142, 160. *The Giant Behemoth* (Allied Artists, 1959). *Providence Journal*, 19 July 1959, as quoted in Mazuzan and Walker, *Controlling the Atom*, 358.

5. "Crapped up": Paul Loeb, *Nuclear Culture: Living and Working in the World's Largest Atomic Complex* (New York: Coward, McCann & Geoghegan, 1982), 12. Association test on *déchets* ("wastes"): Christine Blanchet in *Colloque sur les implications psycho-sociologiques du développement de l'industrie nucléaire*, ed. M. Tubiana (Paris: Société Française de Radioprotection, 1977), 225.

6. Melanie Klein, *The Psycho-Analysis of Children*, trans. Alix Strachey, rev. with H. A. Thorner (New York: Dell, 1975), 129, 145, 239. For anality and aggression see Seymour Fisher and Roger P. Greenberg, *The Scientific Credibility of Freud's Theories and Therapy* (New York: Basic Books, 1977), 154–158. Leo

Szilard, *Leo Szilard: His Version of the Facts*, ed. Spencer Weart and Gertrud Weiss Szilard (Cambridge, Mass.: MIT Press, 1978), 185.

7. World Health Organization, "Mental Health Aspects of the Peaceful Uses of Atomic Energy: Report of a Study Group," Technical Report Series no. 151 (Geneva: WHO, 1958), 14.

8. Rep. Craig Hosmer, 17 Feb. 1969, in David Okrent, "On the History of the Evolution of Light Water Reactor Safety in the United States" (ca. 1979), copy at Niels Bohr Library, American Institute of Physics, College Park, Md., p. 2-411.

9. David E. Lilienthal, *Atomic Energy: A New Start* (New York: Harper & Row, 1980), 73. Alvin M. Weinberg, "Social Institutions and Nuclear Energy," *Science* 177 (1972): 27–34.

10. Glenn T. Seaborg, "Large-Scale Alchemy: Twenty-fifth Anniversary at Hanford-Richland" (1968), in Seaborg, *Nuclear Milestones: A Collection of Speeches* (San Francisco: W. H. Freeman, 1972), 166.

11. Peter Bradford, 7 Oct. 1982, speech to Environmental Defense Fund Associates, New York, as quoted by John Byrne and Steven M. Hoffman, eds., *Governing the Atom: The Politics of Risk* (New Brunswick, N.J.: Transaction, 1996), 53.

17. The Debate Explodes

1. See Ernest J. Yanarella, *The Missile Defense Controversy: Strategy, Technology, and Politics, 1955–1972* (Lexington: University Press of Kentucky, 1977); Anne Hessing Cahn, *Eggheads and Warheads: Scientists and the ABM* (Cambridge, Mass.: MIT Science and Public Policy Program, 1971).

2. "Anywhere except": quoted in Joel Primack and Frank von Hippel, *Advice and Dissent: Scientists in the Political Arena* (New York: Basic Books, 1974), 194n26.

3. Student quoted by Edwin S. Shneidman, *Deaths of Man* (New York: Quadrangle, 1973), 185. NYT, 24 March 1969, p. 12.

4. McNamara address, 18 Sept. 1967, in Ralph E. Lapp, *The Weapons Culture* (New York: W. W. Norton, 1968), app. 12. Volker R. Berghahn, *Militarism: The History of an International Debate, 1861–1979* (Cambridge: Cambridge University Press, 1984), chap. 5.

5. Ernest J. Sternglass, "The Death of All Children," *Esquire*, Sept. 1969, 1a–1d.

6. Ralph E. Lapp, *The Radiation Controversy* (Greenwich, Conn.: Reddy, 1979).

7. Leslie J. Freeman, *Nuclear Witnesses: Insiders Speak Out* (New York: W. W. Norton, 1981), 76.

8. Ibid., 114.

9. John Gofman and Arthur Tamplin, *"Population Control" through Nuclear Pollution* (Chicago: Nelson-Hall, 1970).

10. "Powers That Be" (KNBC-TV, 18 May 1971), transcript and other materials in folder "TV," box 7809, AEC Secretariat Files, Nuclear Regulatory Commission archives, Washington, D.C.

11. Ronnie D. Lipschutz, *Radioactive Waste: Politics, Technology, and Risk* (Cambridge, Mass.: Ballinger, 1980), chap. 4.

12. "Tas de merde," quoted in Colette Guedeney and Gérard Mendel, *L'Angoisse atomique et les centrales nucléaires* (Paris: Payot, 1973), 234. Dieter Rucht, *Von Wyhl nach Gorleben: Bürger gegen Atomprogramme und nukleare Entsorgung* (Munich: Beck, 1977), 127.

13. "Infection": William S. Maynard et al., "Public Values Associated with Nuclear Waste Disposal," Report BNWL-1997 (Seattle, Wash.: Battelle Human Affairs Research Centers, 1976), 173. Peter Faulkner, ed., *The Silent Bomb: A Guide to the Nuclear Energy Controversy* (New York: Random House, 1977), ix, 118–119, and chap. 9.

14. *Doomwatch* (British Film Productions, 1972); *Empire of the Ants* (AIP, 1977).

15. "The Plutonium Connection," PBS, 9 March 1975.

16. David E. Lilienthal, *Atomic Energy: A New Start* (New York: Harper & Row, 1980), 22–23.

17. Amitai Etzioni and Clyde Nunn, "The Public Appreciation of Science in Contemporary America," *Daedalus* 130, no. 3 (Summer 1974): 191–205. "Unspoken fear": Hillier Krieghbaum, *Science, the News and the Public* (New York: New York University Press, 1958), 30.

18. Walter A. Rosenbaum, *The Politics of Environmental Concern* (New York: Praeger, 1973), 63–71.

19. Barry Commoner, *The Closing Circle: Nature, Man, and Technology* (New York: Knopf, 1971), 52; see also 65. "Nature was forever": Frank Graham Jr., *Since Silent Spring* (Boston: Houghton Mifflin, 1970), 13–14. Rachel Carson, *Silent Spring* (New York: Houghton Mifflin, 1962), 14, 18; see also 187–190.

20. "Grim flavor": McKinley C. Olson, "Reacting to the Reactors," *The Nation* 220 (11 Jan. 1975): 15–17.

18. Energy Choices

1. Robert Heinlein, "Let There Be Light," in Heinlein, *The Man Who Sold the Moon* (New York: New American Library, 1951).

2. "Ideally suited": Amory Lovins, "Energy Strategy: The Road Not Taken?" *Foreign Affairs* 55 (Oct. 1976): 89.

3. A summary of what was understood at the time is Richard Wilson et al., *Health Effects of Fossil Fuel Burning: Assessment and Mitigation* (Cambridge, Mass.: Ballinger, 1980).

4. National Academy of Sciences, *Hidden Costs of Energy* (Washington, D.C.: National Academies Press, 2009), online at http://www.nap.edu/catalog.php ?record_id=12794.

5. U.S. Nuclear Regulatory Commission, "Reactor Safety Study: An Assessment of Accident Risks in U.S. Commercial Nuclear Power Plants" (Rasmussen Report), WASH-1400, NUREG 75/014 (Washington, D.C.: NRC, 1975).

6. See Spencer Weart, *The Discovery of Global Warming*, 2d ed. (Cambridge, Mass.: Harvard University Press, 2008); and for more details Weart, "The Public and Climate Change," http://www.aip.org/history/climate/public.htm.

7. "Far cleaner": Harvey Wasserman, *Energy War: Reports from the Front* (Westport, Conn.: Lawrence Hill, 1979), 225.

8. "Il est politique": Alain Touraine et al., *La Prophétie antinucléaire* (Paris: Seuil, 1980), 71; see also 70–79, 153.

9. "Elders . . . unreliable": Robert K. Musil, "Growing Up Nuclear," *BAS* 38, no. 1 (Jan. 1982): 19. Survey: Sybille K. Escalona, "Children and the Threat of Nuclear War," in *Behavioral Science and Human Survival*, ed. Milton Schwebel (Palo Alto, Calif.: Science and Behavior Books, 1965), 201–209.

10. Glenn Seaborg in *NYT*, 10 June 1969, p. 63.

11. E. F. Schumacher, *Small Is Beautiful: Economics As If People Mattered* (London: Blond & Briggs, 1973).

12. Amory Lovins, "Energy Strategy: The Road Not Taken?," *Foreign Affairs* 55 (Oct. 1976): 65–96, on 93. Robert Jungk, *The New Tyranny: How Nuclear Power Enslaves Us*, trans. Christopher Trump (New York: Grosset and Dunlap, 1979).

13. "Choice": Wasserman, *Energy War*, xii. On all this see Dorothy Nelkin and Michael Pollak, *The Atom Besieged: Extraparliamentary Dissent in France and Germany* (Cambridge, Mass.: MIT Press, 1981).

14. Dieter Rucht, *Von Wyhl nach Gorleben: Bürger gegen Atomprogramme und nukleare Entsorgung* (Munich: Beck, 1977).

19. Civilization or Liberation?

1. Roger E. Kasperson, "The Social Amplification of Risk: Progress in Developing an Integrative Framework," in *Social Theories of Risk*, ed. S. Krimsky and D. Golding (Westport, Conn.: Praeger, 1992), 153–178; Nick Pidgeon et al., eds., *The Social Amplification of Risk* (Cambridge: Cambridge University Press, 2003).

2. Todd R. La Porte and Daniel Metlay, "Technology Observed: Attitudes of a Wary Public," *Science* 188 (1975): 121–127; Mark P. Lovington and Robert G. Horne, "Project on Public Images of Nuclear Power and Technical Advance," Thesis no. 78C0051 (Worcester, Mass.: Worcester Polytechnic Institute, 1978);

John A. Meyer III, "1986 Nuclear Opinion Study," Thesis no. 122JMW0321 (Worcester, Mass.: Worcester Polytechnic Institute, 1986); and many other polls. Affective response: Paul Slovik, *The Perception of Risk* (London: Earthscan, 2000), 405–406.

3. "Revolutionary": Joe Shapiro, "The Anti-nuclear Movement," *Science for the People* 12, no. 4 (July–Aug. 1980): 16–21.

4. "Earth raped": Suzanne Gordon, "From Earth Mother to Expert," *Nuclear Times* 1, no. 7 (May 1983): 13–16. "Patriarchy": Mary Daly, *Gyn/Ecology* (Boston: Beacon Press, 1978), as quoted in Dorothy Nelkin, "Nuclear Power as a Feminist Issue," *Environment* 23 (1981): 8.

5. *The Plutonium Incident*, produced by Time-Life (CBS-TV, 1980).

6. Shelly Chaiken and Yaacov Trope, eds., *Dual-Process Theories in Social Psychology* (New York: Guilford Press, 1999).

7. Reuben M. Baron and Stephen J. Misovich, "On the Relationship between Social and Cognitive Modes of Organization," ibid., 586–605.

8. See Richard Olson, *Science Deified and Science Defied: The Historical Significance of Science in Western Culture*, vol. 1: *From the Bronze Age . . . to ca. A.D. 1640* (Berkeley: University of California Press, 1982).

9. Alain Touraine et al., *La Prophétie anti-nucléaire* (Paris: Seuil, 1980), 40–41, 66–67, 73, 93–95, 201, 206–207.

10. William R. Freudenberg, "Risk and Recreancy: Weber, the Division of Labor, and the Rationality of Risk Perceptions," *Social Forces* 71 (1993): 909–932.

11. Sarah Lichtenstein et al., "Judged Frequency of Lethal Events," *Journal of Experimental Psychology: Human Learning and Memory* 4 (1978): 551–578; Baruch Fischhoff, Paul Slovic, and Sarah Lichtenstein, "'The Public' vs. 'The Experts': Perceived vs. Actual Disagreements about Risks of Nuclear Power," in *The Analysis of Actual versus Perceived Risks*, ed. Vincent T. Covello et al. (New York: Plenum, 1983), 235–249.

12. Steven A. Sloman, "The Empirical Case for Two Systems of Reasoning," *Psychological Bulletin* 119 (1996): 3–22. A recent summary: Elke U. Weber, "Experience-Based and Description-Based Perceptions of Long-Term Risk: Why Global Warming Does Not Scare Us (Yet)," *Climatic Change* 77 (2006): 103–120.

13. William L. Rankin and Stanley M. Nealey, *The Relationship of Human Values and Energy Beliefs to Nuclear Power Attitudes* (Seattle: Battelle Memorial Institute Human Affairs Research Centers, 1978), 18.

14. Slovik, *Perception of Risk*, xxxiii and chap. 25.

15. Stephen C. Whitfield et al., "The Future of Nuclear Power: Value Orientations and Risk Perception," *Risk Analysis* 29 (2009): 425–437.

16. William R. Freudenberg, "Risky Thinking: Facts, Values and Blind Spots in Societal Decisions About Risks," *Reliability Engineering and System Safety* 72 (2001): 125–130, citing Freudenberg, "Risk and Recreancy."

17. Jon D. Miller and Kenneth Prewitt, *A National Survey of the Non-Governmental Leadership of American Science and Technology* (De Kalb: Northern Illinois University Public Opinion Laboratory, 1982), 61–64.

18. On songwriters and nuclear weapons in the 1950s and 1960s see the essays by Richard Aquila and Joseph C. Ruff in Alison M. Scott and Christopher D. Geist, eds., *The Writing on the Cloud: American Culture Confronts the Atomic Bomb* (Lanham, Md.: University Press of America, 1997).

19. Richard Curtis and Elizabeth Hogan, *Perils of the Peaceful Atom: The Myth of Safe Nuclear Power Plants* (Garden City, N.Y.: Doubleday, 1969), xiii.

20. As seen in *NYT Magazine*, 5 July 1981, 2.

21. Hans Heinrich Ziemann, *The Accident* (New York: St. Martin's Press, 1979); Ron Kytle, *Meltdown* (New York: McKay, 1976); Bett L. Pohnka and Barbara C. Griffin, *The Nuclear Catastrophe* (Port Washington, N.Y.: Ashley, 1977).

22. Lauriston S. Taylor, "Some Nonscientific Influences on Radiation Protection Standards and Practice," *Health Physics* 39 (1980): 868.

23. S. Robert Lichter, Stanley Rothman, and Linda S. Lichter, *The Media Elite* (Bethesda, Md.: Adler & Adler, 1986), 166–167, 178–184, 216–217.

24. Roger E. Kasperson et al., "Public Opposition to Nuclear Energy: Retrospect and Prospect," in *Sociopolitical Effects of Energy Use and Policy*, ed. Charles T. Unseld et al. (Washington, D.C.: National Academy of Sciences/National Research Council, 1979), 261–292. See Slovik, *Perception of Risk*, 397 and chap. 25.

25. *Superman and Spiderman* (New York: Warner, 1981).

26. Christine Blanchart in *Colloque sur les implications psycho-sociologiques du développement de l'industrie nucléaire*, ed. M. Tubiana (Paris: Société Française de Radioprotection, 1977), 226–227.

27. Philip L. Cantelon and Robert C. Williams, *Crisis Contained: The Department of Energy at Three Mile Island* (Carbondale: Southern Illinois University Press, 1982), chaps. 3–5, Cronkite quoted on 58.

28. *NYT*, 23 April 1980, p. 1; 19 April 1981, p. 6E.

29. http://news.google.com/archivesearch, searching on "nuclear reactor" or "nuclear power" and applying normalization to account for the archives' greater coverage of complex news items as time moves forward (although a search on "dog or cat" shows fairly constant coverage from the 1940s, a search on some arbitrarily chosen longer words, "economics or phenomenon or technology or biology or crisis," shows a rise from ca. 1980 forward). Searches conducted in 2008–2010.

20. Watersheds

1. *Hiroshima Nagasaki 1945* (Erik Barnouw, 1970); see Jack Gould, *NYT*, 4 Aug. 1970.

2. "Finite pool": Elke U. Weber, "Experience-based and Description-based Perceptions of Long-Term Risk: Why Global Warming Does Not Scare Us (Yet)," *Climatic Change* 77 (2006): 111. See Patricia W. Linville and Gregory W. Fischer, "Preferences for Separating or Combining Events," *Journal of Personality and Social Psychology* 60 (1991): 5–23; on evidence for individual mental "space" for worries, e.g., Joseph I. Constans et al., "Stability of Worry Content in GAD Patients: A Descriptive Study," *Journal of Anxiety Disorders* 16 (2002): 311–319.

3. Dorothy Nelkin, "Anti-nuclear Connections: Power and Weapons," *BAS* 37, no. 4 (April 1981): 36–40. Helen Caldicott, *Nuclear Madness: What You Can Do!* (New York: Bantam, 1980), 61. For many aspects of the 1980s see Allan M. Winkler, *Life under a Cloud: American Anxiety about the Atom* (Urbana: University of Illinois Press, 1993), chap. 8.

4. Alistair MacLean, *Goodbye California* (Garden City, N.Y.: Doubleday, 1978). Larry Collins and Dominique Lapierre, *The Fifth Horseman* (New York: Simon & Schuster, 1980). Robert Ludlum, *The Parsifal Mosaic* (New York: Random House, 1982), 616.

5. Jonathan Schell, *The Fate of the Earth* (New York: Knopf, 1982). Dr. Seuss [Theodor Geisel], *The Butter Battle Book* (New York: Random House, 1984).

6. For all events of this period see Lawrence Freedman, *The Evolution of Nuclear Strategy*, 3d ed. (New York: Palgrave Macmillan, 2003).

7. James E. Dougherty and Robert L. Pfaltzgraff Jr., eds., *Shattering Europe's Defense Consensus: The Antinuclear Protest Movement and the Future of NATO* (Washington, D.C.: Pergamon-Brassey's, 1985).

8. Janet Morris, ed., *Afterwar* (New York: Baen, 1985), 8, 12.

9. William Beardslee and John Mack, "The Impact on Children and Adolescents of Nuclear Developments," in *Psychosocial Aspects of Nuclear Developments* (Arlington, Va.: American Psychiatric Association, 1982), 64–93.

10. "The Defense of the United States" (CBS-TV, 14 June 1981); "The Day After" (ABC-TV, 20 Nov. 1983). For more see Kim Newman, *Apocalypse Movies: End of the World Cinema* (New York: St. Martin's Griffin, 2000).

11. On the novels see Paul Brians, *Nuclear Holocausts: Atomic War in Fiction, 1895–1984* (Kent, Ohio: Kent State University Press, 1987), revised online at http://www.wsu.edu/~brians/nuclear/.

12. Lawrence Badash, *A Nuclear Winter's Tale: Science and Politics in the 1980s* (Cambridge, Mass.: MIT Press, 2009). For current understanding: Alan

Robock and Owen Brian Toon, "Local Nuclear War, Global Suffering," *Scientific American* 302 (2010): 74–81.

13. Another example, *Testament* (Paramount, 1983), was frequently compared with *On the Beach* but in fact avoided the latter's shallow, romantic resignation. For the early 1980s see Paul Boyer, *By the Bomb's Early Light: American Thought and Culture at the Dawn of the Atomic Age* (New York: Pantheon, 1985), 361–367.

14. Churchill in Great Britain, Commons, *Debates* 537 (1 March 1955), 1902.

15. "PR splash": Alexander Haig as quoted in E. P. Thomson, ed., *Star Wars* (New York: Pantheon, 1985), 12; see chap. 1 passim.

16. David E. Hoffman, *Dead Hand: The Untold Story of the Cold War Arms Race and Its Dangerous Legacy* (New York: Doubleday, 2009), 90–92 and passim.

17. Interview with Pat Perkins by Orville Butler, 13 Dec. 2005, in Niels Bohr Library and Archives, American Institute of Physics, College Park, Md.; Harrison Brown, "Draw the Line at Star Wars," *BAS* 43, no. 1 (Jan.–Feb. 1987): 3.

18. Edward Tabor Linenthal, *Symbolic Defense: The Cultural Significance of the Strategic Defense Initiative* (Urbana: University of Illinois Press, 1989), 115.

19. Chernobyl Forum, *Chernobyl's Legacy: Health, Environmental and Socio-Economic Impacts*, 2d rev. ed. (Vienna: International Atomic Energy Agency, 2005), online at http://www.iaea.org/Publications/Booklets/Chernobyl/chernobyl.pdf.

20. *Newsweek* 107, no. 19 (12 May 1986): 40.

21. *New Yorker* 62, no. 12 (12 May 1986): 29; *NYT*, 15 May 1986, p. A10.

22. The Worldwatch Institute *Vital Signs* annual is a good source for reactor statistics.

23. James Mahaffey, *Atomic Awakening: A New Look at the History and Future of Nuclear Power* (New York: Pegasus, 2009), 320.

24. John Lewis Gaddis, *The Long Peace: Inquiries into the History of the Cold War* (New York: Oxford University Press, 1987), 231. See also Richard Ned Lebow, "The Long Peace, the End of the Cold War, and the Failure of Realism," *International Organization* 48 (1994): 249–277.

21. The Second Nuclear Age

1. Joseph V. Rees, *Hostages of Each Other: The Transformation of Nuclear Safety since Three Mile Island* (Chicago: University of Chicago Press, 1994), 1, 44, 117.

2. Constance Perin, *Shouldering Risks: The Culture of Control in the Nuclear Power Industry* (Princeton: Princeton University Press, 2005).

3. Statement at 2003 Berlin meeting of World Association of Nuclear Operators, quoted in ibid., x.

4. A 2009 poll found 70 percent of scientists in favor and 27 percent opposed (compared with 42 percent of the public opposed). Pew Research Center for the People & the Press, "Public Praises Science; Scientists Fault Public, Media," 9 July 2009, http://people-press.org/report/?pageid=1550.

5. An estimated 10,000 premature deaths per year in the United States alone. National Academy of Sciences, *Hidden Costs of Energy* (Washington, D.C.: National Academies Press, 2009), online at http://www.nap.edu/catalog.php ?record_id=12794.

6. World Meteorological Organization, *The Changing Atmosphere: Implications for Global Security, Toronto, Canada, 27–30 June 1988: Conference Proceedings* (Geneva: Secretariat of the World Meteorological Organization, 1989), online at http://www.cmos.ca/ChangingAtmosphere1988e.pdf. See Spencer Weart, *The Discovery of Global Warming*, 2d ed. (Cambridge, Mass.: Harvard University Press, 2008), expanded online at http://www.aip.org/history/climate /Internat.htm.

7. "catastrophe" + "global warming" 11,900 articles; + "nuclear reactor," 859; + "nuclear war," 1,730. Accessed 26 June 2010.

8. James E. Hansen et al., "Dangerous Human-Made Interference with Climate: A GISS Model Study," *ArXiv*, 2006, online at http://arxiv.org/abs /physics/0610115.

9. Google accessed 14 Oct. 2010. Stephen Ansolabehere, "Energy Options: Insights for Nuclear Energy," MIT Center for Advanced Nuclear Energy Systems MIT-NES-TR-008 (June 2007), online at http://web.mit.edu/canes /pdfs/nes-008.pdf.

10. Surveys: Paul Slovik, James H. Flynn, and Mark Layman, "Perceived Risk, Trust, and the Politics of Nuclear Waste," *Science* 254 (1991): 1603–1607. Allison M. Macfarlane and Rodney C. Ewing, *Uncertainty Underground: Yucca Mountain and the Nation's High-Level Nuclear Waste* (Cambridge, Mass.: MIT Press, 2006).

11. News accounts of protests, etc. are readily available on the Web; see also Luther J. Carter, *Nuclear Imperatives and Public Trust: Dealing with Radio-active Waste* (Washington, D.C.: Resources for the Future, 1987), 401–402. Acceptance by nearby residents: e.g., Ann Stouffer Bisconti, " 'Not in My Back Yard' Is Really 'Yes! In My Back Yard,' " *Natural Gas & Electricity*, 2010, pp. 23–28.

12. Sir John Hill as told to Carter, *Nuclear Imperatives*, 9.

13. *The Economist*, 11 Nov. 2006, 72. For updates see occasional articles in *Business Week, NYT*, etc.

14. Chernobyl Forum, *Chernobyl's Legacy: Health, Environmental and Socio-Economic Impacts*, 2d ed. (Vienna: International Atomic Energy Agency,

2005), online at http://www.iaea.org/Publications/Booklets/Chernobyl
/chernobyl.pdf, 36; see also 14, 20–21, 41.

15. David P. McCaffrey, *The Politics of Nuclear Power* (Boston: Kluwer, 1991);
Kenneth F. McCallion, *Shoreham and the Rise and Fall of the Nuclear Power
Industry* (Westport, Conn.: Praeger, 1995).

16. Unpublished study by Anthony Leiserowitz, who kindly shared his data.

17. Gallup and other polls at http://www.pollingreport.com/energy.htm. Another
consistent series found even higher favorables: Ann Bisconti, *Record High 70
Percent Favor Nuclear Energy* (Washington, D.C.: Nuclear Energy Institute,
2005), online at http://www.nei.org/documents/PublicOpinion_05–07.pdf. The
trend toward acceptance continued to 2010; for the latest see http://www.nei.org.

18. University of Maryland Program on International Policy Attitudes, "Current
Energy Use Seen to Threaten Environment, Economy, Peace," 2 July 2006, at
www.worldpublicopinion.org/pipa/articles/btenvironmentra/227.php?nid=&id=
&pnt=227.

19. Mick Broderick, "Releasing the Hounds: The Simpsons as Anti-Nuclear
Satire," in *Leaving Springfield: The Simpsons and the Possibility of Opposi-
tional Culture*, ed. John Alberti (Detroit: Wayne State University Press, 2004),
244–272. Episode: "Treehouse of Horror XV," 7 Nov. 2004.

20. Anthony Lane, "Life and Death Matters," *New Yorker* 85, no. 48 (8 Feb. 2010):
76–77; Kurt Loder, "'Edge of Darkness': Dad Reckoning," at http://www.mtv
.com/movies/news/articles/1630736/story.jhtml.

21. Danny Boyle quoted in Jamie Russell, *The Book of the Dead: The Complete
History of Zombie Cinema* (Godalming, Surrey, UK: FAB Press, 2005), 179.

22. James Lovelock, *The Vanishing Face of Gaia: A Final Warning* (New York:
Basic Books, 2009), 6, 116.

23. "The chances": *The Economist*, 26 March 2011, p. 84. Eric Bellman, "Japan's
Farmers Confront Toxins from the Tsunami," *Wall Street Journal*, 6 April 2011,
p. A9.

24. Richard K. Lester, "Why Fukushima Won't Kill Nuclear Power," *Wall Street
Journal*, 6 April 2011, p. A19.

22. Deconstructing Nuclear Weapons

1. T. Milne et al., "An End to UK Nuclear Weapons," British Pugwash Group,
n.d., at http://www.pugwash.org/uk/documents/end-to-uk-nuclear-weapons.pdf,
on p. 35, drawn from *MORI Public Opinion Newsletter*.

2. Compiled January 2006 from http://brain.gallup.com.

3. First reported by William J. Broad, *NYT*, 8 Oct. 1993. David E. Hoffman,
*Dead Hand: The Untold Story of the Cold War Arms Race and Its Dangerous
Legacy* (New York: Doubleday, 2009).

4. Respectively 1,168; 2,903; 10,319 on 10 Oct. 2010 (but the numbers change greatly from month to month).

5. For "radioactive Rambos" among other matters see Paul Brians, "Nuke Pop," http://www.wsu.edu/~brians/nukepop/index.html.

6. Mick Broderick, "Is This the Sum of Our Fears? Nuclear Imagery in Post–Cold War Cinema," in *Atomic Culture: How We Learned to Stop Worrying and Love the Bomb*, ed. Scott C. Zeman and Michael A. Amundson (Boulder: University Press of Colorado, 2004), 127–149, at 144; Jerome F. Shapiro, *Atomic Bomb Cinema: The Apocalyptic Imagination on Film* (New York: Routledge, 2002), 14–15.

7. *Terminator 2: Judgment Day* (Tristar, 1991), *Terminator 3: Rise of the Machines* (Warner Brothers, 2003), *Terminator Salvation* (Warner Brothers, 2009), plus a television series, *The Sarah Connor Chronicles* (Fox, 2007–2009).

8. *The West Wing* (Warner Brothers-NBC), "Galileo," first aired 29 Nov. 2000; *Jericho* (CBS-Paramount, 2006–2008); *24* (Fox, 6th season, winter–spring 2006–2007); *Heroes* (NBC, first season, winter–spring 2006–2007).

9. Robert Coles, *The Moral Life of Children* (Boston: Atlantic Monthly Press, 1986), chap. 7.

10. Paul S. Boyer, "Sixty Years and Counting: Nuclear Themes in American Culture, 1945 to the Present," in *The Atomic Bomb and American Society: New Perspectives*, ed. Rosemary B. Mariner and C. Kurt Piehler (Knoxville: University of Tennessee Press, 2009), 3–18, at 12, 14.

11. *Nuclear War* series (New World Computing, 1989), concluding in 1997 with *Ground Zero*, a fan-made remake. Reviews from http://www.thehouseofgames .net/index.php?t=10&id=49 and http://www.the-underdogs.org/hame.php?id= 3857, accessed 7 Jan. 2011. Given the mutability of the Web, readers may do better to search for their own references for these topics.

12. Atomic toys in Oak Ridge museum, http://www.orau.org/ptp/collection /atomictoys/atomictoys.htm, accessed 15 Sept. 2006.

13. Noi Sawaragi, "On the Battlefield of 'Superflat': Subculture and Art in Postwar Japan," in *Little Boy: The Arts of Japan's Exploding Subculture*, ed. Takashi Murakami (New York: Japan Society; New Haven: Yale University Press, 2005), 187–207, at 204; see 203–205; also Alexandra Munroe, "Introducing Little Boy," ibid., 240–261, at 247.

14. Jacques Derrida, "No Apocalypse, Not Now: Full Speed Ahead, Seven Missiles, Seven Missives," *Diacritics* 14 (1984): 20–31, as quoted by J. Fisher Solomon, *Discourse and Reference in the Nuclear Age* (Norman: University of Oklahoma Press, 1988), 22–23.

23. Tyrants and Terrorists

1. Tom Clancy, *The Sum of All Fears* (New York: Putnam, 1991).

2. Among the few who notice two separate post–Cold War periods without specifying the 9/11/2001 events as a crucial division is William J. Kinsella, "One Hundred Years of Nuclear Discourse: Four Master Themes and Their Implications for Environmental Communication," in *The Environmental Communication Yearbook*, vol. 2, ed. Susan L. Senecah (Mahwah, N.J.: Lawrence Erlbaum Associates, 2005), 49–72. For novels 1960s–early 1980s featuring terrorist attempts at bombing or extortion see Paul Brians, *Nuclear Holocausts: Atomic War in Fiction, 1895–1984* (Kent, Ohio: Kent State University Press, 1987), 37–38, 151, 178, 189, 201, 267, 275, 293; online version at http://www.wsu.edu/~brians/nuclear/.

3. John Mueller, *Atomic Obsession: Nuclear Alarmism from Hiroshima to Al-Qaeda* (New York: Oxford University Press, 2009), 95; see 90–95.

4. Pew Research Center for the People & the Press, search of data on the terms "nuclear," "atomic," "weapon," "reactor." http://people-press.org/nii/bydate.php.

5. 1996 poll: Pew Research Center for the People & the Press, "Public Apathetic about Nuclear Terrorism," 11 April 1996, http:// people-press.org/reports/display .php3?ReportID=128, accessed 8 Jan. 2007. 1998 poll: Keating Holland/CNN, "Poll: Many Americans Worry about Nuclear Terrorism," http://www.cnn.com /ALLPOLITICS/1998/06/16/poll/, accessed 3 Jan. 2007. Jessica Stern, *The Ultimate Terrorists* (Cambridge, Mass.: Harvard University Press, 1999); Gary Ackerman and William C. Potter, "Catastrophic Nuclear Terrorism: A Preventable Peril," in *Global Catastrophic Risks*, ed. Nick Bostrom and Milan M. Cirkovic (New York: Oxford University Press, 2008), 402–449; see 404–406. Nadine Gurr and Benjamin Cole, *The New Face of Nuclear Terrorism: Threats from Weapons of Mass Destruction* (London: I. B. Tauris, 2002), 3–8.

6. Gallup polls from http://brain.gallup.com. Pew Research Center for People & the Press, "Two Years Later, the Fear Lingers," 4 Sept. 2003, people-press.org /reports/print.php3?PageID=735. World survey: http://www.globescan.com/rf _gi_first_01.htm. Further polls can be readily be turned up with Web searches.

7. J. F. O. McAlister, "The Spy Who Knew Too Much," *Time* 168, no. 25 (18 Dec. 2006): 32; Walter Litvinenko, quoted in *New Scientist* 192 no. 2581 (9 Dec. 2006): 9.

8. The large literature on these issues is itself a sign of their prominence. One good summary is Stern, *The Ultimate Terrorists*. Another example: Graham T. Allison, *Nuclear Terrorism: The Ultimate Preventable Catastrophe* (New York: Henry Holt–Times Books, 2004).

9. The Google News Archive, http://news.google.com/archivesearch/, finds a rapid rise for the terms "child molester" and "drug criminal," both starting around 1985 and leveling off in the 1990s.

10. Joseph Conrad, *The Secret Agent* (1907), a masterpiece; Emile Zola, *Paris* (1898), a failure; more recently, e.g., Doris Lessing, *The Good Terrorist* (1985); John Updike, *Terrorist* (2006).

11. Paul Forman, "(Re)Cognizing Postmodernity: Helps for Historians—of Science Especially," *Berichte zur Wissenschaftsgeschichte* 33 (2010): 157–175.

12. Noah Feldman, "Islam, Terror and the Second Nuclear Age," NYT *Magazine*, 29 Oct. 2006, p. 53.

13. Muhammad el-Baradei (Director-General of the International Atomic Energy Agency), quoted in John Tagliabue, "A Nation Challenged: Atomic Anxiety. Threat of Nuclear Terror Has Increased, Official Says," *NYT*, 2 Nov. 2001. "Traumatizing": James Carroll on "Morning Edition," National Public Radio, 30 May 2006.

14. Tom Engelhardt, "9/11 in a Movie-Made World," *The Nation*, 25 Sept. 2006, online at http://www.thenation.com/doc/20060925/engelhardt/2, gives these quotes from *NYT*.

15. The *Los Angeles Times* used the term on September 12; by the weekend it was universally used for what some had called, e.g., the "collapse site." My thanks to Bo Jacobs and Los Alamos archivist Roger Meade for clarifying the origins of the term.

16. Google News Archive; see John Mueller, *Atomic Obsession*, 11.

17. Allison, *Nuclear Terrorism*, 129. Similarly see Ron Suskind, *The One Percent Doctrine: Deep Inside America's Pursuit of Its Enemies since 9/11* (New York: Simon & Schuster, 2006), 62. Speech rewrite: Timothy Noah, "Moderation equals suicide," Slate.com, 21 May 2009, http://www.slate.com/id/2218837/.

18. Google search conducted 11 Sept. 2006. The methodology here is complex because I had to remove overlaps by searching on different combinations of the terms. The first group associated with "terrorists" (Google conveniently included "terrorism" and "terrorists" in "terrorist" searches) included "nuclear weapon" (2.8 million all by itself), "nuclear bomb," "atomic bomb," and "dirty bomb." The second group included "anthrax" (1.0 million), "smallpox," "biological weapon," "germ," and "virus," which got 9 million combinations, but most of these seemed to be computer viruses or descriptions of terrorism itself as a virus. "Chemical weapon" and "poison gas" minus overlaps yielded only about 730,000 hits. Also, the search found about 12.2 million pages for "terrorist AND bomb AND (nuclear OR atomic)," almost as many as the roughly 13.8 million for "terrorist AND bomb NOT nuclear NOT atomic." Finally, searches conducted by the public on Google

in 2004–2009 for "dirty bomb" consistently outnumbered searches for the much broader term "biological weapon," as seen in http://www.google.com/insights/search/.

19. Stern, *The Ultimate Terrorists*, 3.

20. For these and more quotes: Mueller, *Atomic Obsession*, 19–21.

21. Richard C. Leone and Greg Anrig Jr., eds., *The War on Our Freedoms: Civil Liberties in an Age of Terrorism* (New York: Public Affairs, 2003).

22. Bush: Allison, *Nuclear Terrorism*, 38; Cheney: Noah, "Moderation Equals Suicide," see also Suskind, *The One Percent Doctrine*.

23. Frank Rich, *The Greatest Story Ever Sold: The Decline and Fall of Truth from 9/11 to Katrina* (New York: Penguin, 2006).

24. Paul Wolfowitz (deputy secretary of defense), interview by Sam Tanenhaus, *Vanity Fair*, July 2003, as cited by Allison, *Nuclear Terrorism*, 123. "Not substantively": Commission on the Intelligence Capabilities of the United States Regarding Weapons of Mass Destruction, *Report to the President of the United States* (Washington, D.C., 2005), 75, online at http://www.gpoaccess.gov/wmd/index.html.

24. The Modern Arcanum

1. Fred Kirby quoted by Charles Wolfe, notes to the recording *Atomic Café* (Archives Project, 1981). See Michael J. Yavendetti, "American Reactions to the Use of Atomic Bombs on Japan, 1945–1947," Ph.D. diss., University of California, Berkeley, 1970, 255–256.

2. "Basic power": for example, John Cockcroft in British Broadcasting Corporation, *Atomic Challenge: A Symposium* (London: Winchester, 1947), 2.

3. Gerald J. Ringer, "The Bomb as a Living Symbol: An Interpretation," Ph.D. diss., Florida State University, 1966, 116–117.

4. Lowell Blanchard, "Jesus Hits Like an Atom Bomb" (1949–1950), on recording *Atomic Café*. Prophecy: Revelation 6–16.

5. Hal Lindsey with C. C. Carlson, *The Late Great Planet Earth* (Grand Rapids, Mich.: Zondervan, 1970). Tim LaHaye and Jerry B. Jenkins, *Left Behind: A Novel of the Earth's Last Days* (Wheaton, Ill.: Tyndale House, 1996).

6. Arthur C. Clarke, "If I Forget Thee, O Earth . . ," reprinted in Clarke, *Across the Sea of Stars* (New York: Harcourt, Brace & World, 1959), 63–67.

7. Arthur C. Clarke, *Childhood's End* (New York: Ballantine, 1953); see also Clarke, *2001: A Space Odyssey* (New York: New American Library, 1968), which also sold millions of copies.

8. C. G. Jung, *Flying Saucers: A Modern Myth of Things Seen in the Skies*, trans. R. F. C. Hull (1964; reprint, Princeton: Princeton University Press, 1978), 14, 22–33, 55–62.

9. Michael J. Carey, "Psychological Fallout," *BAS* 38, no. 1 (Jan. 1982): 23. *Pravda*, 7 Aug. 1964, p. 3, in *Current Digest of the Soviet Press* 16, no. 32 (1964), p. 17.

10. Here as elsewhere for more detailed references see Weart, *Nuclear Fear* (Cambridge, Mass.: Harvard University Press, 1988); a new reference (Japanese) is Masuji Ibuse, *Black Rain* (Tokyo: Kodansha International, 1969), 54.

11. Robert G. Wasson and V. P. Wasson, *Mushrooms, Russia and History*, 2 vols. (New York: Pantheon, 1957).

12. Robert G. Wasson, *Soma, Divine Mushroom of Immortality* (New York: Harcourt Brace Jovanovitch, 1971). Richard Evans Schultes and Albert Hofmann, *Plants of the Gods: Origins of Hallucinogenic Use* (New York: McGraw-Hill, 1979).

13. Anatole France, *Penguin Island*, trans. A. W. Evans (New York: Dodd, Mead, 1909), 336.

14. Americo Favale to Roland Anderson, 1 Aug. 1956. My thanks to Richard Hewlett and Roger Anders for digging this letter out of the AEC Patent Branch records, Department of Energy, Germantown, Md.

15. Carl G. Jung, *Mandala Symbolism*, trans. R. F. C. Hull (Princeton: Princeton University Press, 1972). Heinrich Zimmer, *Myths and Symbols in Indian Art and Civilization* (New York: Harper & Brothers, 1962), 139–148.

16. Carl G. Jung, *Mysterium Coniunctionis: An Inquiry into the Separation of Psychic Opposites in Alchemy*, trans. R. F. C. Hull, 2d ed. (Princeton: Princeton University Press, 1970), 463.

17. Clarence P. Hornung, *Hornung's Handbook of Designs and Devices*, 2d rev. ed. (New York: Dover Publications, 1946), 39, fig. 349, as discussed by Bill Geerhart, "An Indelible Cold War Symbol: The Complete History of the Fallout Shelter Sign," at http://knol.google.com/k/bill-geerhart/an-indelible-cold-war-symbol/1uefuvb7s5ifz/12# (accessed 6 May 2010).

18. See also Ira Chernus, *Dr. Strangegod: On the Symbolic Meaning of Nuclear Weapons* (Columbia: University of South Carolina Press, 1986).

25. Artistic Transmutations

1. W. H. Auden, *The Age of Anxiety: A Baroque Eclogue* (New York: Random House, 1947).

2. Robert Jay Lifton, *The Broken Connection: On Death and the Continuity of Life* (New York: Simon & Schuster, 1979). A good collection is Jim Schley, ed., *Writing in a Nuclear Age* (Hanover, N.H.: University Press of New England, 1984).

3. Kurt Vonnegut, *Palm Sunday: An Autobiographical Collage* (New York: Laurel, 1984), 69.

4. John Braine, "People Kill People," in *Voices from the Crowd: Against the H-Bomb*, ed. David Boulton (Philadelphia: Dufour, 1964), 181. Directors: *Show* (June 1962): 78–81.

5. Pat Frank, "Hiroshima: Point of No Return," *Saturday Review*, 24 Dec. 1960, p. 25. Mordecai Roshwald, *Level 7* (New York: McGraw-Hill, 1960).

6. Lifton, *Broken Connection*; Lifton and Greg Mitchell, *Hiroshima in America: A Half Century of Denial* (New York: Avon, 1996).

7. Paul Brians, *Nuclear Holocausts: Atomic War in Fiction, 1895–1984* (Kent, Ohio: Kent State University Press, 1987) (updated online at http://www.wsu.edu /~brians/nuclear/), 22.

8. Richard Martin, "Detonating on Canvas: The Abstract Bomb in American Art," in *The Writing on the Cloud: American Culture Confronts the Atomic Bomb*, ed. Alison M. Scott and Christopher D. Geist (Lanham, Md.: University Press of America, 1997), 73.

9. Salvador Dali with André Parinaud, *The Unspeakable Confessions of Salvador Dali* (New York: William Morrow, 1976), 202. Klein quoted by Stephen Petersen, "Explosive Proposition: Artists React to the Atomic Age," *Science in Context* 17 (2004): 579–609, q.v. for all 1950s art. Jean Tinguely, *Hommage à New York* (1960); see Harold Rosenberg, *The De-Definition of Art: Action Art to Pop to Earthworks* (New York: Horizon, 1972), 156–166.

10. Albert E. Stone, *Literary Aftershocks: American Writers, Readers, and the Bomb* (New York: Twayne; Maxwell Macmillan, 1994), 128–129.

11. Cai Guo-Qiang, *The Century with Mushroom Clouds: Project for the Twentieth Century* (1995–1996), partly replicated in solo retrospective, Guggenheim Museum, New York (2008), from which the quote is taken. Gathering artifacts: James Acord. Reproducing: Gregory Green, Jim Sanborn. For some other works 1945–2000 see Jonathan Jones, "Magic Mushrooms," *The Guardian*, 6 Aug. 2002, at http://www.guardian.co.uk/artanddesign/2002/aug /06/art.artsfeatures.

12. See Paul Fussell, *The Great War and Modern Memory* (London: Oxford University Press, 1975), 8, 34–35, 203. Kurt Vonnegut, *Cat's Cradle* (New York: Holt, Reinhart & Winston, 1965); Joseph Heller, *Catch-22* (New York: Simon & Schuster, 1961). "I wrote it" quoted in Lawrence H. Suid, *Guts and Glory: Great American War Movies* (Reading, Mass.: Addison-Wesley, 1978), 271.

13. Takashi Murakami painting *Time Bokan* (2001). Murakami, ed., *Little Boy: The Arts of Japan's Exploding Subculture* (New York: Japan Society; New Haven: Yale University Press, 2005).

14. Kurt Vonnegut, *Slaughterhouse-5: or the Children's Crusade* (New York: Delta, 1969), 17. See also Paul Boyer, *By the Bomb's Early Light: American Thought*

and Culture at the Dawn of the Atomic Age (New York: Pantheon, 1985), chap. 20.

15. For Brecht's *Galileo* see Gerhard Szczesny, *Das Leben des Galilei and der Fall Bertolt Brecht* (Frankfurt: Ullstein, 1966).

16. Heinar Kipphardt, *In the Matter of J. Robert Oppenheimer*, trans. Ruth Speirs (New York: Hill and Wang, 1968), 146–164. Cf. Robert O. Butler, *Countrymen of Bones* (New York: Horizon, 1983); Thomas McMahon, *Principles of American Nuclear Chemistry: A Novel* (Boston: Little, Brown, 1981).

17. *Dr. Atomic* libretto by Peter Sellars. John Joseph Adams, *Hallelujah Junction: Composing an American Life* (New York: Farrar, Straus and Giroux, 2008), 271 ff.

18. Michael Frayn, *Copenhagen* (1998; reprint, New York: Anchor, 2000). Broadway run of 326 performances in 2000; BBC film adaptation, 2002.

19. William Golding, *Lord of the Flies* (London: Faber & Faber, 1954). See Bernard S. Oldsey and Stanley Weintraub, *The Art of William Golding* (New York: Harcourt, Brace & World, 1965), chap. 2.

20. Carl G. Jung, "On the Nature of the Psyche" (1947), in Jung, *Collected Works*, vol. 8, trans. R. F. C. Hull (New York: Pantheon, 1960), 159–234; see 218, 222.

21. Frank Sullivan, "The Cliché Expert Testifies on the Atom," in Sullivan, *A Rock in Every Snowball* (Boston: Little, Brown, 1946), 31. Karel Čapek, *An Atomic Phantasy: Krakatit*, trans. Lawrence Hyde (1924; reprint, London: Allen & Unwin, 1948), 287.

22. Russell Hoban, *Riddley Walker* (New York: Summit, 1980); Denis Johnson, *Fiskadoro* (New York: Knopf, 1985).

23. Cormac McCarthy, *The Road* (New York: Knopf, 2006).

24. E.g., Denise Levertov, *The Poet in the World* (New York: New Directions, 1973), 121–122; Levertov, *Candles in Babylon* (New York: New Directions, 1982), 73; Gary Snyder, "LMFBR," in *The Postmoderns: The New American Poetry Revised*, ed. Donald Allen and George F. Butterick (New York: Grove, 1982), 281.

25. For this and the following see Spencer Weart, *The Discovery of Global Warming*, 2d ed. (Cambridge, Mass.: Harvard University Press, 2008); and the expanded website of the same name at http://www.aip.org/history/climate/.

26. Ian McEwan, *Solar* (New York: Doubleday, 2010), 15–16.

27. "More compelling": Richard Hamblyn, "Message in the Wilderness," *Times Literary Supplement* no. 5389 (14 July 2006): 18.

28. Interview with Nicholas Wroe, "Ian McEwan: 'It's Good to Get Your Hands Dirty a Bit,'" *The Guardian*, 6 March 2010, at http://www.guardian.co.uk/books/2010/mar/06/ian-mcewan-solar.

29. E.g., Bill McKibben, *Eaarth: Making a Life on a Tough New Planet* (New York: Henry Holt, 2010).

30. James Hansen, "Is There Still Time to Avoid 'Dangerous Anthropogenic Interference' with Global Climate?," address to American Geophysical Union, San Francisco, 6 Dec. 2005, at http://www.columbia.edu/~jeh1/2005/Keeling _20051206.pdf.

A Personal Note

1. David Lilienthal, *Change, Hope, and the Bomb* (Princeton: Princeton University Press, 1963), 20; Lilienthal, *Atomic Energy: A New Start* (New York: Harper & Row, 1980).
2. Spencer Weart, *Never at War: Why Democracies Won't Fight One Another* (New Haven: Yale University Press, 1998).

FURTHER READING

On the "First Nuclear Age," through 1987, for much more detail and documentation see my earlier work, *Nuclear Fear: A History of Images* (Cambridge, Mass.: Harvard University Press, 1988). Unlike the present shorter book, it attempts a world survey rather than concentrating on the United States. The *Alsos Digital Library for Nuclear Issues* online at http://alsos.wlu.edu/ is a large, annotated historical bibliography covering all aspects.

Attitudes to nuclear imagery have not been explored far beyond the pioneering works, notably Robert Jay Lifton, *The Broken Connection: On Death and the Continuity of Life* (New York: Simon & Schuster, 1979); and Paul Boyer, *By the Bomb's Early Light: American Thought and Culture at the Dawn of the Atomic Age* (New York: Pantheon, 1985). Other histories discussing public reactions, mostly American, include Michael Scheibach, *Atomic Narratives and American Youth: Coming of Age with the Atom, 1945–1955* (Jefferson, N.C.: McFarland, 2003); Allan M. Winkler, *Life under a Cloud: American Anxiety about the Atom* (Urbana: University of Illinois Press, 1993); Margot W. Henriksen, *Dr. Strangelove's America: Society and Culture in the Atomic Age* (Berkeley: University of California Press, 1997); Rosemary B. Mariner and G. Kurt Piehler, eds., *The Atomic Bomb and American Society: New Perspectives* (Knoxville: University of Tennessee Press, 2009), including in particular Piehler, "Bibliographic Essay," pp. 407–425; Michael J. Hogan, ed., *Hiroshima in History and Memory* (New York: Cambridge University Press, 1996); Robert Lifton and Greg Mitchell, *Hiroshima in America: A Half Century of Denial* (New York: Avon, 1996); John Mueller, *Atomic Obsession: Nuclear Alarmism from Hiroshima to Al-Qaeda* (New York: Oxford University Press, 2009). Also Michael Ortiz Hill, *Dreaming the End of the World: Apocalypse as a Rite of Passage*, 2d ed. (Putnam, Conn.: Spring Publications, 2004); Lawrence Wittner, *Confronting the Bomb: A Short History of the World Disarmament Movement* (Stanford, Calif.: Stanford University Press, 2009), a condensation of *The Struggle against the Bomb*, 3 vols. (Stanford, Calif.: Stanford University Press, 1993–2003).

On transmutational and apocalyptic thinking I was influenced by the works of Freud and Jung as well as, in particular, Mircea Eliade, *The Forge and the Crucible: The Origins and Structure of Alchemy*, trans. Stephen Corrin (New York: Harper & Row, 1971); Joseph Campbell, *The Masks of God*, vol. 1: *Primitive Mythology* (New

York: Viking, 1969); Ernest Becker, *The Denial of Death* (New York: Free Press, 1973); John Bowlby, *Attachment and Loss*, vol. 2, *Separation: Anxiety and Anger* (London: Hogarth, 1973), and vol. 3, *Sadness and Depression* (New York: Basic Books, 1980). More recent sociological and psychological discoveries are not all summarized in books. For introductions see Dan Ariely, *Predictably Irrational: The Hidden Forces That Shape Our Decisions*, rev. ed. (New York: Harper, 2009); and Antonio R. Damasio, *Descartes' Error: Emotion, Reason, and the Human Brain* (New York: Avon, 1994). On risk see Paul Slovik, *The Perception of Risk* (London: Earthscan, 2000); Nick Pidgeon et al., eds., *The Social Amplification of Risk* (Cambridge: Cambridge University Press, 2003); on community trauma, Kai T. Erikson, *A New Species of Trouble: Explorations in Disaster, Trauma, and Community* (New York: Norton, 1994). Millenarian history is summarized in Frederic J. Baumgartner, *Longing for the End: A History of Millennialism in Western Civilization* (New York: St. Martin's Press, 1999); see also the classic Norman Cohn, *The Pursuit of the Millennium*, 3d ed. (New York: Oxford University Press, 1970).

The many works on nuclear films, fiction, and other arts include Jack G. Shaheen, ed., *Nuclear War Films* (Carbondale: Southern Illinois University Press, 1978); Mick Broderick, *Nuclear Movies: A Critical Analysis and Filmography of International Feature Length Films Dealing with Experimentation, Aliens, Terrorism, Holocaust, and Other Disaster Scenarios, 1914–1990* (Jefferson, N.C.: McFarland, 1991); Jerome F. Shapiro, *Atomic Bomb Cinema: The Apocalyptic Imagination on Film* (New York: Routledge, 2002); Kim Newman, *Apocalypse Movies: End of the World Cinema* (New York: St. Martin's Griffin, 2000); Joyce A. Evans, *Celluloid Mushroom Clouds: Hollywood and the Atomic Bomb* (Boulder, Colo.: Westview, 1998); John Canaday, *The Nuclear Muse: Literature, Physics, and the First Atomic Bombs* (Madison: University of Wisconsin Press, 2000); Paul Brians, *Nuclear Holocausts: Atomic War in Fiction, 1895–1984* (Kent, Ohio: Kent State University Press, 1987), updated online at http://www.wsu.edu/~brians/nuclear/; Bruce H. Franklin, *War Stars: The Superweapon and the American Imagination* (New York: Oxford University Press, 1988); Albert E. Stone, *Literary Aftershocks: American Writers, Readers, and the Bomb* (New York: Twayne; Maxwell Macmillan, 1994); Takashi Murakami, ed., *Little Boy: The Arts of Japan's Exploding Subculture* (New York: Japan Society; New Haven: Yale University Press, 2005); Robert Jacobs, ed., *Filling the Hole in the Nuclear Future: Art and Popular Culture Respond to the Bomb* (Lanham, Md.: Lexington, 2010). On the background of futuristic writings see I. F. Clarke, *The Pattern of Expectation, 1644–2001* (New York: Basic, 1979); W. Warren Wagar, *Terminal Visions: The Literature of Last Things* (Bloomington: Indiana University Press, 1982).

The actual history of events is scarcely covered in this book, as it is readily available in many places. A very brief survey of weapons history and issues is Joseph Cirincione, *Bomb Scare: The History and Future of Nuclear Weapons* (New York: Columbia University Press, 2007); and for the crucial core period see Lawrence Badash,

Scientists and the Development of Nuclear Weapons: From Fission to the Limited Test Ban Treaty, 1939–1963 (Amherst, N.Y.: Prometheus, 1995). For comprehensive coverage (though less so on the 1960s and 1970s) see Richard Rhodes: *The Making of the Atomic Bomb* (New York: Simon & Schuster, 1996); *Dark Sun: The Making of the Hydrogen Bomb* (New York: Simon & Schuster, 1996); *Arsenals of Folly: The Making of the Nuclear Arms Race* (New York: Knopf, 2007); and *The Twilight of the Bombs: Recent Challenges, New Dangers, and the Prospects for a World without Nuclear Weapons* (New York: Knopf, 2010). A few of the countless works on special topics: Lawrence Freedman, *The Evolution of Nuclear Strategy*, 3d ed. (New York: Palgrave Macmillan, 2003); P. D. Smith, *Doomsday Men: The Real Dr. Strangelove and the Dream of the Superweapon* (New York: St. Martin's Press, 2007) (from the 1800s to the 1960s); Barton C. Hacker, *Elements of Controversy: the Atomic Energy Commission and Radiation Safety in Nuclear Weapons Testing, 1947–1974* (Berkeley: University of California Press, 1994); Robert A. Divine, *Blowing on the Wind: the Nuclear Test Ban Debate, 1954–1960* (New York: Oxford University Press, 1978); Guy Oakes, *The Imaginary War: Civil Defense and American Cold War Culture* (New York: Oxford University Press, 1994); Dee Garrison, *Bracing for Armageddon: Why Civil Defense Never Worked* (New York: Oxford University Press, 2006); Nadine Gurr and Benjamin Cole, *The New Face of Nuclear Terrorism: Threats from Weapons of Mass Destruction* (London: I. B. Tauris, 2002); Charles D. Ferguson and William C. Potter, *The Four Faces of Nuclear Terrorism* (New York: Routledge, 2005).

Stephanie Cooke, *In Mortal Hands: A Cautionary History of the Nuclear Age* (New York: Bloomsbury, 2009), is a popularized history of the connection between weapons and civilian power; for a more sympathetic history of the industry see James Mahaffey, *Atomic Awakening: A New Look at the History and Future of Nuclear Power* (New York: Pegasus, 2009); and for recent matters, W. J. Nuttall, *Nuclear Renaissance: Technologies and Policies for the Future of Nuclear Power* (Bristol, UK: Institute of Physics, 2005). I have found no up-to-date comprehensive historical survey of the U.S. or world nuclear industry; for recent decades one must resort to industry journals and business magazines. Older works worth consulting include Irvin C. Bupp and Jean-Claude Derian, *Light Water: How the Nuclear Dream Dissolved* (New York: Basic, 1978); Bertrand Goldschmidt, *The Atomic Complex: A Worldwide Political History of Nuclear Energy* (La Grange, Ill.: American Nuclear Society, 1982), which also includes weapons history and should be compared with the equally partisan, more complete, but less uniformly accurate Peter Pringle and James J. Spigelman, *The Nuclear Barons* (New York: Holt, Rinehart & Winston, 1981); Steven L. Del Sesto, *Science, Politics, and Controversy: Civilian Nuclear Power in the United States, 1946–1974* (Boulder, Colo.: Westview, 1979). Concentrating on specific topics but of general value are four books by the Nuclear Regulatory Commission's historian, J. Samuel Walker: *Containing the Atom: Nuclear Regulation in a Changing Environment, 1963–1971*; *Permissible Dose: A History of Radiation Protection in the*

Twentieth Century; Three Mile Island: A Nuclear Crisis in Historical Perspective; The Road to Yucca Mountain: The Development of Radioactive Waste Policy in the United States (Berkeley: University of California Press, 1992, 2000, 2004, 2009, respectively); as well as George T. Mazuzan and J. Samuel Walker, *Controlling the Atom: The Beginnings of Nuclear Regulation, 1946–1962* (Berkeley: University of California Press, 1984). For other special topics: Kenneth F. McCallion, *Shoreham and the Rise and Fall of the Nuclear Power Industry* (Westport, Conn.: Praeger, 1995); Luther J. Carter, *Nuclear Imperatives and Public Trust: Dealing with Radioactive Waste* (Washington, D.C.: Resources for the Future, 1987); John Byrne and Steven M. Hoffman, eds., *Governing the Atom: The Politics of Risk* (New Brunswick, N.J.: Transaction, 1996); Allison M. Macfarlane and Rodney C. Ewing, *Uncertainty Underground: Yucca Mountain and the Nation's High-level Nuclear Waste* (Cambridge, Mass.: MIT Press, 2006); Constance Perin, *Shouldering Risks: The Culture of Control in the Nuclear Power Industry* (Princeton: Princeton University Press, 2005).

Summaries and reviews of popular culture items are easily found by Web searches (even if inaccurate, these are significant as reactions to the item). Conelrad at http://www.conelrad.com looks at popular culture of the early nuclear decades. For contemporary issues, Physicians for Social Responsibility http://www.psr.org and the Union of Concerned Scientists http://ucsusa.org address antinuclear weapons and power issues; the Nuclear Energy Institute http://www.nei.org offers a pro-nuclear power viewpoint plus statistics, etc.

INDEX